Chromosomes: Eukaryotic, Prokaryotic, and Viral

Volume I

Repetitive Human DNA: The Shape of Things to Come. Sequence-Specific DNA-Binding Proteins Involved in Gene Transcription. Physical and Topological Properties of Closed Circular DNA. Structure of the 300A Chromatin Filament. Three-Dimensional Computer Reconstructions of Chromosomes in Human Mitotic Cells. The Kinetochore and its Role in Cell Division. X Inactivation in Mammals, An Update. The Y Chromosome of *Drosophila*.

Volume II

Meiosis. Chromosome Structure and Function During Oogenesis and Early Embryogenesis. Chromatin Organization in Sperm. Structure and Function of Polytene Chromosomes. *Saccharomyces cerevisiae:* Structure and Behavior of Natural and Artificial Chromosomes. DNA Replication in Higher Plants. Chromatin Structure of Plant Genes. Ploidy Manipulations in the Potato. The Chloroplast Genome and Regulation of its Expression.

Volume III

Bacterial Chromatin (A Critical Review of Structure-Function Relationships). The Chromosomal DNA Replication Origin, *oriC,* in Bacteria. Replication and Segregation Control of *Escherichia coli* Chromosomes. Termination of Replication in *Bacillus subtilis, Escherichia coli* and R6K. Polyoma and SV40 Chromosomes. The Genome of Cauliflower Mosaic Virus: Organization and General Characteristics. Reinitiation of DNA Replication in Bacteriophage Lambda. *In Vivo* Fate of Bacteriophage T4 DNA. Double-Stranded DNA Packaged in Bacteriophages: Conformation, Energetics and Packaging Pathway. Bacteriophage P1 DNA Packaging. Bacteriophage P22 DNA Packaging.

Chromosomes: Eukaryotic, Prokaryotic, and Viral

Volume II

Editor

Kenneth W. Adolph, Ph.D.
Associate Professor
Department of Biochemistry
University of Minnesota Medical School
Minneapolis, Minnesota

CRC Press, Inc.
Boca Raton, Florida

Library of Congress Cataloging-in-Publication Data

Chromosomes: eukaryotic, prokaryotic, and viral.

  Includes bibliographies and index.
  1. Chromosomes. I. Adolph, Kenneth W., 1944-
QH600.C498   1989   574.87'322   88-35340
ISBN 0-8493-4397-6 (v. 1)

  Direct all inquiries to CRC Press, Inc., 2000 Corporate Blvd., N.W., Boca Raton, Florida, 33431.

© 1990 by CRC Press, Inc.

International Standard Book Number 0-8493-4397-6 (Volume I)
International Standard Book Number 0-8493-4398-4 (Volume II)
International Standard Book Number 0-8493-4399-2 (Volume III)

Library of Congress Card Number 88-35340
Printed in the United States

PREFACE

All animal cells, plant cells, and viruses capable of reproduction contain genetic material. For each living system, the genetic material exists as a complex of the coding molecules of DNA or RNA with proteins. The protein molecules serve to protect the DNA from degrading enzymes and shearing, and also control expression of the genetic information. Eukaryotic chromosomes are organized by the association of DNA with histones: these small, highly charged proteins coat the DNA to form the bead-like structures of the nucleosomes. Chromatin fibers are produced by the helical coiling of the beads-on-a-string filament of nucleosomes, and folding of the chromatin fiber results in the recognizable morphology of mitotic chromosomes. Nonhistone proteins further contribute to the structure and function of eukaryotic chromosomes, particularly in the regulation of gene expression.

The characteristic structures of mitotic chromosomes are observed during the process of normal cell division. But other processes and activities can influence DNA packaging in eukaryotic cells. The reduction of the diploid to haploid complement of chromosomes in meiosis is accompanied by special DNA packaging in the formation of eggs and sperm. In addition, chromosomes are active in replication and transcription, and this requires changes in the organization of chromosomes.

Prokaryotes, included bacteria and cyanobacteria (blue-green algae), have much smaller genomes than eukaryotes and their chromosomes are correspondingly less complex. The intestinal bacterium *E. coli* has 1000 × less DNA than human cells, lacks the histone proteins, and doesn't undergo mitosis or meiosis. Yet the DNA is contained in a defined nucleoid and is organized as supercoiled loops. This arrangement permits the efficient expression of genes and the segregation of replicated chromosomes to daughter cells. The lower complexity of prokaryotic chromosomes is an advantage for studies of the molecular biology and biochemistry of transcription and replication.

Viral chromosomes have even smaller genomes as a consequence of their parasitic reproduction cycles. For example, animal viruses such as polyoma and SV40 have DNA molecules that are 1000 × shorter than those of *E. coli*. Basic differences in the replication of animal, plant, and bacterial viruses have produced a variety of means of packaging viral genomes. Some animal viruses, included polyoma and SV40, have true minichromosomes with histones and nucleosomes. The heads of bacterial viruses such as T7 and P22 contain tighly folded or wound DNA strands that are free of bound protein. And most plant viruses (cauliflower mosaic virus is a notable exception) have RNA genomes enmeshed in protective coat protein subunits.

Major factors which determine the structures of chromosomes therefore include: the biochemical nature of DNA as a linear molecule composed of A, T, C and G subunits, the size of the genome and extent of DNA coiling and interaction with proteins, the nature of the histone proteins as globular and highly-charged, the activity of the genome in replication and transcription, and the requirement for special DNA packaging in mitosis and meiosis. These and additional topics will be examined in the following chapters, which will hopefully convey the variety and fascination of eukaryotic, prokaryotic, and viral chromosomes.

The chapters are divided into five sections; the first three of these are concerned with eukaryotic chromosomes, while the remaining two sections are devoted to prokaryotic and viral chromosomes. Each section begins with an introduction to give a brief overview of the subject matter and to relate it to the topics in other sections. Reviewing the results of research on different systems is important because, although chromosomes possess vastly different degrees of complexity, all chromosomes share similar features. Genetic information is encoded in the same DNA (or RNA) molecules and the sequence of genes must be protected and compacted by interacting with proteins. However, even though all chromosomes have the underlying unity of being protein-nucleic acid complexes, the special features observed for chromosomes of different sources makes studying chromosomes particularly challenging.

Chromosomes: Eukaryotic, Prokaryotic, and Viral should be a valuable resource for readers with a variety of interests and backgrounds. It is hoped that the information presented will be useful and that a sense is imparted of the excitement of research on chromosomes.

THE EDITOR

Kenneth W. Adolph, Ph.D., is presently a faculty member in the Department of Biochemistry of the University of Minnesota Medical School in Minneapolis. His research concerns two fundamental aspects of chromosomes: the structure of chromosomes determined by analysis of electron micrographs and the roles of nonhistone proteins in chromosome organization. He also maintains an interest in virus assembly. Kenneth W. Adolph has been a faculty member at the University of Minnesota since 1978 and is currently an Associate Professor. Postdoctoral training at Princeton University in the Department of Biochemical Sciences preceded this appointment; metaphase chromosome substructure and nuclear substructure were investigated in the laboratory of U.K. Laemmli. Earlier postdoctoral and graduate research was concerned with the assembly of viruses, protein-nucleic acid complexes much simpler than eukaryotic chromosomes and equally interesting. The structure of an icosahedral plant virus was studied during a postdoctoral year working with D. L. D. Caspar at the Rosenstiel Center, Brandeis University. Prior to this, the editor had a postdoctoral position at the Medical Research Council Laboratory of Molecular Biology in Cambridge, England. Research in the laboratory of Aaron Klug involved the *in vitro* reassembly of another simple, icosahedral plant virus. Kenneth W. Adolph received his Ph.D. from the Department of Biophysics, University of Chicago. His thesis concerned the isolation and characterization of cyanophages, and his advisor was R. Haselkorn. B.S. and M.S. degrees were received from the Department of Physics at the University of Wisconsin, Milwaukee.

CONTRIBUTORS

Kenneth W. Adolph, Ph.D.
Associate Professor
Department of Biochemistry
University of Minnesota Medical School
Minneapolis, Minnesota

Aimée Hayes Bakken, Ph.D.
Associate Professor
Department of Zoology
University of Washington
Seattle, Washington

Peta C. Bonham-Smith, Ph.D.
Department of Biological Sciences
University of Calgary
Calgary, Alberta, Canada

Don P. Bourque, Ph.D.
Associate Professor
Department of Biochemistry
University of Arizona
Tucson, Arizona

Gordon Cannon, Ph.D.
Assistant Professor
Department of Chemistry
University of Southern Mississippi
Hattiesburg, Mississippi

Robert Ferl, Ph.D.
Associate Professor
Institute for Food and Agricultural
  Sciences
University of Florida
Gainesville, Florida

Sabine Heinhorst, Ph.D.
Research Associate Professor
Department of Chemistry
University of Southern Mississippi
Hattiesburg, Mississippi

Pamela J. Hines, Ph.D.
Department of Medical Genetics
University of Washington
Seattle, Washington

Milan Alexander Jamrich, Dr.Rer.Nat.
Senior Staff Fellow
Laboratory of Molecular Genetics
National Institute of Child Health and
  Human Development
National Institutes of Health
Bethesda, Maryland

Anna-Lisa Paul, M.S.
Research Assistant
Department of Botany
University of Florida
Gainesville, Florida

S. J. Peloquin, Ph.D.
Campbell Bascom Professor
Departments of Horticulture and Genetics
University of Wisconsin
Madison, Wisconsin

Michael S. Risley, Ph.D.
Associate Professor
Department of Biological Sciences
Fordham University
Bronx, New York

Herbert Stern, Ph.D.
Professor
Department of Biology
University of California at San Diego
La Jolla, California

Arthur Weissbach, Ph.D.
Associate Director
Roche Institute of Molecular Biology
Nutley, New Jersery

Joanna E. Werner, Ph.D.
Department of Horticulture
University of Wisconsin
Madison, Wisconsin

Georgia L. Yerk, Ph.D.
Department of Horticulture
University of Wisconsin
Madison, Wisconsin

Virginia A. Zakian, Ph.D.
Department of Basic Sciences
Fred Hutchinson Cancer Research Center
Seattle, Washington

TABLE OF CONTENTS

Volume 2

SECTION I. EUKARYOTIC CHROMOSOMES: MEIOSIS

SECTION II. EUKARYOTIC CHROMOSOMES: SPECIAL, PLANT, AND ORGANELLE

Section I. Eukaryotic Chromosomes: Meiosis

INTRODUCTION

The process of meiosis and the production of oocytes and sperm involve major changes in chromosome organization within specialized cells. The reduction of the diploid number of chromosomes to the haploid number includes divisions of meiosis which have stages similar to those of mitosis (prophase, metaphase, anaphase, telophase). But the packaging of DNA in sperm and oocytes has unique features. This is because gametes are not simply cells with half the normal number of chromosomes, but are structures designed to effectively bring about union of parental chromosomes during fertilization. Their proper functioning is essential to successfully initiate development of the embryo. Meiosis, oogenesis, and DNA packing in sperm are clearly important topics and are therefore reviewed in this section.

A fundamental challenge for studies of meiosis is to explain the process in terms of molecular cell biology. Light and electron microscopy have been applied to the problem for many years and have provided useful and detailed descriptions of the cell biology of the phenomenon. But understanding the molecular interactions which underlie the visible manifestations of meiosis is crucially important and much remains to be learned. The meiotic behavior of cells requires a developmental switch from the mitotic behavior of normal cell division, a switch controlled by a precise pattern of molecular regulation. Certain mechanisms, particularly those involving chromosome disjunction, are common to both mitosis and meiosis, although with modifications. Other events, namely the pairing of homologous chromosomes and crossing-over or recombination, give meiosis its distinctive features. Clarifying the molecular mechanisms and regulatory features of these events constitutes a central goal of chromosome research.

The changes in genome organization which accompany meiosis are coordinated with cellular changes to bring about fertilization and embryogenesis. In oogenesis and spermatogenesis, not only do chromosomes enter a developmental pathway, but the cells differentiate to produce the structures of oocytes and sperm which unite to form a diploid zygote. The entire process from gamete production to growth of the multicellular embryo involves a sequence of mitotic, modified mitotic, and meiotic cellular divisions. In addition, oocytes and sperm are distinguished by unique biochemical compositions, besides their special morphologies. Oocytes are characterized by pools of RNA and protein molecules and yolk materials, all needed for early development of the embryo. Sperm of many animal species undergo a major alteration of chromosome structure with the replacement of the histone proteins by the protamines. Meiosis is a subject that calls for attention in this book, and the following chapters are therefore devoted to a discussion of the topic.

Chapter 1

MEIOSIS

Herbert Stern

TABLE OF CONTENTS

I. INTRODUCTION

The biological role of meiosis is simple enough. It is the exclusive gateway to sexual reproduction through which the chromosome number per cell is halved and recombination occurs. Recombination between chromosome sets is virtually universal; recombination by chromosome exchange is nearly so. Beyond the gateway, the pathways leading to gamete development are diverse but all culminate in a cell-cell fusion, thereby restoring diploidy. In strong contrast to the simplicity of its biological role, the mechanisms underlying the process are complex enough to pose a challenge that has eluded the research grasps of molecular and cellular biology. Whereas the structural aspects of meiosis, as analyzed by light and electron microscopy, are well studied and soundly based,[1,2] the molecular events that are characteristic of and essential to the process are barely understood. The intellectual challenge offered by the phenomenon of meiosis is readily perceived even on cursory examination; the thin response to that challenge is readily accounted for by the paucity of materials suitable for biochemical and molecular analysis. Possibly, the molecular-genetic studies of meiosis in *Saccharomyces cerevisiae* that are being increasingly pursued[3] may alter this picture in the near future. At present, however, the picture is nearly one-sided, the thin side being provided by the very limited studies of meiosis in lilies, mice, and, as already mentioned, yeast.

II. THE ISSUES

Textbooks on general and cell biology almost always couple meiosis and mitosis in a single treatment. There are indeed fundamental features that are common to both. Chromosome reproduction and disjunction are the centerpiece of the two processes which, undoubtedly, are evolutionarily linked. But the single treatment frequently conveys the impression that they differ only in details of chromosome behavior. The profundity of meiosis thus tends to be overlooked. Four issues need to be addressed in analyzing the meiotic process but only one of these truly represents a modification of mitosis. The first major issue concerns the mechanism involved in switching chromosome behavior from the mitotic to meiotic pattern. The switch occurs in cells that are the cloned end-products of a series of mitoses. These mitotically dividing precursors may be set aside very early in development, as occurs in most animals, or close to maturation as occurs in higher plants. It is a switch that applies to one cell generation only; none of the progeny from a meiotic division can repeat the process unless a fertilization intervenes. The factors that regulate meiocyte behavior persist no longer than the duration of meiosis. They are not transmitted except as a genetic potential. In many cases, the pattern of meiocyte metabolism is partly determined by the anticipation of postmeiotic events, but that component of the pattern will not be considered here. It comprises a group of diverse differentiative processes.

The second distinctive issue that is central to analysis of meiosis concerns the pairing of homologous chromosomes. Meiotic pairing is a transient event, the pairs being rendered stable for part of meiotic prophase. In higher eukaryotes the process of pairing (zygotene) may extend for several days. The paired state (pachytene), which may also extend for several days in higher eukaryotes, is associated with the third issue, crossing-over. The issue of crossing-over, or recombination, is not confined to molecular mechanisms even though these mechanisms are a critical part of the meiotic process. As will be seen, the regulatory features of meiotic recombination are more puzzling and probably more complex than the molecular mechanisms by which such recombination is achieved. The fourth issue provides for a direct linkage to mitosis—chromosome disjunction. Centromere behavior is the key factor in governing chromosome disjunction at the two anaphase stages. Although that behavior is a prominent modification of the mitotic one, it is not the sole chromosomal factor essential to meiotic disjunction. In addition to spindle body formation, stability of homologous pairing, crossing-over, and sister-chromatid adhesion are equally important items.

III. THE PREMEIOTIC PERIOD: COMMITMENT TO MEIOSIS

The conditions under which cells become committed to a meiotic division are almost as varied as the life cycles of different species. For *S. cerevisiae* the transfer of diploid cells to a medium deficient in ammonia (or a ready source such as glutamine) is sufficient to initiate an altered developmental pathway that leads to meiosis and sporulation. For flowering plants, length of day may be critical in generating floral primordia within which a germline is differentiated and from which meiocytes (often referred to as "microsporocytes" for the male and "megasporocytes" for the female) are ultimately formed. Mammals, like most other animals, differentiate a germline early in embryogenesis, a line that is destined to give rise ultimately to spermatocytes or oocytes which, following meiosis, develop into gametes, sperm or egg. In mammals, however, analysis of the mechanism of meiotic initiation appears to be complicated by the difference in germline development between females and males. In females, the entire meiotic prophase occurs in the unborn fetus; in humans this occurs between the 4th and 6th months of pregnancy. In the case of males, pubertal development is required to set the stage for meiosis. Despite these differences in developmental settings, which undoubtedly involve cumulative intracellular changes prior to meiotic initiation, there appears to be a common feature to the process of commitment. This generalization is based on observations of yeast[4] and of lily.[5] In both cases, an irreversible commitment to meiosis occurs in cells closely after completion of the premeiotic S-phase. Prior to that period, cells can be induced to revert to mitosis, but whether changes relevant to meiosis occur during the precommitment interval is unclear. What is clear is that presumptive meiotic cells can be induced to revert to mitosis only during the short interval before meiosis is initiated.

A. PROPERTIES OF THE PREMEIOTIC S-PHASE

The S-phase preceding meiosis is longer than that in mitotic cycles among a broad range of species. It is about five times as long in meiocytes of *Triturus,*[6] 2 to 3 times as long in

S. cerevisiae,[7] and about 5 times as long in *Lilium*.[8] The function of such attenuation is unclear but is believed to be due to the rate at which replicons are activated. The sharp change in functional properties of premeiotic cells at the time of meiotic commitment is illustrated by experiments on blue light inhibition of mitosis in *Trillium*.[9] Brief treatment with blue light of presumptive meiotic cells induced to revert to mitosis inhibits the process, but the same treatment of cells just beyond the point of reversion has no inhibitory effect on the course of meiosis. Clearly, there is a deep-seated change in the metabolic pattern of these cells that occurs over the very brief interval of time on either side of premeiotic S-phase termination.

A highly important and distinguishing characteristic of premeiotic S-phase is the suppressed replication of a group of DNA sequences which, for reasons that will later become apparent, is referred to as zygotene DNA (zygDNA). The phenomenon has been directly demonstrated in lilies[10] and indirectly in mice.[11] It would thus appear that cells entering meiosis do so with an incompletely replicated genome. In lilies the unreplicated portion constitutes about 0.2% of the genome. If premeiotic cells are induced to revert to a mitotic division, zygDNA is replicated before mitosis occurs. The ameiotic mutation described in *Zea mays* may be related to zygDNA behavior. Presumptive meiocytes homozygous for the mutation undergo mitosis following what would otherwise be a premeiotic S-phase and then degenerate.[12] The significance of zygDNA sequences to meiosis will be considered in relation to chromosome synapsis. From the standpoint of meiotic commitment, the mechanism responsible for suppression of zygDNA replication is a critical factor.

B. SUPPRESSION OF ZygDNA REPLICATION
1. The L-Protein

A protein, referred to as "L-protein", has been identified as the agent responsible for suppression of zygDNA replication.[13] It is mainly, if not entirely, localized in the nuclear membrane. It is not extractable with non-ionic detergents but is extractable with deoxycholate. The extraction procedure has no significant effect on activity. The fidelity with which nuclei isolated from lily microsporocytes mimic the *in vivo* replication pattern has made the finding possible. Isolated premeiotic nuclei from S-phase cells can, under *in vitro* conditions, replicate their DNA except for zygDNA sequences. G2 or leptotene nuclei synthesize negligible levels of DNA. The same result is obtained if the nuclei are treated with non-ionic detergent prior to incubation. Treatment of these same nuclei with deoxycholate, however, permits zygDNA replication, while addition of the protein to such treated nuclei restores the suppression. No significant inhibitory action by the L-protein on the remainder of the genome has been demonstrable. The behavior of the protein could be related to the irreversible commitment of the cells to meiosis. L-protein is not detectable in the cells during premeiotic S-phase nor is it present during the interval when the cells can be reverted to mitosis. It appears abruptly in premeiotic cells when they become irreversibly commited to meiosis. This coincidence in time points to the probability that L-protein is a factor responsible for irreversible commitment. The timing of L-protein formation invites another significant speculation. Its absence during all or nearly all of the premeiotic S-phase suggests that zygDNA does not normally replicate in mitotic cycles until the termination of S-phase, possibly during the G2 period. If this is so, it may be supposed that, in mitotic cells, replication of zygDNA sequences signals the completion of sister chromatid formation. ZygDNA would thus appear to have a distinctive and general structural function in chromosomes. If so, zygDNA should have similar but not identical roles in both mitosis and meiosis. The speculation begs experimental testing.

2. L-Protein: A Factor in Meiotic Commitment

The probable association of zygDNA replication with chromosome pairing (to be discussed in the next section) makes possible a reasonable scheme of regulation. Although direct evidence is lacking, it is at least consistent with a few key observations. The scheme is simply one in which the suppression of zygDNA replication effectively halts the completion of chromosome replication and, thus, normal mitotic progression. Coordinated with this suppression in premeiotic cells is the activation of a mechanism that makes possible the synthesis of certain meiosis-specific components essential to the pairing structure ("synaptonemal complex"). This scheme is reasonable if it is assumed that the initiation of pairing requires the presence of unreplicated zygDNA. It is therefore relevant that perturbations of premeiotic cells during the interval between S-phase and leptotene cause abnormalities in the pairing process and in chiasma formation. This has been demonstrated in lilies by explanting the meiocytes at different developmental stages,[14] in *Ornithogalum* pollen mother cells by cold treatment,[15] and in *Chlamydomonas* by treatment of cells with phenethyl alcohol.[16] These effects are most probably related to events that are essential to the components of synapsis rather than synapsis itself; in lilies at least, where stage of treatment is unambiguous, the effects are only observed in premeiotic treatments. One long-held view that initiation of homolog alignment occurs during premeiosis is not sustained by the evidence accumulated.[1,12] Although a very precise assignment of developmental stage is unlikely in these different studies, it is clear that the interval between the premeiotic S-phase and the entry into meiosis encompasses processes that are essential to chromosome synapsis and crossing-over. Such evidence is consistent with the speculative model that a key to the irreversible commitment of cells to meiosis lies in the suppression of premeiotic zygDNA replication by L-protein. Obviously, that suppression must be accompanied by the activation of other essential processes.

IV. CHROMOSOME SYNAPSIS: STRUCTURAL ASPECTS

A. SOME GENERAL COMMENTS

In the normal course of meiosis, homologous chromosomes somehow find each other by the start of zygotene and begin the extended process of synapsis, a process that requires 2 to 3 days in organisms as diverse as lily and mouse. When observed through the light microscope, homologs appear to be perfectly aligned at the termination of the pairing process. Cytogenetic evidence points to an effectively precise juxtapositioning of corresponding genes within the chromosome pairs. With some notable exceptions, translocations or inversions in one member of a pair are accommodated by appropriate structural configurations that align homologous regions. Diagrammatic representation of the situation is straightforward; the mechanism by which this occurs raises difficult questions. Precise alignment of homologous genes implies correspondingly precise DNA sequence matching. It is clear, however, that pairing is initiated when the chromosomes are already in a partially compacted state. In *Zea mays* the length of the paired chromosomes is 0.015% of the DNA length and in humans it is 0.01%.[17] Most of the compacted chromatin in each chromosome is distant from its homologous partner and only a small fraction of the chromosome can be involved in homologous alignment. Electron microscope (EM) studies reinforce this conclusion. This aspect of alignment is illustrated by the electron micrograph in Figure 1. If, indeed, the precision of synapsis depends upon DNA-DNA matching, then only a correspondingly small fraction of the DNA can be involved in the matching process. Although chromosomes do alter length between the initiation and completion of pairing, the extent of shortening is trivial when compared with the percentages of shortening just listed. Chromosome pairing needs to be recognized as a highly complex event that is most probably affected by a large number of factors. What follows will touch upon some of these factors but it will not adequately account for the nature of the synaptic process.

B. THE LEPTOTENE STAGE

This first recognized stage of meiosis, which begins after a relatively short G2 interval,

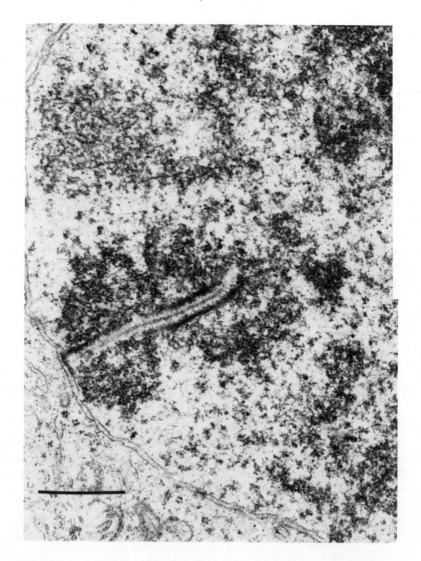

FIGURE 1.   A transmission EM photograph of a section of a pachytene spermatocyte. The major components of the SC are clearly visible: lateral elements, central region, and central element. The bulk of the chromatin lies outside the complex and illustrates the point made in the text that only a very small fraction of the DNA is involved in homologous alignment. The association of telomeres with the nuclear membrane, a common but not universal phenomenon, is also discussed in the text. (Magnification: bar = 1 μm.) (From Chandley, A. C., Hotta, Y., and Stern, H., *Chromosoma*, 62, 243, 1977. With permission.)

is the stage at which a uniquely meiotic modification of chromosome structure occurs in anticipation of synapsis. A ribbon-like, possibly tubular, structure measuring about 35 nm in width forms along the presumed border between the two sister chromatids. The structure is referred to as the "lateral element" (LE) or "lateral component" (LC), and it extends the full length of the partially compacted leptotene chromosome (Figure 2). At this stage, but not at later ones, it is also called the "axial element" because it would appear to delineate the axis of the chromosome as organized for the pairing process. Unlike mitotic chromosomes at the prophase stage, the doubleness resulting from prior replication is not structurally evident. The singleness may well reflect the suppression of zygDNA replication which would prevent the separation of sister chromatids. Alternatively, the very close appositioning of

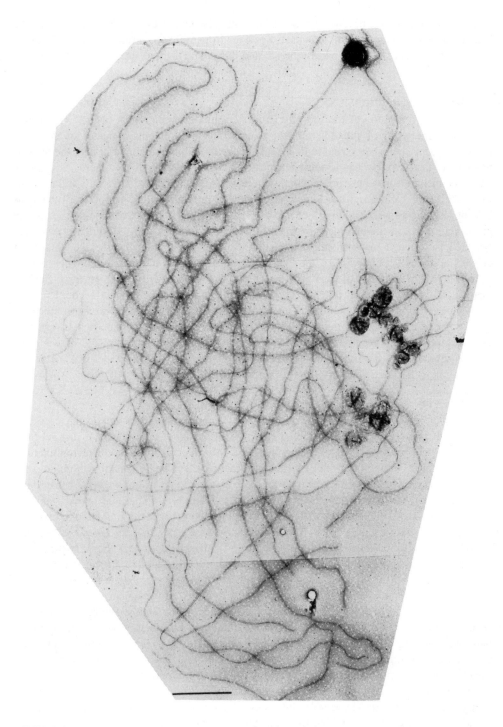

FIGURE 2.   A chromosomal spread of an entire pachytene nucleus from a lily. The lateral elements (LEs), each of which is associated with 2 sister chromatids, are selectively stained by this technique. Most of the chromatin, which is barely visible, is subtended from the lateral elements. The central region lying between the LEs is also unstained and is best seen in Figure 1. The cluster of circles remains unexplained. This spread was prepared and photographed by Dr. Clare Hasenkampf who is currently in the author's laboratory. The preparation represents a distinguished victory over a formidable technical challenge inasmuch as the total DNA content of this nucleus is 12×10^{10} bp. (From Hasenkampf, C., *J. Heredity*, 80, 197, 1989. With permission.)

sister chromatids could be due to protein binding. The nature of the binding between the LE and chromatin is unknown. It is clear however from still early immunochemical studies that specific polypeptides are part of the LE and that some of these are unique to the pairing structure.[18,19]

C. THE PAIRING PROCESS

Following formation of the LEs, the process of homolog alignment begins. The occurrence of some form of prealignment, if it occurs at all, remains a matter of debate.[20] Commonly, telomeres are found to be loosely concentrated in a region at or near the nuclear membrane and in many cases pairing appears to be initiated at the telomeres.[1] Precisely what role the nuclear membrane might play in facilitating homolog alignment is still a matter for speculation. Colchicine has long been known to interfere with the process of synapsis but not with the paired structure once it is formed.[21] The action of colchicine on the pairing process suggests an involvement of microtubules in alignment but no evidence has been obtained for such a mechanism. A role of the nuclear membrane in the mechanism of colchicine action is pointed to by the evidence that there is a colchicine-binding component in the nuclear membranes of meiocytes[22] and also that there is colchicine-sensitive fibrillar material in prezygotene nuclei.[23] Whether or not these two components account for the colchicine effect, there is no doubt that initiation of pairing may occur in several chromosomal regions and not purely at the membrane-associated telomeres. As soon as particular segments of the chromosomes become aligned, a synaptonemal complex is completed between the aligned segments by the formation of a central region, about 100 nm wide, that includes a "central element" (CE), about 30 nm in diameter.

D. ROLE OF THE SYNAPTONEMAL COMPLEX (SC)

One generalization about the role of the SC can be made; meiocytes that display crossing-over also display SCs between the paired chromosomes (Figure 1). That role is distinct from the one assigned to synapsis between segments of DNA molecules undergoing recombination, a process that is common to both meiotic and somatic cells. The SC positions homologs for recombination but it does not appear to be a direct participant in the process. It has other functions, a few of which have been defined. The SC does stabilize the paired state and thus makes possible the reductional disjunction of chromosomes at the first meiotic division. Fulfillment of that role, however, requires the occurrence of at least one crossover between members of a pair. When prophase ends, the SCs disappear so that the only structures holding the pairs intact is the crossover. Unpaired homologs lack the critical structure and thus lack a mechanism for positioning themselves at the metaphase plate in proper alignment for reductional disjunction. Such disjunction also requires a mechanism to maintain sister chromatid adhesion following dispersal of the complex at the end of prophase.[24] Without sister chromatid adhesion a crossover could not maintain the integrity of the chromosome pair. Inasmuch as SC formation is a precondition for crossing-over and possibly provides for sister chromatid adhesion, it can be stated that the SC plays a critical role in regulating meiotic chromosome disjunction, an indispensable requirement for sexual reproduction. There are situations in which reductional disjunction occurs without crossing-over. The female silkworm is a good example. Its oocytes have dispensed with the process of recombination, and instead, modify but maintain the SCs until the chromosomes are secured to the metaphase I spindle.[25]

A central question about the role of the SC, its relation to chromosome pairing, remains unanswered. It is known that morphologically normal SCs can bridge nonhomologous regions.[1] It has also been observed that paired heterologous regions spanned by an SC may be replaced in favor of homologous pairing.[17] The conclusion may be drawn that the SC is an unlikely source of factors that sense DNA sequences and effect homologous alignments.

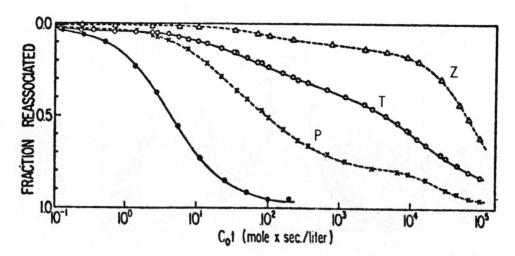

FIGURE 3.   Cot curves of DNA components that are active in meiosis. The solid curve with solid circles is a reference curve of *E. coli* DNA. The other curves are for total lily nuclear DNA (T), zygDNA (Z), and pachytene DNA (P). Note that the zygDNA sample consists of few repeats, whereas the PDNA curve consists of many moderate repeats. These curves represent the first set of fractions analyzed. (From Hotta, Y. and Stern, H., *The Eukaryote Chromosome*, Peacock, W. J. and Brock, R. D., Eds., The Australian University Press, Canberra, 1975, 281. With permission.)

It seems more likely that the juxtapositioning of homologous regions occurs by direct DNA-DNA interaction and, as discussed below, it is likely that specific regions of chromosomal DNA function in the alignment process. Such a conjecture, however, invites other questions. There is no evidence for the presence of DNA filaments that span the complex, although it is conceivable that methods of detection are not sufficiently sensitive to demonstrate the presence of what may be very few filaments crossing the complex. Whether or not the segments of what may be termed "synaptic DNA" persist or withdraw following alignment, a mechanism is required to coordinate the alignment of DNA sequences with completion of complex formation. It may be that coordination alone is sufficient to ensure homologous pairing in the presence of lateral elements that line the pairing regions. The nature of the binding between the LEs and chromatin or DNA is highly relevant to this issue but, as yet, we have no information about its nature.

V. METABOLIC EVENTS CORRELATED WITH SYNAPSIS

A. REPLICATION OF ZygDNA
A distinctive feature of the meiotic process is the semiconservative replication during zygotene of the chromosomal DNA fraction that was suppressed in its replication during the premeiotic S-phase.[10] That event was discussed earlier in connection with premeiotic DNA replication. The DNA in question is referred to as zygDNA (zygotene DNA) because its replication occurs during the zygotene interval and is thus correlated with the pairing process. The phenomenon has been directly demonstrated in only two genera of liliaceous plants, *Trillium* and *Lilium*. Indirect evidence, discussed below, points to a similar phenomenon in mouse spermatocytes. This evidence is reinforced by the radioautographic studies of Moses et al.[26] who demonstrated DNA replication at the zygotene stage by pulse-labeling mouse spermatocytes and oocytes and analyzing the grain distribution in nuclear spreads. The plant materials permit a more extensive molecular analysis because the meiocytes develop more or less synchronously and can be cultured through most of meiosis *in vitro*. The principal pieces of experimental evidence that reveal the properties to be described are diagrammed in Figures 3 and 4.

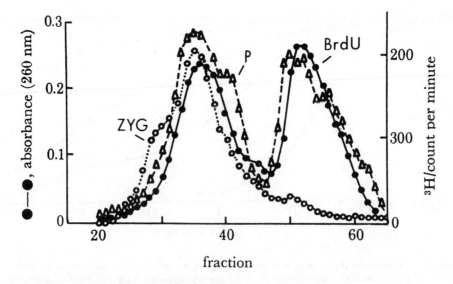

FIGURE 4. Evidence for semi-conservative replication of zygDNA (zyg) and repair-replication of pachytene DNA (P) during meiotic prophase of lily microsporocytes. The meiocytes were exposed to BrdU during the premeiotic S-phase (BrdU). Samples of the cells thus treated were pulse labeled either during zygotene (zyg) or during pachytene (P). Culture was continued until the termination of meiosis and DNA prepared from each group of cells. The DNA samples were then centrifuged to equilibrium in alkaline CsCl gradients. The profile of the S-phase labeled cells displays two peaks. The one at the left is in the heavy end of the gradient; it contains the newly synthesized DNA strands that have incorporated the brominated uridine. The peak to the right contains the old strands. The labeled zygDNA profile is entirely in the heavy peak because it is ligated to the newly synthesized strands, as is to be expected because it is completing its replication. The labeled PDNA is present in both strands because it is a product of repair-replication. (From Stern, H. and Hotta, Y., *Phil. Trans. R. Soc. London Ser. B.*, 277, 277, 1977.)

DNA prepared from cells pulse-labeled at zygotene and centrifuged to equilibrium in a CsCl gradient displays a satellite-like profile and thus is quite different from that of total nuclear DNA. ZygDNA has none of the properties of a typical satellite. Unlike the common examples of late-replicating DNA, it consists of no highly repeated sequences but rather of low copy number sequences as is evident from its Cot profile (Figure 3). Its satellite behavior is due to yet another distinctive meiotic property. Newly replicated zygDNA strands remain unligated to their flanking sequences until after pachytene, at about the time when the first meiotic division occurs.[27] The discontinuities render the ends of the zygDNA segments susceptible to mechanical shearing during DNA purification. Treatment of isolated pachytene nuclei with S1 micrococcal nuclease is also sufficient to release the zygDNA sequences. Zonal sedimentation of sequences thus released behave like fragments that are 5 to 10 kbp in length.[28] The evidence that the pulse-labelled zygDNA is a product of delayed semicon-servative replication is provided by two different sets of analyses. Labeled zygDNA se-quences hybridized to nuclear DNA prepared from successive developmental stages display a profile that indicates the saturation level at pachytene to be about twice that at leptotene, the transition occurring during zygotene. It is also possible to demonstrate that the zygDNA strands synthesized are part of those strands of total DNA that were synthesized during premeiotic S-phase. The graph in Figure 4 was obtained by labeling meiocytes with BrdU during the premeiotic S-phase and incubating the cells on reaching zygotene with tritiated deoxyadenosine. Following the zygotene labeling, the cells were allowed to develop until close to the end of meiosis, and the DNA was then extracted. When the DNA from such a preparation is centrifuged in an alkaline CsCl gradient, the UV absorbance profile displays one light and one heavy peak, the latter being heavy because of the BrdU incorporated.

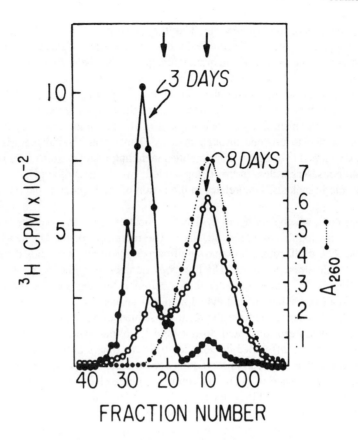

FIGURE 5. Isopycnic profiles of zygDNA labelled with BrdU and [³H]deoxyadenosine and centrifuged to equilibrium in an alkaline CsCl gradient. The 3-day profile is of cells that were harvested immediately after the labeling period. The 8-day profile is of cells harvested near the end of meiosis. The 3-day peak is slightly to the left of where that of unsubstituted zygDNA would be (leftmost arrow). Almost all of the 8-day profile tracks the optical density profile of total nuclear DNA. There is no significant displacement because the total BrdU incorporated by the zygDNA which is now ligated to the bulk of nuclear DNA is too small to cause an appreciable change in buoyant density. The peak of unsubstituted total nuclear DNA is indicated by the second arrow. (From Hotta, Y. and Stern, H., *Chromosoma*, 55, 171, 1976. With permission.)

Traced against the UV absorbance profile, the zygotene label may be seen to be confined to the heavy peak, thus indicating that it is exclusively associated with strands synthesized during premeiotic S-phase. The eventual ligation of the newly synthesized zygDNA strands to their neighboring strands can be demonstrated by pulse-labeling zygotene cells with BrdU and [³H]deoxyadenosine and, as in the previous experiment, harvesting the cells toward the end of meiosis. If the extracted DNA is centrifuged to equilibrium in a CsCl gradient, the zygDNA no longer behaves as a satellite but tracks the optical density profile of total nuclear DNA, thereby demonstrating its ligation to the rest of the genome (Figure 5). Relevant to the role of zygDNA is the demonstration by *in situ* hybridization of pachytene cells, using purified zygDNA as a probe, that the sequences are distributed more or less uniformly along pachytene chromosomes.[31]

B. ACTIVITIES OF ZygDNA DURING THE PAIRING INTERVAL

Relevant to the regulation of the meiotic process is the fact that zygDNA comprises a group of sequences of high complexity distributed as segments, several kbp in length, among all the chromosomes of the lily genome. That they undergo replication in coordination with chromosome pairing suggests their role in the process; their delayed ligation might be coordinated with some aspect of meiotic disjunction. It has been found that administration of physiologically high levels of deoxyadenosine that reversibly inhibit zygDNA synthesis inhibit the process of pairing, particularly at the time of pairing initiation.[29] For reasons not yet understood, treatment with hydroxyurea does not elicit the same effect, but the cytological studies were unaccompanied by a measurement of the inhibition. Cells treated during G2 and leptotene failed to proceed through meiosis. Those that entered meiosis did undergo pairing but developed cytological abnormalities including meiotic arrest.[30] Pairing may have been heterologous. Regardless of this complication, two features of zygDNA behavior point to a special role in meiosis. One relates to the action of the L-protein, the other to zygDNA transcription.

The L-protein, which suppresses zygDNA replication, also, as would be expected, binds specifically to ds(double-stranded)zygDNA.[13] The remainder of the DNA, which accounts for more than 99% of the total, shows no significant competition for binding if the competition is between double-stranded forms. The low copy number of zygDNA, as deduced from its Cot curve (Figure 3), makes it likely that a common binding site for the protein would represent a small percentage of individual zygDNA sequences. Whether or not the binding sites in all zygDNA segments have the identical sequence, or nearly so, most of the cloned fragments that have been generated do not bind L-protein. The cloned fragments that do bind display a striking difference in behavior from the others in the presence of L-protein. No kind of interaction has been observed between the nonbinding fragments and L-protein in the presence or absence of ATP. By contrast, supercoiled plasmids that bind L-protein undergo relaxation when ATP is added to the mixture. The relaxation is not a result of topoisomerase activity. When the relaxed plasmids are denatured and the products resolved by gel electrophoresis, only two new bands are observed, one in the position of a ss(single-stranded)DNA circle and the other in that of a linear ssDNA strand. The length of each of the strands is equal to that of the original duplex. Thus, L-protein behaves like a nickase on binding to zygDNA in the presence of ATP. It is easy to visualize the provision of single-strand tails as a mechanism whereby zygDNA can form heteroduplexes from the zygDNA tails of the juxtaposed homologs in order to align them. There is, of course, no evidence that this is so and the nicking activity may partly or entirely relate to transcriptional activities of zygDNA segments. Nevertheless, the model is especially attractive in the absence of any other molecular evidence on the chromosomal synaptic process. A diagram of this speculative model is shown in Figure 6. Because we do not know the precise timing of transcription and pairing (except that they are approximately simultaneous), the model assumes that heteroduplex formation precedes replication and that only one of the sister chromatids is directly involved in the process. The model can be made even more appealing by further assuming that zygDNA segments function as chromosomal axes, such axes being essential from the standpoint of the structural organization of chromosomes at meiotic prophase as revealed by electron microscopy.

The transcription of some zygDNA sequences is not an expected requirement of the pairing process. It was observed because the low copy number of zygDNA made it seem likely that it would be transcribed, especially so in a genome whose haploid value is of the order of 3×10^{10} bp. Using mixed zygDNA sequences as a probe, no hybridizations were detected in poly(A+)RNA preparations from rapidly-dividing somatic tissues. The only positive results were obtained with such preparations made from meiocytes primarily at the zygotene stage. The transcripts appear in very early zygotene, perhaps even late leptotene, and decline very quickly at the end of zygotene.[11] At their peak level, and only then, the "zygRNA" sequences constitute about 40% of the total poly(A+)RNA pool. They may turn over more rapidly than any other group of meiotic transcripts, but this is purely con-

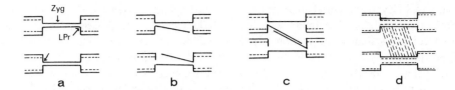

FIGURE 6.   Diagram to illustrate how zygDNA and L-protein might be involved in aligning chromosomes for homologous synapsis. Only one segment of zygDNA flanked by the ends of adjoining S-phase replicated DNA is shown. The replicated regions account for about 99.8% of the nuclear DNA. The G2-leptotene stages are drawn in (a). The location of L-protein (Lpr) is indicated for the two sites the nickase is imagined to act prior to initiation of replication. This results in formation of single-strand tails from only one of the two sister chromatids in each chromosome (b). The segments become aligned by heteroduplex formation between the homologous zygDNA segments (c). No model is prosposed for the events that follow alignment except for the ultimate outcome which includes the near-completion of replication (the short vertical lines remain single) and the stabilization of the alignment by completion of SC formation (d). The SC is represented by the diagonal lines.

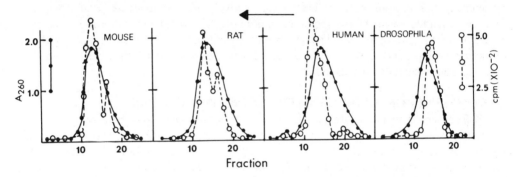

FIGURE 7.   The conservation of *Lilium* zygRNA sequences across the phylogenetic spectrum: an illustration. Each of the DNA samples was centrifuged to equilibrium in a CsCl gradient and individual fractions tested for hybridization with labelled lily zygRNA under moderately stringent conditions. The radioactivities do not track the optical density profiles except for murine DNA. The relatively high GC content of lily zygDNA is clearly not a general characteristic of the transcribed zygDNA sequences. (From Stern, H. and Hotta, Y., *Aneuploidy: Etiology and Mechanisms*, Dellarco, V., Voytek, P. E., and Hollaender, A., Eds., 1985, 305.)

jecture. Because the lifespan of individuals in this presumably complex array of transcripts is unknown, the proportion of zygDNA sequences that are transcribed is uncertain. It is conceivable that transcripts of any particular sequence are quickly destroyed so that there is a rapid succession of zygRNA pools through zygotene. It is also conceivable that only a select subset of zygDNA sequences is transcribed. Either possibility is consistent with the observation that at saturation levels only about 1 to 2% of zygDNA sequences are hybridized. Mixtures of early and late zygRNA extracts did not display saturation levels above 2%; selective transcription is the likelier possibility, one that is also more reasonable from the standpoint of zygRNA sequence conservation as discussed below.

By constructing zygDNA-Sepharose columns it is relatively easy to check other meiotic systems for the presence of zygRNA if sequences are evolutionarily conserved. Conservation seems probable because of their meiotic specificity. General evidence for the conservation of zygRNA is demonstrated by its hybridization to DNA from a variety of eukaryotic species. This is illustrated in Figure 7 by the hybridization of zygRNA to total nuclear DNA from mouse, rat, human, and *Drosophila* under moderately stringent conditions. The DNA samples were centrifuged in CsCl gradients. The profiles of hybridization make it evident that the transcribed zygDNA sequences from the different species do not have a uniform GC composition. A more detailed analysis of zygRNA conservation has been made with mouse spermatocytes. Zygotene-enriched fractions of mouse spermatocytes are readily obtained and from these poly(A+) extracts were made. When these were passed through a lily zygDNA column a significant fraction of poly(A+)RNA was retained. Only extracts from prophase spermatocytes behaved in this way. The highest levels obtained were about 5% of the total poly(A+)RNA pool,[11] a value significantly lower than what was obtained with lily. This is to be expected; meiotic stage homogeneity is very much greater in lily than in mouse preparations. Moreover, transcriptional activities during meiosis related to postmeiotic events are nearly absent in lily, whereas transcription related to spermiogenesis is probably high in the mouse. Two other properties of mouse zygRNA provide additional evidence for a functional correspondence between mouse and lily zygRNA. Mouse zygRNA hybridized to DNA prepared from lily meiocytes at successive stages of meiosis display the same profile of saturation values as described for labeled zygDNA. The sequences to which mouse zygRNA hybridizes replicate during zygotene as does zygDNA. If mouse meiocytes have sequences similar to lily zygDNA, such sequences should fail to complete ligation when replicated during zygotene, and should be excisable from prophase spermatocytes by treating isolated nuclei with micrococcal nuclease. When DNA from nuclei thus treated is resolved by centrifugation in a glycerol gradient, a quantitatively minor component, well separated from the remainder of the DNA, should be present that hybridizes with either mouse or lily zygRNA. This is indeed the case. The observation not only attests to the presence of zygRNA transcripts during mouse meiosis but also to a fraction of mouse DNA that corresponds in sequence and behavior to lily zygDNA.

C. COMMENTS ON A MECHANISTIC MODEL OF SYNAPSIS

The conserved sequence organization of transcribed zygDNA and the correspondence in meiotic behavior between lily and mouse point strongly to a fundamental role of this group of sequences in meiosis. As already discussed, transcribed zygDNA sequences have also been found to be conserved in the DNA of *Drosophila*, rats, and humans.[31] The most likely meiotic role of zygDNA is one that is associated with synapsis. Allowing for the paucity of direct experimental evidence on zygDNA function, it remains attractive to propose that zygDNA is a chromosomal component that behaves like an axis and thus provides an array of low copy number sequences that serve to align chromosomes in precise fashion. If the distribution of zygDNA segments is constant with respect to the remainder of the genome, the apparently accurate and effective positioning of genes for recombination can be rationalized despite the extremely small fraction of DNA that is presumed to be involved in alignment on the basis of EM observations. As mentioned earlier, it would be conceptually gratifying if zygDNA could be shown to function as a general signal for the completion of chromosome replication. If so, it could be asserted that whereas chromosome replication is completed in somatic cells before the prophase of mitosis, it is completed after the termination of prophase in meiotic cells. This assertion is vaguely reminiscent of an intuitive assertion made by C. D. Darlington[32] in his "Precocity Theory" of meiosis, a theory that gave the process conceptual coherence without the benefit of any existing evidence, direct or indirect, that cytology could provide at the time. The theory asserted that chromosome pairing preceded chromosome replication and that such precocity made possible crossing-over after the entry of cells into meiosis. The evidence now indicates that cells enter meiosis not prior to DNA replication but rather before completion of replication. The point that needs to be emphasized is the presence in all chromosomes of specific DNA sequences that function to serve the structural needs of the meiotic process. Those needs probably lie in the meiotic phenomenon of chromosome synapsis. Even so, any model of the synaptic process suffers from a major gap in our understanding of mechanism. How an alignment of chromosomes by DNA matching coordinates with the formation of a stabilizing synaptonemal complex is yet to be elucidated.

VI. FROM SYNAPSIS TO RECOMBINATION: THE PROBLEMS

A. GENERAL

The simplest description of the relation between pairing and recombination is that homologous chromosomes pair and, having been accurately paired, undergo crossing-over, one or more chiasmata being formed as result of the process. Yet, even this description brings with it some difficult questions. The distribution of chiasmata in a chromosome pair is not random. If two chiasmata are present in a chromosome arm their relative positions indicate the operation of a factor that prevents a random distribution. In some species the telomeric regions are favored as sites of crossing-over; in other cases more central regions are favored. Generally, crossing-over is extremely rare in constitutive heterochromatic regions which house highly repeated sequences. Consideration must therefore be given to properties of the chromosome that are responsible for preferred sites of crossing-over under conditions where synapsis is uniform along the chromosome length.

B. RELEVANCE OF CHROMOSOMAL SYNAPSIS TO MOLECULAR SYNAPSIS

As mentioned earlier, two types of synapsis occur in meiosis, one addressed to whole chromosome pairing and the other to the molecular sites of recombination. They are quite different if only because the proximity of paired components at DNA recombining sites is of molecular dimension, whereas that of paired chromosomes is resolvable by electron if not light microscopy. It would appear that chromosome synapsis facilitates but does not ensure molecular synapsis. As discussed earlier, apparently normal synaptonemal complexes may form between heterologous chromosome segments, but these are unaccompanied by crossing-over. Occasionally, the view has been expressed that recombination occurs during the S-phase and thus in advance of general synapsis. The link between the pairing of chromosomes and crossing-over is not self-evident even though it certainly occurs. The site of occurence is most probably in the "recombination nodule".[17] The nodule is a dense structure, spherical or ellipsoidal in shape, that is regularly associated with the synaptonemal complex.[33] The strong correlation between the number of nodules observed at pachytene and the number of chiasmata (or crossovers) leads to the conclusion, broadly accepted, that these structures represent the ultimate sites of exchange. The mechanism by which the sites are formed is unclear. The number of nodules found at zygotene is 1.5 to 2 times that found at pachytene.[17] The reduction may represent failures in the occurrence of crossovers in those nodules that disappear, but evidence for this conjecture is lacking. It is apparent, however, that crossovers are somehow directly coupled with recombination, and that they are also coupled with the synaptonemal complex which is a precondition for meiotic crossing-over.

C. MOLECULAR MECHANISMS: CONCERNS

The most intensive molecular genetic studies that are relevant to meiosis have been in the area of recombination. Although the phenomenon of recombination by itself receives major research attention, there have been many efforts to address the specific features of meiotic recombination.[3,34] The mechanisms themselves are not yet adequately resolved and they therefore furnish a limited but important approach to the meiotic process. The total number of different enzymes involved in recombination is large, many of which, but not all, operate in meiotic systems. This is probably best revealed in genetic studies of *Saccharomyces cerevisiae* in which a mutation affecting meiotic recombination may not have identical effects in meiosis.[3] Such comparisons are less common in the more complex eukaryotic organisms, particularly mammals, because the spontaneous occurrence of mitotic recombination is rare. What may be uniquely relevant to meiotic recombination is not the particulars of the enzyme pathways, of which there are at least several, but rather the regulatory factors that govern their activities. The organization of the various factors governing the various aspects of DNA metabolism that have thus far been identified is the substance of the discussion that follows. Most of the discussion has a narrow experimental base. *Lilium* remains the most fruitful experimental material supplemented by analyses of mouse spermatocytes.

VII. DNA METABOLISM FOLLOWING SYNAPSIS

A. DNA SYNTHESIS DURING PACHYTENE

The character of DNA synthesis changes sharply after meiocytes have completed zygotene and entered the pachytene stage.[10] Geneticists have long considered pachytene to be the interval of crossing-over. Repair replication rather than semiconservative replication is the pattern of DNA synthesis at this stage. Various tests made on DNA from pulse-labeled pachytene cells uniformly indicate repair activity. In one particularly informative set of experiments that has already been described in relation to zygDNA behavior, cells are exposed to BrdU during premeiotic S-phase so as to render the buoyant density of the newly synthesized strand heavy relative to the old complementary strand. If cells thus treated are pulse-labeled at pachytene, the label is found in both heavy and light strands when the DNA is centrifuged to equilibrium in a CsCl gradient. By contrast, zygDNA label is found, as expected, only in the heavy strands (Figure 4). A similar result for pachytene labeling has been obtained for mouse spermatocytes.[35] Unlike zygDNA, pachytene-labeled DNA (PDNA) hybridized to successive stages of meiosis displays no change in saturation levels. A distinctive feature of PDNA labeling is also seen with respect to satellite DNA regions. In neither mouse nor man are the satellites labeled during pachytene.[36] A more striking example of satellite DNA exclusion from pachytene labeling has been found in the Arabian oryx *(Oryx lencoryx)* in which close to 50% of the DNA is satellite.[37] This latter behavior is directly related to the genetic evidence that crossing-over in regions of constitutive heterochromatin is very rare. Even more direct evidence for a relation between PDNA repair-replication and crossing-over is the absence or near absence of such replication in situations where homologous pairing is absent or much reduced. This occurs in one of the infertile lily hybrids and in meiocytes treated with colchicine to suppress pairing.[38] In some way, homologous pairing regulates the occurrence of programmed repair-replication so that crossing-over is correspondingly reduced. The one point deserving emphasis is that the process of PDNA repair-replication is a major component of the meiotic program. Its programmed occurrence in lily and mouse indicates that it is central rather than accessory to the meiotic process. The additional features to be discussed will underline this opinion because of the relatively profound organizational changes that pachytene cells undergo in sustaining the program. The physiology of meiotic tissues provides significant pointers to relevant molecular mechanisms.

B. NATURE OF REPAIR-REPLICATION

DNA samples extracted under alkaline conditions from cells at successive stages of meiosis and subjected to zonal sedimentation in alkaline gradients display a characteristic set of profiles. At all meiotic stages except pachytene a single but broad sedimentation peak is observed having a mean value greater than 100S.[39] The sedimentation profile of DNA prepared from pachytene cells has two peaks, one at about 63S and the other at greater than 100S. If DNA extraction and sedimentation are carried out under neutral conditions, a single and broad peak with a modal value of about 250S is observed at all meiotic stages. The 63S component in the alkaline gradient accounts for about 50% of the total DNA. The conclusion drawn from these observations is that about half the nuclear DNA is subjected to nicking at intervals having a mean value of 160 kb (kilobases). By contrast, double-strand cuts are comparatively rare. There is thus a clear program for introducing single-strand discontinuities in chromosomal DNA following the completion of synapsis, but there is no such program with respect to double-strand discontinuities. Although the latter have been proposed as the mechanism of meiotic recombination,[40] the pattern of DNA behavior indicates otherwise. The principal argument in support of the double-strand model is the observation in yeast that rad52 mutants are incapable of healing double-strand breaks and that meiotic recombination fails in such mutants. Behavior of yeast meiocytes do not quite support that model;[3] single-strand interruptions do occur in coordination with meiotic recombination in yeast and the action of rad52 appears to be related to a nuclease activity. As would be expected, the consequences of deficient homologous synapsis for nicking are similar to those already described for repair replication; the 63S component that is normally present in ssDNA from pachytene cells is lacking in asynaptic cells.[38] Clearly, nicking, like repair-replication, does not occur in the absence of homologous pairing.

The nicking process is mediated by an endonuclease that begins to appear at zygotene and reaches a peak of activity during early pachytene.[41] Its properties differ from other nucleases present in somatic tissues and it is presumed to be a meiosis-specific endonuclease. In one respect the difference between the lily somatic and the meiotic endonucleases is of interest in relation to the discussion of double-strand breaks. Both have acidic optima, pH 5.6 for the somatic and 5.2 for the meiotic. The somatic enzyme produces double-strand breaks; the meiotic one produces only single-strand interruptions. Unlike the meiotic endonuclease, there is no sharp rise and fall in activity in cells adjacent to the meiocytes. Moreover, in one strain of lily the somatic enzyme is present in meiocytes but its activity does not display the profile characteristic of the meiotic component. In any case, the presence of a double-strand cutting enzyme is not essential to meiosis, whereas the nicking enzyme has been found in all varieties tested. One additional property of the meiotic enzyme indicates the probable complexity of controls that relate to the repair-replication process. The activity of the enzyme gives rise to 5'OH and 3'OH termini. A 3'OH terminus is required for DNA polymerase activity. That requirement can be satisfied by the acid phosphatase present in the meiocytes (see Figure 9 for the activity profile of the enzyme).

Most models of recombination that are based on experimental data require an interruption in one strand of the DNA duplex. The Holliday model and the Aviemore model of Meselson and Radding have been widely discussed.[42] It is therefore reasonable to propose that the programmed nicking which occurs during pachytene is designed to serve as potential initiation sites for recombination. This proposal is much strengthened by the fact, described above, that similar but not identical programs of nicking have been observed in mouse and yeast. In yeast, however, it would appear that the number of nicks is similar to the number of recombinations,[3] whereas in lily and mice that number is well in excess of the number of recombinations.[10] The enormous differences in chromosome size may account for the difference in ratios of nicks/recombinations.

C. CHARACTERISTICS OF NICKED CHROMOSOME SITES

An undetermined but high proportion of the nicks that are introduced by meiotic endonuclease are extended by an unidentified exonuclease to form gaps,[10] an event in line with the requirements of recombination models. The nicks, and thus the gaps, are not distributed randomly (Figure 3). Pulse-labeled PDNA, if denatured, reassociates with kinetics characteristic of moderately repeated sequences.[43,44] Roughly calculated, there are about 700 families of these sequences, each repeated about 1000 times. Moreover, in contrast with the divergence commonly encountered among moderately repeated sequences, the PDNA families exhibit extremely low levels of divergence. In effect, they are highly conserved sequences and they even display some sequence conservation across a broad spectrum of plant species.[45]

Taken together, the programmed timing of nicking, the regulated activation of a meiosis-specific endonuclease, and the selective introduction of nicks by the endonuclease into moderately repeated sequences add further emphasis to the view that the organization of meiotic cells for recombination is far more elaborate than are the requirements for affecting the immediate molecular events in recombination itself. This very brief summary has some obvious deficiencies. The principal one is the lack of any information on the structural relationship between the PDNA sequences and the genes that might undergo crossing-over. If the PDNA sequences define all potential crossover sites, every gene with the potential for being involved in a crossover should be contiguous with a PDNA segment. No such evidence has yet been obtained and, indeed, may be very difficult to obtain. In the studies of Moses et al.[26] on pachytene DNA labeling in spermatocytes and oocytes of mice, they found approximately 40% of the total label concentrated in the region of the synaptonemal complex. In view of the low concentration of DNA in the SC region, the evidence points to relatively intense repair-replication in that region and thus to the direct involvement of PDNA regions in crossing-over. Nevertheless, this statistical assessment points to a probability; it does not provide the information needed for an unambiguous description of the process.

VIII. ORGANIZATION OF PACHYTENE SITES

A. THE ISOLATION OF PDNA CHROMATIN

Serendipity made possible the isolation of chromatin fragments housing PDNA sequences.[46] During late zygotene and pachytene the chromatin regions housing PDNA are hypersensitive to DNase II. Moreover, the fragments released by the enzyme in the presence of Mg^{2+} remain in suspension like those containing transcribing DNA segments.[47] Two types of chromatin fragments from pachytene nuclei can be separated by treatment with ribonuclease: PDNA-containing fragments remain in suspension, while the chromatin fragments that contain transcribing DNA precipitate. The deproteinized PDNA fragments have modal lengths of about 2000 bp or 200 bp depending upon the duration of DNase II treatment. The smaller fragments account for virtually all of the PDNA label, as do the larger fragments which contain the smaller ones. If the large fragments are treated with exonuclease, the radioactivity is lost well before the entire fragment is digested. It may be inferred that the large fragments are tripartite in organization, both ends of the fragments consisting of the labeled PDNA, the middle portion not being labeled. These labeled components of the PDNA are referred to as "PsnDNA" because of their relationship to PsnRNA, a component discussed in section C.

B. PATTERN OF PACHYTENE DNA NICKING

An important question that has been answered is the structural relation between the PDNA segments and the 63S ssDNA fragments found in the DNA prepared from pachytene cells. A modal length of 160 kb is considerably longer than that of the large PDNA fragment which is 2 kb in length. The location and number of PDNA segments in the 63S strands provide significant information about the organization of pachytene repair-replication activity. If 63S strands isolated either from pulse-labeled pachytene cells or from pachytene cells previously labeled during the premeiotic S-phase are treated with exonuclease I, about 30% of the pachytene label is lost at the initiation of hydrolysis and 65% is lost at the completion of digestion.[46] The loss of uniformly labeled DNA in the 63S fragments from pachytene cells is linear. Because the exonuclease digests in a 3' to 5' direction, it may be inferred that there is a polarity in the nicking pattern of the endonuclease such that about $2/_3$ of nick repair activity occurs in the 5' ends of the 160 kb ssDNA strands. The mechanism responsible for the polarity is unknown, but it is clear that the PDNA segments flank the 63S strands.

Moreover, this pattern must apply to both complementary strands so that the nicks introduced in the duplex must be at opposite ends of the PDNA segments, presumably in one or the other of PsnDNA termini within the PDNA segments. Duplex 50S DNA fragments isolated from pachytene labeled cells have been shown by exonuclease digestion to be labeled only at their ends. The significant feature of these observations is the occurrence of an organized pattern of nicking; such organization is most unlikely in the absence of a relevant and specific function.

The extent to which the population of PsnDNA sequences in the genome undergoes nicking has been determined by using a mixed PsnDNA probe to track the distribution of the sequences along the alkaline sedimentation profile of pachytene DNA. The evidence is unambiguous. In DNA extracts made from early- and mid-pachytene cells, hybridization with the probe occurs almost entirely in the 63S region where approximately 50% of the DNA is found. The conclusion to be drawn is that all potential nicking sites are nicked. The number of such nicks exceeds the number of crossovers by several orders of magnitude. This strongly suggests that one of the mechanisms for achieving a regularity of recombination during meiosis is to provide a huge excess of potential initiation sites for recombination. This principle of excess is commonplace in reproductive systems, as for example in pollen and sperm production; it is no more distinctive in ensuring recombination than it is in ensuring fertilization.

C. CHARACTERISTICS OF PDNA CHROMATIN IN RELATION TO SYNAPSIS

Particles isolated by the DNase II procedure behave similarly to nucleosomes when centrifuged in a glycerol or sucrose gradient.[48] Limited digestion of isolated nuclei at various meiotic stages, so as to provide a DNA nucleosome ladder, demonstrates striking differences in the digestibility of the DNA spacers. At pachytene, a PsnDNA probe hybridizes almost entirely to the mononucleosome DNA band. At earlier meiotic stages, the probe hybridizes more or less in proportion to the DNA content of each band.[49] The relatively high sensitivity of PsnDNA chromatin to nucleases would appear to be general. This sensitivity is attributable to the composition of the chromatin. These chromatin particles have three major components — RNA, protein, and DNA. The DNA is, of course, the PsnDNA and its known properties have been described. The RNA is a small nuclear RNA (hence, PsnRNA) that is about 125 nucleotides in length, but this may be an underestimate because of the RNase digestion used in isolating the chromatin particles. Northern blots of total nuclear RNA probed with PsnDNA indicate a length closer to 140 nucleotides (Figure 8). Moreover, such blots of the total RNA taken at different meiotic stages indicate the absence of PsnRNA at all stages except for zygotene and pachytene. PsnRNA is homologous with PsnDNA; in hybridization experiments at least 50% of PsnDNA is hybridized with PsnRNA at saturation. The number of different PsnRNA sequences is at least equal to the number of PsnDNA families.

The protein components of PDNA chromatin do not include significant amounts of histones. When suspended in 0.4 M NaCl, the particles disperse, a behavior that would not occur if histones were part of the structure. Only one major protein component ("Psn-protein") has been found in protein preparations from the chromatin particles. The resistance of PsnRNA in PDNA chromatin is attributable to the Psn-protein. Moreover, it has been found that the protein selectively binds to PsnRNA, and when thus complexed it is resistant to RNase attack.[68] It would be interesting to know whether any of the recombinogenic enzymes are present among the quantitatively minor components of these chromatin particles. This has not yet been determined but, obviously, if they are present PDNA chromatin could be regarded as a functional unit of meiotic recombination.

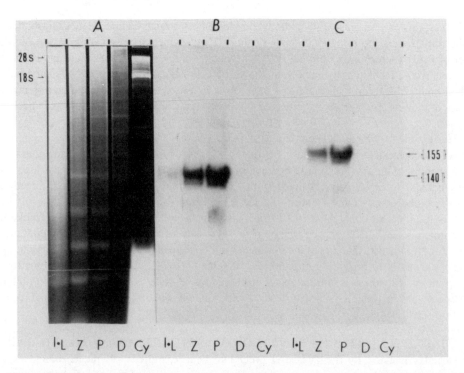

FIGURE 8.   Northern blot of nuclear RNA from sequential meiotic stages and of cytoplasmic RNA (Cy) from zygotene/pachytene mixture. Preparations were run in 6% acrylamide gels and probed with labeled PsnDNA. The total RNA in the extracts is shown in A, and the radiogram of the probed gel is shown in B. Gel C is a variant of B in which a slightly different condition for gel separation was used. The stages analyzed are interphase/leptotene (IL), zygotene (Z), pachytene (P), and diplotene/diakinesis (D). (Hotta, Y. and Stern, H., unpublished result.)

IX. BEHAVIOR OF PDNA CHROMATIN IN RELATION TO SYNAPSIS

A. THE EFFECTS OF ASYNAPSIS

As briefly discussed earlier, a lily hybrid deficient in homologous pairing displays a corresponding deficiency in DNA nicking. A similar effect is observed if synaptic varieties are treated prior to zygotene with colchicine, a treatment that generates asynapsis.[38,50] The viability of the meiocytes is unaffected in these situations although the gametes are infertile because of the resulting aneuploidy. The question that arises concerns the mechanism responsible for coupling asynapsis with a failure to activate the nicking process. That mechanism does not involve a suppression of the synthesis of meiotic endonuclease. The meiotic profile of endonuclease activity measured in extracts from the hybrid cells is identical with corresponding profiles from chiasmatic cells. No genic defect is responsible for the behavior displayed by the asynaptic hybrid. By appropriate treatment of developing flower buds with colchicine, the premeiotic germ cells can be converted to tetraploids.[51] The tetraploid cells contain pairs of identical chromosomes so that homologous pairing is possible and, as a result, chiasmata are formed. Because the genotype of the tetraploid is identical to that of the diploid hybrid, it is evident that the absence of crossing-over in the diploid is not due to a genic defect but must be due to the absence of homologous synapsis. Biochemical analyses of the tetraploid pachytene cells indicate a normal pattern of DNA nicking and repair-replication activity.[38] It would therefore appear probable that the defect lies in the composition of PDNA chromatin.

B. FUNCTIONAL PROPERTIES OF PDNA CHROMATIN

Isolated nuclei from leptotene and early zygotene cells do not release chromatin particles when briefly treated with DNase II. This is consistent with the observation mentioned earlier that the PsnDNA chromatin regions from these early prophase stages are not differentially sensitive to micrococcal nuclease. If isolated pachytene nuclei are thus treated, Psn-chromatin particles are released as expected. This, however, is not the case if the pachytene nuclei are isolated from the achiasmatic hybrid.[48] Recent unpublished analyses also indicate that neither Psn-protein nor PsnRNA are present at normal levels. The analytical data thus point to an absence of chromatin modification at what must be loosely defined as the pachytene stage.

The role of PsnRNA has been demonstrated by incubating the isolated achiasmatic nuclei with a preparation of PsnRNA, an uncharacterized fraction of proteins from pachytene cells, and ATP. Following incubation, treatment of the nuclei with DNase II releases chromatin particles containing PsnDNA.[48] Similar pretreatment of normal pachytene nuclei followed by DNase II increases the quantity of PsnDNA released. Corresponding experiments with meiotic endonuclease rather than DNase II yield parallel results. Omission of any one of the three components added renders the treatments ineffective. Two of the critical components are PsnRNA and Psn-protein, but except for there being an ATP requirement, the mechanism affecting the change in chromatin components is undetermined. It is apparent that under normal conditions pachytene nicking is regulated by governing the accessibility of the relevant PDNA sites to endonuclease, accessibility being determined by the composition of chromatin.

C. CONCLUDING CONSIDERATIONS

In describing the salient features of meiotic recombination, two of those stressed were regularity and selectivity. The behavior of DNA, as discussed in this section, touched on both features although hardly in a definitive way. The pattern of nicking is viewed as contributing to the regularity by providing a large excess of initiating sites such that the limiting factor in recombination is assured of effective action. The fact that all the presumed potential initiation sites (PDNA segments) undergo nicking makes it probable that chromosomes of higher eukaryotes have a prefixed number of meiotic recombination sites built into their structure. Such an organization implies, of course, that selectivity is required in both the positioning of the sites and in the mechanism that limits the nicking to those sites. There is no information on positioning at the chromosome level except for the evidence that few if any sites are present in satellite DNA components. Selectivity in positioning of potential nicking sites is also indicated by the fact that about 50% of lily DNA, which is isolated as single-strand fragments of lengths greater than 350 kb (the upper limit of the 63S fraction), contains no PsnDNA sites and thus adds to the evidence for selectivity in potential nicking sites. The mean number of reciprocal crossovers per meiotic cell is of the order of 30, so that exclusion of half genome (haploid value: 3×10^{10} bp) from nicking is hardly adequate to account for the ultimately small number of recombinations. It may be recalled that the satellite DNA of the Arabian Oryx, which accounts for 50% of the animal's genomic DNA, also excludes nick-repair activity at pachytene. Mechanisms more restrictive than those imposed by the distribution of PDNA sites must function in limiting the number of crossovers within the genome as a whole. The recombination nodule very likely represents part or all of the ultimate restriction process, but as yet the link between ultimate and initial steps in meiotic recombination is lacking.

The role played by chromatin structure in determining selectivity is of particular interest. It might have been supposed that the endonuclease itself selects sites of nicking but the only selectivity found is the requirement for a dsDNA substrate. The precise timing of nicking is also not determined by the enzyme except for the extreme situation in which its mutated form is inactive. In this hypothetical situation there would be no nicking and no recombination. Unlike the endonuclease, the L-protein nicks DNA selectively, but this protein is active during early- to mid-zygotene and not during pachytene; it relates almost certainly to synapsis. Regulation of nicking by altering chromatin composition allows for a much wider diversity of selected sites without diminishing the capacity for regulation. In the absence of any information about the location of these segments with respect to transcribed genes, meaningful speculation is impossible. What begs some discussion is the manner in which chromatin alterations are made at PDNA sites. It is conceivable, but unlikely, that each PsnDNA sequence is separately transcribed and that the resulting PsnRNAs complex with Psn-protein and replace the histones in the presence of ATP. This would require highly synchronized transcription at several thousand sites. It would seem more likely that a relatively small number of "master genes" transcribe the PsnRNA sequences which then spread through the nucleus and, after complexing with Psn-protein, replace the histones. PsnRNA would thus provide the sequence information for siting and the protein would effect the replacement and provide for accessibility.

An important link must exist between the process of synapsis and the synthesis of components essential to the generation of nuclease accessibility. There appear to be two types of mechanisms for regulating synapsis-associated events. In one type, coordination is achieved by each of the component events having a fixed time of appearance. The behavior of meiotic endonuclease in the absence of homologous synapsis is an excellent, but not the only, illustration. Certain recombinogenic proteins will be shown later to fall into this category. In all such cases activities reach a peak at the time when synapsis is or would have just been completed; their respective activity profiles are normally coordinated with, but are indifferent to, the actual occurrence of synapsis. The components essential to accessibility are clearly not in that grouping. The low level of R-protein (to be discussed) in asynaptic situations is another example. That their formation is linked to the process of synapsis and not to its completion is evident from the fact that, at least in the case of PsnRNA, formation begins during zygotene, while in the case of R-protein it begins no later than the start of zygotene. Synapsis does have functions other than the juxtapositioning of chromosomes; among these are functions related to metabolic activities of the meiocytes. This aspect of chromosome behavior has not been generally considered except, perhaps, in relation to somatic pairing in the Diptera.

X. THE WAVE OF RECOMBINOGENIC PROTEINS

A. GENERAL

Even a superficial survey of the metabolic events accompanying the meiotic process in *Lilium* reveals a striking display of enzyme regulation with respect to recombination. The condition that permits such display is the high, though not perfect, degree of meiotic synchrony and the duration of each of the relevant prophase stages, approximately 2 to 4 days per stage. The period intervening between completion of the premeiotic S-phase and the initiation of synapsis at zygotene is sufficiently long so that there is very little ambiguity about the stages at which particular proteins are formed and then removed. No other system provides such rich advantages for biochemical analysis of the process; it deserves broader study despite the near impossibility of a genetic characterization and the labor required to obtain sufficient amounts of material for biochemical analysis.

The metabolic pattern of the prophase stages being considered reveals two different regulatory groupings of components that are directly involved in one aspect or another of DNA metabolism, involvements that can at least be visualized as being related to recombination. The components of one group which may, perhaps correctly, be referred to as the "DNA Housekeeping" group have broad temporal activity profiles. They are already at a relatively high level during the premeiotic interval and more or less remain so until the interval following termination of pachytene when they undergo various rates of decline

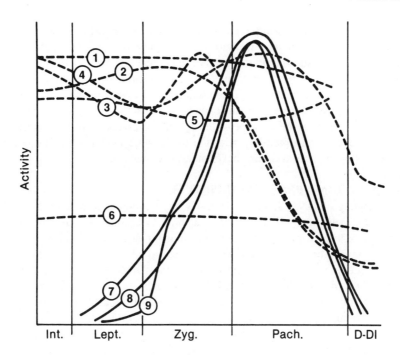

FIGURE 9.   Profiles of housekeeping (dashed lines) and meiosis-specific (solid lines) enzyme activities during microsporocyte development in *Lilium*. 1: DNA polymerase B. 2: Polynucleotide kinase. 3: Polynucleotide ligase. 4: Topoisomerase II. 5: Acid phosphatase. 6: Topoisomerase I. 7: R-protein. 8: U-protein. 9: Endonuclease. (From Stern, H. and Hotta, Y., *Stadler Symp.*, 15, 25, 1983. With permission.)

(Figure 9). The latter decay in metabolic activity is not unique to pathways in which DNA is the substrate; metabolic activity in general falls after pachytene and rises again in the postmeiotic generation of haploid cells, the microspores. This feature of metabolism in lily meiocytes is not general inasmuch as most animal species have no postmeiotic resurgence of metabolic activity until after fertilization. Not surprisingly, the respective metabolic patterns in meiocytes of plants and animals diverge widely with respect to activities required for postmeiotic development.

There is one partial exception to the statement made above about housekeeping enzymes—DNA polymerase. There is a major switch in the type of DNA polymerase that predominates at meiosis. In mouse spermatocytes, DNA polymerase β virtually replaces the α form and in lily meiocytes the substitution is similar except that DNA polymerase B almost replaces DNA polymerase A.[52] The switch would appear to be directed at the intense repair-replication that occurs during the pachytene interval. A relatively high capacity for DNA repair during pachytene has been shown to be the case in both mouse and lily meiocytes.[53,54] Among the other housekeeping enzymes traced, some of which are also related to repair, are polynucleotide kinase, DNA ligase, topoisomerase I and II, the II being much more active than the I during meiotic prophase.[55] Undoubtedly, the list of DNA housekeeping enzymes could, with a little effort, be considerably lengthened. The relevant point is that mutations affecting the activity of any of these enzymes are likely to affect not only somatic tissues but also meiotic ones. This is also evident from studies of *Saccharomyces cerevisiae* in which mutations affecting radiation-repair in vegetatively growing cells were found to affect the meiotic process.[3]

The second regulatory grouping of enzymes is of more immediate interest to this description of the meiotic process than is the housekeeping group. All these enzymes are inactive, or nearly so, during the premeiotic S-phase, and all rise abruptly during late leptotene and zygotene (or during zygotene alone). A maximum level is achieved during late zygotene or early pachytene (Figure 9). This pattern of activity provides emphatic evidence that the enzymatic activities related to recombination are a characteristic of meiotic prophase. Models that prefer to link meiotic recombination with S-phase replication are inconsistent with the behavior of the recombinogenic enzymes during meiosis. Unless the meiotic profiles of activity are grossly misleading, it is clear that, at least in higher eukaryotes, recombinational activities are at a peak during late zygotene and early pachytene. The biochemical evidence is in harmony with the cytogenetic evidence for the occurrence of crossing-over after the chromosomes have undergone synapsis.

B. THE UNWINDING PROTEIN: U-PROTEIN

The unwinding of DNA duplexes is clearly an essential step in recombination and, as discussed earlier, also in the homologous matching of chromosome regions for synapsis. The U-protein[56] can perform that function and, thus far, it is the only identified property of that protein. This contrasts with the *Escherichia coli* recA protein which not only unwinds duplex DNA but can also effect strand transfer to an ssDNA receptor. The U-protein has a molecular weight of approximately 130 kDa and binds strongly to ssDNA and less so to dsDNA. It acts on dsDNA only if the duplex is linear; if circular, a nick or gap in one of the strands is required. Moreover, a 3′OH terminus is essential to unwinding action, a process that also requires ATP. In effect, the protein is an ATP-dependent helicase. The unwinding distance is limited to about 400 bp in the case of an internal nick and to 50 bp at duplex ends. Internal gaps allow for a slightly greater unwinding distance. The extent of unwinding at nicks or gaps is sufficient to provide for stable duplex formation between each of the ssDNA strands from the interacting homologous chromatids. The protein does not unwind RNA-DNA hybrids.

A similar protein has not been identified in mitotic tissues, but no direct evidence is available concerning the role of U-protein in synapsis and/or recombination. The profile of its activity is consistent with either or both roles but unless the chromosomes are much closer together at the initiation of alignment than they are when paired (about 200 μm), the single strands formed are too short for interaction. Activity is very low during the premeiotic S-phase, begins to rise during leptotene, and is at its maximum during early pachytene. The relatively intense activity during meiotic prophase and the apparent absence of the protein in somatic cells point to its being a meiosis-specific factor and also points to the zygotene-pachytene stages as important intervals in DNA metabolic events related to pairing and recombination. It is of interest that the profile of U-protein activity is indifferent to the occurrence of homologous synapsis, a feature that is shared with some but not all of the recombinogenic enzymes. In such cases the respective profiles of activity during meiosis in achiasmatic cells are indistinguishable from those in chiasmatic ones.

C. THE R-PROTEIN

From the standpoint of meiosis, the R-protein displays many more interesting features than the U-protein. Its most relevant property is the capacity to catalyze the reannealing of denatured DNA *in vitro*, regardless of source.[57] In some respects it resembles the gene-32 protein.[58] It binds to DNA cooperatively and does not require ATP for activity. However, the meiotic protein, whether derived from lily or mouse, would appear to be more effective in reannealing denatured DNA and less effective in unwinding duplex DNA than the gene-32 protein.[59] Unwinding of duplex DNA by R-protein only occurs in media lacking Mg^{2+}, such situations being somewhat unstable for the duplex configuration; in "standard" DNA suspension media, R-protein is ineffective as an unwinding agent. In theory, at least, the actions of U-protein and R-protein are complementary. Unlike the U-protein, the R-protein is localized in the nuclear membrane in both lily and mouse meiocytes, a localization similar

Table 1
Binding of Phosphorylated and
Dephosphorylated R-Protein to DNA[60]

Treatment	μg Protein bound/μg DNA	
	Single strand	Double strand
Native form	5.6	0.07
Dephosphorylated	22.76	20.8
Rephosphorylated	6.0	0.15

to that of the L-protein. The latter protein, it should be recalled, probably functions in synapsis. The R-protein, like the L-protein, is present at a low level in asynaptic situations.[38] The mechanism regulating the absence of a rise in R-protein activity is unknown. The protein is synthesized during early prophase, a process that may be tied to synapsis. Alternatively, synapsis may be required to stabilize the protein so that in its absence, the half-life of the protein may be much reduced. Either possibility carries the important implication that synapsis not only involves a homologous positioning of chromosomes to facilitate crossing-over but also regulates certain metabolic events that are essential to the process.

One regulatory property governing the capacity of R-protein to catalyze duplex DNA formation is its state of phosphorylation.[60] The native protein, as extracted, appears to have two phosphorylated sites. If the protein is treated with phosphatase, the capacity to reanneal is lost. The loss may be assigned to the striking change in binding affinity; dephosphorylated R-protein binds equally to ss- and dsDNA. The level of binding in the protein thus treated is significantly higher than that of the native protein as shown in Table 1. If the native protein, or the dephosphorylated derivative, is treated with a cAMP protein kinase, the number of phosphorylated residues per protein molecule is much greater than two and it no longer binds to either ss- or dsDNA. The loss of activity by the dephosphorylated protein is reversible if an endogenous protein kinase is used; an example of such recovery is included in Table 1.

Phosphorylation affects the affinity of R-protein for ss- and dsDNA. Where binding to both forms is equal and relatively strong, the protein does not favor either form. Where binding to both forms is weak, there is no tendency to duplex formation. The meiocyte does have the potential to alter the helicase properties of the R-protein, a potential that may be utilized in either synapsis or recombination. The probability of such involvement appears likely because of the interaction of the R-protein with DNA polymerase β.[52] As discussed earlier, the β-form predominates over the α-form during meiosis; it is about 50 times more active in spermatocytes and 10 times more active in lily meiocytes. The R-protein has a stimulatory effect on the activity of the β-polymerase but not on the α-form. It may be supposed that the effect is directed at the repair-replication activity related to recombination. Whether or not this occurs, the important feature of R-protein behavior is its temporal correlation with other activities that are both meiosis specific and prophase specific.

D. THE RecA-LIKE PROTEIN

Two of the principal questions that have concerned students of meiosis are the "when" and the "how" of recombination. The first of these questions—and the easier one to answer—has already been briefly discussed; the second one can at best be given a partial answer. The evidence in both cases is still circumstantial, but the evidence on the issue of "when" is now overwhelming. The discovery of proteins in meiocytes that function directly in recombination makes it possible to infer unequivocally from their temporal behavior the timing of the molecular events in the recombination process. The discovery also provides some insights, limited of course, as to how recombination might be effected. In the case of

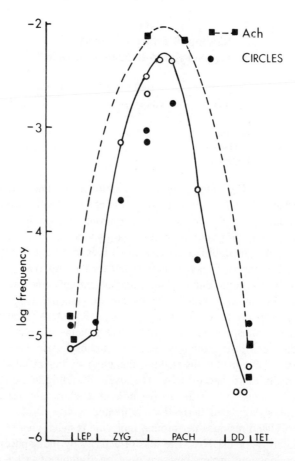

FIGURE 10.   Profile of m-rec activity in microsporocytes of *Lilium* at sequential stages of meiosis. The activities and protein concentrations are expressed in terms of 100 mg microsporocyte DNA. In this experiment, DNA-dependent ATPase was used to determine m-rec activity (1 unit = 1 nmol of P_i released in 30 min at 23°C). The activities are proportional to D-loop activities at all stages. The stages analyzed were premeiotic interphase including S-phase (I), leptotene (L), zygotene (Z), pachytene (P), diplotene/diakinesis (D), and cell tetrads (T). (From Hotta, Y., Tabata, S., Bouchard, R. A., Pinon, R., and Stern, H., *Chromosoma*, 93, 140, 1985. With permission.)

the recA-like protein, its virtual absence during the premeiotic S-phase and its steep rise during zygotene to a peak in early pachytene provides compelling evidence on the timing of recombination (Figure 10). The genetic evidence for the recombinogenic role of recA in *E. coli* is fully established,[61] and it is therefore reasonable to assume that its counterpart in meiocytes plays a similar role. Two lines of consideration need attention in making this assumption: the degree to which the properties of the meiotic recA-like protein are similar to those of recA and the extent to which the protein identified in meiotic cells is specific to meiosis.

The recA-like protein found in meiotic cells[62] is referred to as the "m-rec" protein, the "m" referring to meiotic. The protein has been prepared from lily meiocytes and mouse spermatocytes; the properties of the two products are virtually identical except for temperature optima. Lily anthers and mouse testes are adapted to operate at rather different temperatures. A brief description of the properties tested follows. In each case no substantial difference from recA behavior has been noted. M-rec protein binds to ssDNA, but not to RNA. The binding is higher in the presence of ATP. If complementary strands are present, the m-rec protein catalyzes their reassociation. Although no direct comparisons between m-rec and R-protein have been made, rough comparison would indicate the R-protein to be more efficient in such catalysis. In a mixture containing a 10:1 ratio of protein: φX174 DNA for mouse m-rec[62] and the same ratio of mouse R-protein: T7 DNA, the respective levels of reassociation after a 1 h incubation are about 40% for m-rec and 80% for R-protein. It should be noted that T7 DNA is a much larger molecule than φX174 so that in terms of bp reassociated the difference in rate is much greater. The relevance of the comparison to meiosis is unclear in the absence of any evidence for the exclusive involvement of R-protein in chromosome synapsis. The significant point is the capacity of m-rec to displace one of the strands in a circular duplex by a ssDNA fragment whose sequence is the same as that of the displaced strand—the phenomenon of "D-loop formation". The presence of ATP is required for this activity, a characteristic of recA but not of a comparable protein of *Ustilago*[63] which retains 40% of its D-loop activity in the absence of ATP. The topological properties of a closed circular duplex limit the extent to which the invading strand can displace the resident one. This is not the case for combinations of linear duplex with single-strand circle or linear ssDNA and nicked circular duplex. Under suitable conditions, m-rec can catalyze the transfer of an entire strand to either circular structure. It may be concluded that, with respect to recA type activity, meiotic cells contain a recombinational mechanism very similar to the much studied *E. coli*. The conclusion in no way implies that m-rec adequately accounts for the recombinase enzymes in meiocytes. We have too little knowledge of the composition of meiotic cells to make such judgments; there may indeed be more than one recA-type protein. One example of this is discussed below.

Three lines of evidence point to the meiotic specificity of the m-rec protein. The first of these is its profile of activity during meiotic development (Figure 10). As with the U- and R-proteins, m-rec activity rises sharply during zygotene, reaching its maximum during early pachytene, and then declines. In lily meiocytes the increase in activity from preleptotene to pachytene is at least 150- to 200-fold, initial activity being difficult to determine. A second line of evidence concerns the effect of defective synapsis on the protein. Like the R-protein, failure in homologous chromosome pairing is accompanied by a failure in the rise of m-rec activity.[69] There is, however, no ready explanation for m-rec or R-protein responding in one way to synaptic failure and for U-protein or endonuclease responding in the opposite manner. It is nevertheless clear that the responsiveness of m-rec to synaptic failure reflects its intimate involvement with one aspect or another of chromosomal events during meiotic prophase.

The third line of evidence relates to the differences in recombinase activities between meiotic and mitotic cells. These differences are concerned with either physical and catalytic characteristics or with temperature optima for activity. In both lilies and mice, recA-like activities have been found in mitotic as well as meiotic cells. The findings in mice are particularly informative because of the several types of tissues that can be examined for activity, especially those somatic tissues in which recombination is known to be active. One striking feature relates to molecular weight. In both lilies and mice, proteins with recA-like activities that have been isolated from somatic or vegetative tissues display molecular weights of about 70 to 75 kDa when measured in SDS-acrylamide gels. By contrast, those isolated from meiotic tissues display molecular weights of 43 to 45 kDa.[62] The proteins have not been sufficiently purified to account for the nature of the difference, but the fact that organisms as evolutionarily distant as lilies and mice display nearly identical differences in molecular size of rec proteins between meiotic and mitotic cells makes it very likely that the proteins meet somewhat different functional needs. Whatever the differences in function, no apparent differences exist with respect to the catalytic properties of the enzymes. The characteristics described above for m-rec apply equally to those of "s-rec" (somatic-rec).

The only observed difference concerns specific catalytic rates. Assuming the purity of meiotic and mitotic enzymes to be similar, the assays indicate a much higher specific activity for the meiotic enzymes. For "equal" amounts of enzyme, D-loop formation is 6 times more rapid with the meiotic enzyme as with the spleen enzyme. Approximately the same difference has been observed with respect to the strand transfer reaction. The most impressive evidence for the existence of intrinsic differences between somatic and meiotic enzymes is in their temperature optima.

It is generally recognized that meiosis in male mammals becomes defective if testes are subjected to a temperature of 37°C. Plant cytogeneticists are aware that for many species relatively low temperatures are required for effective fertilization. The optimal temperature for meiosis in the male mouse is 33°C and for lily it is 23 to 25°C. It is significant that this property is reflected in the *in vitro* activities of the m-rec enzymes. In the case of mouse, the optimal temperature for s-rec, regardless of tissue of origin, is close to 37°C. D-loop formation by mouse m-rec is 3 times more rapid at 33°C than at 37°C. In lily, the rate is 4 times more rapid at 25°C than at 33°C. The latter temperature results in the higher rate in the case of s-rec. A similar situation occurs with respect to strand transfer. In mouse, m-rec is 4 times more rapid at 33°C than at 37°C; in lily, it is 3 times more rapid at 25° than at 33°C. Similar differences have been found with respect to DNA binding. It is doubtful that this temperature differential between meiotic and mitotic cells is general. Meiocytes in fetal ovaries of mammals very probably undergo meiosis at 37°C; there is no accurate information about the behavior of the female meiocytes of plants with respect to temperature. Whether or not there is a deep significance to the lower temperature has not yet been perceived; the important feature of this behavior is the demonstration that very similar activities in mitotic and meiotic cells are carried out by different enzymes. What is unclear is whether m-rec is adapted to meet other physiological requirements of the meiocyte or whether the temperature characteristic of the meiocyte is determined by that of the recombination mechanism.

E. RECOMBINASE SYSTEM: *IN VITRO* ASSAYS

Despite the overwhelming evidence for a key role of recA protein in recombination, an *in vitro* system in which its role could be directly demonstrated is lacking. Several successful attempts have been made to develop such a system in the hope that it might provide an approach toward a better understanding of the recombination process.[64] Although the fruits of that goal are not yet evident, the search for an *in vitro* system in meiotic cells has provided another line of evidence supplementing and reinforcing the assertions made about the prominence of recombinational activities during meiotic prophase. As in the case of m-rec, the activities in somatic and meiotic cells are similar but the similarities in no way obscure the evidence that meiotic cells develop a distinctive set of recombinogenic proteins. The assay system commonly used in measuring recombinase activity consists of two types of plasmids that are each mutant in a different region of the same gene so that one of the recombinant products inherits an active form of the gene. The latter may code for resistance to an antibiotic, such as tetracycline, and the frequency of recombinations is determined by transforming bacteria with a plasmid mixture previously exposed to the cell extracts. When extracts from meiocytes of lily and mouse were thus tested, two features of the results obtained were significant in assessing the organization of the recombination mechanism during meiosis. There was relatively little activity during the premeiotic S-phase, but following the entry of cells into meiosis recombinational activity rose sharply. The respective profiles of activity in extracts prepared from successive stages of meiosis in lily and mouse were similar. A comparable result was obtained when *S. cerevisiae* cells were switched to sporulation medium.[62] As in other assays, lily meiocytes provided the most closely synchronous cell preparations and the increase from leptotene to early pachytene was about 700-fold (Figure

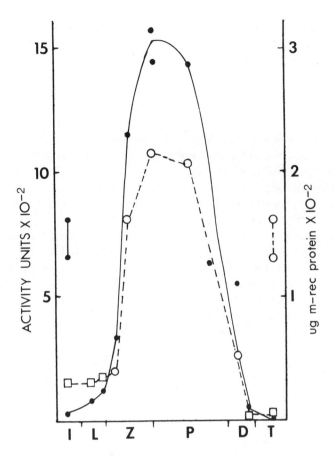

FIGURE 11.   Recombinase activities in meiotic extracts of *Lilium*. The solid line traces activities of chiasmatic cells, and the dashed line traces those of the achiasmatic hybrid, cv. Black Beauty. (Note that activities are expressed on a logarithmic scale.) The solid circles indicate levels of activity if both mutant plasmids are present as supercoiled structures. The open circles represent activities if one of the plasmids is present as a linear duplex. (From Hotta, Y., Tabata, S., Bouchard, R. A., Pinon, R., and Stern, H., *Chromosoma*, 93, 140, 1985. With permission.)

11). One inference is easily drawn: the molecular events that are immediately related to meiotic recombination occur at a time when chromosomes have completed the pairing process.

Some features of the recombination process may be noted even though their relevance to *in vivo* events is unclear. At least one of the mutant plasmids must be supercoiled. The other mutant plasmid may be a supercoiled circle or a linear duplex. The latter gives the higher rate. If one of the plasmid substrates is rendered single-stranded, no recombination occurs. It is clear that the initial and essential step in the recombination process must be affected by the meiotic extract, but how many of the subsequent steps are performed in the bacteria is not known. *E. coli* recA and lily m-rec have a stimulatory effect on the recombinational activity of the extracts, their respective levels of stimulation being the same. Lily extracts are stimulated 1.4 times by recA or m-Rec and mouse extracts, 2.5 times. Among the various enzyme additives tested, lily U-protein has the highest stimulatory effect on the recombinase activity of meiotic extracts—1.8 times for lily extracts and 2.9 times for mouse extracts. The combined presence of rec and U-proteins stimulates only slightly more than U-protein alone. None of these additives, alone or in combination, effects recombination in the absence of meiotic extract.

Similarly prepared extracts from somatic tissues display the same recombinational activities as do meiocyte extracts. Bone marrow extracts from mice have the highest somatic rate, but these are well below the levels reached by the prophase spermatocytes.[62] Somatic and meiotic recombinase extracts show the same difference in temperature response as did the s-rec and m-rec proteins. Extracts of mouse spermatocytes have significantly higher recombinase activity at 33°C than 37°C; lily meiocyte extracts have the higher activity at 25°C than at 30°C.[69] As in the case of the rec proteins, the functional significance of the temperature difference is unknown except for its indication that the protein(s) responsible for recombination in meiotic extracts is different from the protein(s) in somatic extracts.

The profiles of recombinase activity during meiosis are unaffected by an absence of homologous chromosome pairing (Figure 11). Mouse spermatocytes cannot be tested in this way: the evidence was obtained with lilies. However, supportive observations were made with *S. cerevisiae*. Yeast diploids that are α/α or a/a undergo the same 100-fold increase in recombinase activity as meiotically competent a/α diploids 6 to 7 h following immersion in sporulation medium. Thus, whereas R-protein and m-rec fail to undergo their normal meiotic rise in the absence of homologous pairing, U-protein and recombinase activity are indifferent to the absence. One might speculate that the R-protein and m-rec are directly involved in synapsis whereas the other two are involved in recombination, an involvement that is consistent with the stimulatory action of U-protein on the extracted recombinase activity. The speculation has little else to recommend it.

XI. GENERAL CONSIDERATIONS

A. COMMENTS

There are many specific and clearly formulated questions that need to be addressed in order to better understand the mechanisms sustaining the meiotic process. The obvious ones are readily listed. The switch in a line of cells from a history of repeated mitotic divisions to a one-time terminal meiotic division must involve a unique set of signals. No single set of stimuli can be assigned to the activation of those unique meiosis-inducing signals. In yeast and other fungi the signals are activated by ammonia and glucose deprivation. In higher plants a given photoperiod may induce a complex sequence of events producing factors that ultimately lead to flowering. Meiosis is, of course, a component of the flowering process. Among animals, the ultimate event of meiosis is assigned to cells of the germline set aside during embryogenesis. The research challenge is to identify the unique meiosis-inducing signals which, presumably, have been highly conserved. The challenge has been confronted by many but answered by none. In the case of mammals, some preliminary claims have been made about the identification of a meiosis-inducing substance in maturing mouse testes and in fetal ovaries,[65] but there has been no definitive conclusion to the identification efforts. A note of caution needs to be introduced here. The term "meiosis-inducing substance" has frequently been erroneously used in referring to substances that are responsible for the resumption of meiosis from premetaphase I in the course of ovulation. The resumption of an arrested meiosis is distinctly different from an induction of the process.

The entry into meiosis is accompanied by a series of unique chromosomal changes. The first cytologically identifiable prophase stage—leptotene—is readily distinguishable from a mitotic prophase. The morphology of leptotene threads must surely reflect unique features of chromatin composition which have yet to be characterized. From leptotene on through zygotene and pachytene the structural components of the synaptonemal complex have been thoroughly described. Indeed, EM analyses of the meiotic process must be singled out as the most fruitful advance in our understanding of the process over the past 30 years. These advances leave little doubt that the prophase of meiosis, during which pairing and recombination occur, require the formation of an elaborate structure to make the two events possible. It also makes apparent the need for a much longer prophase period than that occurring in mitosis. The essential point is that meiosis is not purely a modified mitosis. It is a modified mitosis only with respect to the regulation of centromere division and the disjunction of chromosomes at metaphase II. Most of the process involves a distinctive pattern of chromosome behavior that requires an elaborate mechanism to effect the homologous alignment of chromosomes and rigorously regulate crossovers. The rich background of information that decorates our knowledge of these different structural developments becomes impoverished when we attempt to translate the developments in biochemical or molecular terms. We have no knowledge of how homologous chromosomes find each other while being partially compacted, nor how their alignment is secured by formation of the synaptonemal complex. We do not know the nature of the forces that bind the lateral elements to the chromatids, nor of those that unite the lateral to the central elements. Moving from chromosome pairing to crossing-over further blurs the picture. The recombination nodule, another fruitful product of EM analyses, is totally lacking in molecular characterization.

Given the indifferent state of our knowledge of the individual mechanisms that sustain the meiotic process, it is probably more useful for the reader to conclude this chapter by sketching the broad outlines of meiosis, particularly as they relate to the biochemical/ molecular events underlying the process. Before doing so it is best to note that the information available for the sketch is derived from the meiotic behavior of a small number of species, a situation that should have become apparent from the various studies discussed earlier. The biochemical characterization of different meiotic stages is feasible in very few species, most of them in the *Liliaceae*. Thus far, however, most of what has been observed is at least consistent with features of meiosis noted in other systems. It is therefore possible to outline the biochemical organization of meiotic cells with some confidence that the outline is not entirely misleading and with the hope that this will stimulate other studies. The experimental grounds for the outline are provided in the preceding sections.

B. THE ORGANIZATION OF MEIOSIS: AN OUTLINE

At some point just prior to or just after the termination of premeiotic DNA synthesis, presumptive meiotic cells become irreversibly committed to meiosis. This commitment precludes the cells from undergoing mitotic divisions. It also involves chromosomal changes, presumably those anticipating the needs of synapsis. How many distinctive proteins are synthesized at the time is unknown. We know of one protein, the L-protein, that has an immediate function at the time of its synthesis, suppression of zygDNA replication. Incompletely replicated chromosomes cannot enter the premitotic G2 phase, and if suppression of zygDNA replication is irreversible, the calls cannot undergo mitosis. Under these conditions, the metabolism of the premeiotic G2 phase becomes modified to accommodate the transcriptional and translational activities essential to the pairing process. The interval spanning the end of S-phase and the start of leptotene is critical to meiosis despite its brevity. That interval must embrace a set of transcriptions that are fundamental to meiotic development. Not surprisingly, perturbation of cells during this brief G2 phase causes abnormalities in the pairing process. This phase and the leptotene stage constitute an interval of relatively intense meiosis-specific syntheses diluted, to be sure, by the many syntheses that serve housekeeping functions. Regardless of the extent of dilution, these syntheses, if identified and characterized, would provide a much needed insight into the organization of the meiotic process. The events occurring in that interval would account for the scattered bits of evidence linking some perturbations of prezygotene cells with abnormalities in chiasma formation. One structural change that is manifest during this interval is the formation of lateral elements along the newly replicated sister chromatids. Modification of the nuclear membrane also occurs during this period, a change that may be related to the observed association of telomeres with the nuclear membrane, an association which some believe to be important for homologous chromosome alignment.

Zygotene marks the initiation of several molecular processes that are at least partly related to synapsis, and partly to activities at pachytene. The most distinctive event that has been observed is the partial replication of the hitherto suppressed zygDNA sequences. It is partial because the replication is incomplete; short gaps of undetermined size remain at the ends of the newly replicated strands. The event is unusual from at least three standpoints. It is the only known case in which an S-phase is programmed to be incomplete and one in which the replication remains incomplete even after its programmed resumption. The unusual pattern of zygDNA behavior is underlined by the occurrence of each of the replication intervals at specific stages of meiosis. Chromosome synapsis seems to be tied to the initiation of zygDNA replication. Its incompleteness until after pachytene suggests a role in chromosome disjunction, perhaps by facilitating the removal of DNA interlocks, or by completing the theoretical axial DNA replication and thus removing sister chromatid attachments which otherwise inhibit chromatid separation, a requirement for normal disjunction. Besides the activities of zygDNA itself, two zygDNA-related activities are almost certainly significant to chromosome synapsis. The L-protein, by virtue of its zygDNA-specific nickase activity, makes possible the availability of ssDNA, which may then actually be provided by the action of U-protein. If this sketch is realistic, L-protein would be directly involved in the homologous juxtapositioning of the chromosomes. The highly conserved zygRNA transcripts are almost certainly involved in the synaptic process but the nature of that involvement is unknown.

What emerges from these various considerations is the view that chromosome pairing consists of several separate but coordinated processes. One of these is the juxtapositioning of specific DNA segments that are broadly distributed among all the chromosomes and serve as alignment sites for homologous pairing. ZygDNA is a candidate for this function, but a different set of segments could fulfill that role. The critical point is that among the comparatively large chromosomes of higher forms only a small fraction of chromosomal DNA can be directly involved in homologous alignment, presumably the fraction that also serves as part of the chromosome axis. The alignment itself does not constitute a stable configuration so that a second process is required to render it secure. This is provided by the formation of the synaptonemal complex. In the absence of contrary evidence it may be assumed that the coupling of the lateral elements by the components of the central element (in effect, SC formation) is indifferent to homologous matching. If so, there must be a mechanism for coordinating DNA matching with SC formation. Inasmuch as SCs may be present between heterologously paired chromosomes, the mechanism for coordinating the two events is beyond the reach of current speculations on meiotic synapsis. The apparent independence of the stabilizing function of the SC from the homology of chromosome alignment is a fundamental issue that begs close attention.

As the zygotene stage flows into pachytene, significant changes occur in chromatin composition and DNA metabolism. Certain DNA regions, again a small fraction of the genome, are programmed to serve as potential sites for recombination. The number of such regions is in very large excess of the number of actual crossovers, a number that ensures the regularity of meiotic recombination. Site selection appears to be a property of chromatin composition, and the selection is effected by molecules of snRNA that are homologous with segments of the DNA at those sites. Their chromatin composition is changed by replacing all or most of the histones by a meiosis-specific protein, tentatively designated as "Psn-protein". The replacement renders the sites susceptible to the action of an endonuclease. Chromatin in all other regions protects the DNA against nuclease activity. Although a meiosis-specific endonuclease is induced at the time of zygotene, the action of that enzyme is regulated by chromatin composition at the PDNA sites. The enzyme is believed to initiate the first of the enzymatic steps leading to recombination. The programmed introduction of nicks at each of the PDNA sites provides fairly compelling evidence that single-strand interruptions in the DNA are at least one of the major processes leading to recombination.

A most impressive event at the zygotene-pachytene stages is the rather abrupt appearance of the recombinogenic enzymes discussed earlier. It is impressive by virtue of the tight correlation between the activities of these enzymes and the time of chromosome pairing. The meiotic profiles of the proteins furnish a diagrammatic illustration of the way in which the meiocyte is organized to effect recombination while also validating the classical cytogenetic view that crossing-over occurs once the chromosomes have paired. Although the necessary calculations have not been made, it is reasonable to consider the levels of enzyme activity as being in large excess over what is required to effect the actual number of recombinations. A limiting factor must be present to maintain the low number of crossovers per chromosome and per meiosis. The recombination nodule may house that factor, but it is yet to be identified. Judging from genetic evidence, there must exist a mechanism specific to each chromosome that controls its capacity to mediate crossing-over.

A concluding and hesitantly optimistic note may be sounded about our progress in understanding meiosis. It is a process for which each of the successive stages has been described in fine detail by light microscopy, and almost in the same detail but at much higher resolution by electron microscopy. To a very limited extent molecular and biochemical counterparts to a number of key stages have been found, and this gives hope that a more dynamic account of the well described process is in the offing. This cheerful vision needs to be balanced by the recognition that nature has not been very generous in its experimental offerings for the molecular/biochemical studies of meiosis. This lack of charity may be bypassed by the introduction of cellular techniques that will render many meiotic systems as attractive targets for research, or better yet, by the discovery of an organism that offers easy access to the genetic, molecular, and structural aspects of meiosis.

ACKNOWLEDGMENTS

I wish to thank Dr. Clare Hasenkampf for her most helpful review of the manuscript. I also wish to express the strongest gratitude to the National Science Foundation, the National Institute for Child Health and Human Development, and the American Cancer Society for support of the many studies that constitute the bulk of this article.

REFERENCES

1. **Gilles, C. B.,** The synaptonemal complex in higher plants, *CRC Crit. Rev. Plant Sci.,* 2, 81, 1984.
2. **von Wettstein, D.,** The assembly of the synaptonemal complex, *Philos. Trans. R. Soc. London, Ser. B,* 277, 235, 1977.
3. **Resnick, M. A.,** Investigating the genetic control of biochemical events in meiotic recombination, in *Meiosis,* Moens, P., Ed., Academic Press, New York, 1987, 157.
4. **Esposito, R. E. and Klapholz, S.,** Meiosis and ascospore development, in *Molecular Biology of the Yeast Saccharomyces cerevisiae. Life Cycles and Inheritance,* Strathern, J. N., Jones, E. W., and Broach, J. R., Eds., Cold Spring Harbor Laboratory, Cold Spring Harbor, New York, 1981, 211.
5. **Ito, M. and Takegami, M. H.,** Commitment of mitotic cells to meiosis during the G2 phase of premeiosis, *Plant Cell Physiol.,* 23, 943, 1982.
6. **Callan, H. G.,** Replication of DNA in the chromosomes of eukaryotes, *Proc. R. Soc. London B,* 181, 19, 1972.
7. **Williamson, D. H., Johnston, L. H., Fennell, D., and Simchen, G.,** The timing of the S-phase and other nuclear events in yeast meiosis, *Exp. Cell Res.,* 145, 209, 1983.
8. **Holm, P. B.,** The premeiotic DNA replication of euchromatin and heterochromatin in *Lilium longiflorum, Carlsberg Res. Commun.,* 42, 249, 1977.
9. **Ninnemann, H. and Epel, B.,** Inhibition of cell division by blue light, *Exp. Cell Res.,* 79, 318, 1973.
10. **Stern, H. and Hotta, Y.,** The biochemistry of meiosis, in *Meiosis,* Moens, P. B., Ed., Academic Press, New York, 1987, 303.
11. **Hotta, Y., Tabata, S., Stubbs, L., and Stern, H.,** Meiosis-specific transcripts of a DNA component replicated during chromosome pairing: homology across the phylogenetic spectrum, *Cell,* 40, 785, 1985.

12. **Palmer, R. G.,** Cytological studies of ameiotic and normal maize with reference to premeiotic pairing, *Chromosoma,* 35, 233, 1971.
13. **Hotta, Y., Tabata, S., and Stern, H.,** Replication and nicking of zygotene DNA sequences: control by a meiosis-specific protein, *Chromosoma,* 90, 243, 1984.
14. **Stern, H. and Hotta, Y.,** Chromosome behavior during development of meiotic tissue, in *The Control of Nuclear Activity,* Goldstein, L., Ed., Prentice-Hall, Englewood Cliffs, NJ, 1967, 36.
15. **Church, K. and Wimber, D. E.,** Meiosis in *Ornithogalum virens.* II. Univalent production by preprophase cold treatment, *Exp. Cell Res.,* 64, 119, 1971.
16. **Chiu, S. M. and Hastings, P. J.,** Premeiotic DNA synthesis and the recombination in *Chlamydomonas reinhardii, Genetics,* 73, 29, 1973.
17. **von Wettstein, D., Rasmussen, S. W., and Holm, P. B.,** The synaptonemal complex in genetic segregation, *Annu. Rev. Genet.,* 18, 331, 1984.
18. **Dresser, M. E.,** The synaptonemal complex and meiosis: an immunochemical approach, in *Meiosis,* Moens, P. B., Ed., Academic Press, New York, 1987, 245.
19. **Moens, P. B., Heyting, C., Dietrich, A. J. J., van Raamsdonk, W., and Chen, Q.,** Synaptonemal complex antigen location and conservation, *J. Cell Biol.,* 105, 93, 1987.
20. **Heslop-Harrison, J. S., Smith, J. B., and Bennett, M. D.,** The absence of somatic association of centromeres of homologous chromosomes in grass mitotic metaphases, *Chromosoma,* 96, 119, 1988.
21. **Shepard, J., Boothroyd, E. R., and Stern, H.,** The effect of colchicine on synapsis and chiasma formation in microsporocytes of *Lilium, Chromosoma,* 44, 423, 1974.
22. **Hotta, Y. and Shepard, J.,** Biochemical aspects of colchicine action on meiotic cells, *Mol. Gen. Genet.,* 122, 243, 1973.
23. **Bennett, M. D. and Smith, J. B.,** The effect of colchicine on fibrillar material in wheat meiocytes, *J. Cell Sci.,* 38, 33, 1979.
24. **Maguire, M. P.,** A possible role for the synaptonemal complex in chiasma maintenance, *Exp. Cell Res.,* 112, 297, 1978.
25. **Rasmussen, S. W.,** The transformation of the synaptonemal complex into the "elimination chromatin" in *Bombyx mori* oocytes, *Chromosoma,* 60, 205, 1977.
26. **Moses, M. J., Dresser, M. E., and Poorman, P. A.,** Composition and role of the synaptonemal complex, in *Symposia of the Society for Experimental Biology,* Vol. 38, Evans, C. W. and Dickinson, H., Eds., University of Cambridge, Cambridge, 1984, 245.
27. **Hotta, Y. and Stern, H.,** Persistent discontinuities in late replicating DNA during meiosis in *Lilium, Chromosoma,* 55, 171, 1976.
28. **Stern, H. and Hotta, Y.,** Biochemistry of meiosis, *Philos. Trans. R. Soc. London Ser. B,* 277, 277, 1977.
29. **Roth, T. F. and Ito, M.,** DNA-dependent formation of the synaptinemal complex at meiotic prophase, *J. Cell Biol.,* 35, 247, 1967.
30. **Takegami, M. H. and Ito, M.,** Effect of hydroxyurea on mitotic and meiotic divisions in explanted lily microsporocytes, *Cell Struct. Funct.,* 7, 29, 1982.
31. **Stern, H. and Hotta, Y.,** Molecular biology of meiosis: synapsis-associated phenomena, in *Aneuploidy: Etiology and Mechanisms,* Dellarco, V., Voytek, P. E., and Hollaender, A., Eds., Plenum Press, New York, 1985, 305.
32. **Darlington, C. D.,** *Recent Advances in Cytology,* Blakiston, Philadelphia, 1937.
33. **Carpenter, A. T. C.,** Recombination nodules and synaptonemal complex in recombination-defective females of *Drosophila melanogaster, Chromosoma,* 25, 259, 1979.
34. **Hastings, P. J.,** Meiotic recombination interpreted as heteroduplex correction, in *Meiosis,* Moens, P. B., Ed., Academic Press, New York, 1987, 107.
35. **Hotta, Y., Chandley, A. C., and Stern, H.,** Biochemical analysis of meiosis in the male mouse. II. DNA metabolism at pachytene, *Chromosoma,* 62, 255, 1977.
36. **Hotta, Y. and Stern, H.,** Absence of satellite DNA synthesis during meiotic prophase in mouse and human spermatocytes, *Chromosoma,* 69, 323, 1978.
37. **Stern, H. and Hotta, Y.,** The organization of DNA metabolism during the recombination phase of meiosis with special reference to humans, *Mol. Cell. Biochem.,* 29, 145, 1980.
38. **Hotta, Y., Bennett, M. D., Toledo, L. A., and Stern, H.,** Regulation of R-protein and endonuclease activities in meiocytes by homologous chromosome pairing, *Chromosoma,* 72, 19, 1979.
39. **Hotta, Y. and Stern, H.,** DNA scission and repair during pachytene in *Lilium, Chromosoma,* 46, 279, 1974.
40. **Szostak, J. W., Orr-Weaver, T. L., Rothstein, R. J., and Stahl, F. W.,** The double-strand break repair model for recombination, *Cell,* 33, 25, 1983.
41. **Howell, S. H. and Stern, H.,** The appearance of DNA breakage and repair activities in the synchronous meiotic cycle of *Lilium, J. Mol. Biol.,* 55, 357, 1971.
42. **Hastings, P. J.,** Models of heteroduplex formation, in *Meiosis,* Moens, P. B., Ed., Academic Press, New York, 1987, 139.

43. **Smyth, D. R. and Stern, H.,** Repeated DNA synthesized during pachytene in *Lilium henryi, Nature New Biol.,* 245, 94, 1973.

44. **Bouchard, R. A. and Stern, H.,** DNA synthesized at pachytene in *Lilium:* a nondivergent subclass of moderately repetitive sequences, *Chromosoma,* 81, 349, 1980.

45. **Friedman, E., Bouchard, R. A., and Stern, H.,** DNA sequences repaired at pachytene exhibit strong homology among distantly related higher plants, *Chromosoma,* 87, 409, 1982.

46. **Hotta, Y. and Stern, H.,** The organization of DNA segments undergoing repair synthesis during pachytene, *Chromosoma,* 89, 127, 1984.

47. **Gottesfeld, J. M. and Butler, P. J. G.,** Structure of transcriptionally active chromatin subunits, *Nucleic Acids Res.,* 4, 3155, 1977.

48. **Hotta, Y. and Stern, H.,** Small nuclear RNA molecules that regulate nuclease accessibility in specific chromatin regions of meiotic cells, *Cell,* 27, 309, 1981.

49. **Hotta, Y., de la Pena, A., Tabata, S., and Stern, H.,** Control of enzyme accessibility to specific DNA sequences during meiotic prophase by alterations in chromosome structure, *Cytologia,* 50, 611, 1985.

50. **Bennett, M. D., Toledo, L. A., and Stern, H.,** The effect of colchicine on meiosis in *Lilium speciosum* cv. Rosemede, *Chromosoma,* 72, 175, 1979.

51. **Toledo, L. A., Bennett, M. D., and Stern, H.,** Cytological investigations of the effect of colchicine on meiosis in *Lilium* hybrid cv. Black Beauty microsporocytes, *Chromosoma,* 72, 157, 1979.

52. **Sakaguchi, K., Hotta, Y., and Stern, H.,** Chromatin-associated DNA polymerase activity in meiotic cells of lily and mouse: its stimulation by meiotic helix destabilizing protein, *Cell Struct. Funct.,* 5, 323, 1980.

53. **Kofman-Alfaro, S. and Chandley, A. C.,** Radiation-initiated DNA synthesis in spermatogenic cells of the mouse, *Exp. Cell Res.,* 69, 33, 1971.

54. **Stern, H. and Hotta, Y.,** Regulatory mechanisms in meiotic crossing-over, *Annu. Rev. Plant Physiol.,* 29, 415, 1978.

55. **Stern, H. and Hotta, Y.,** Meiotic aspects of chromosome organization, *Stadler Symposia,* Vol. 15, University of Missouri Press, Columbia, MO, 1983, 25.

56. **Hotta, Y. and Stern, H.,** DNA unwinding protein from meiotic cells of *Lilium, Biochemistry,* 17, 1872, 1978.

57. **Hotta, Y. and Stern, H.,** A DNA-binding protein in meiotic cells of *Lilium, Dev. Biol.,* 26, 87, 1971.

58. **Alberts, B. M. and Frey, L.,** T4 bacteriophage gene 32: a structural protein in replication and recombination of DNA, *Nature (London),* 227, 1313, 1970.

59. **Mather, J. and Hotta, Y.,** A phosphorylatable DNA-binding protein associated with a lipoprotein fraction from rat spermatocyte nuclei, *Exp. Cell Res.,* 109, 181, 1977.

60. **Hotta, Y. and Stern, H.,** The effect of dephosphorylation on the properties of a helix-destabilizing protein from meiotic cells and its partial reversal by a protein kinase, *Eur. J. Biochem.,* 95, 31, 1979.

61. **Radding, C. M.,** Homologous pairing and strand exchange in genetic recombination, *Annu. Rev. Genet.,* 16, 405, 1982.

62. **Hotta, Y., Tabata, S., Bouchard, R. A., Pinon, R., and Stern, H.,** General recombination mechanisms in extracts of meiotic cells, *Chromosoma,* 93, 140, 1985.

63. **Kmiec, E. B. and Holloman, W. K.,** Synapsis promoted by *Ustilago* rec I protein, *Cell,* 36, 593, 1984.

64. **Symington, L. S., Fogarty, L. M., and Kolodner, R.,** Genetic recombination of homologous plasmids catalyzed by cell-free extracts of *Saccharomyces cerevisiae, Cell,* 35, 805, 1983.

65. **Byskov, A. G.,** Regulation of meiosis in mammals, *Ann. Biol. Animale, Biochem. Biophys.,* 19, 1251, 1979.

66. **Chandley, A. C., Hotta, Y., and Stern, H.,** Biochemical analysis of meiosis in the male mouse. I. Separation and DNA labeling of specific spermatogenic stages, *Chromosoma,* 62, 243, 1977.

67. **Hotta, Y. and Stern, H.,** Zygotene and pachytene-labelled sequences in the meiotic organization of chromosomes, in *The Eukaryote Chromosome,* Peacock, W. J. and Brock, R. D., Eds., The Australian University Press, Canberra, 1975, 281.

68. **Hotta, Y.,** unpublished data.

69. **Hotta, Y. and Stern, H.,** unpublished data.

Chapter 2

CHROMOSOME STRUCTURE AND FUNCTION DURING OOGENESIS AND EARLY EMBRYOGENESIS

Aimée Hayes Bakken and Pamela Jean Hines

TABLE OF CONTENTS

I. INTRODUCTION

Oogenesis and early embryogenesis represent periods of most extreme change in the structure and function of chromosomes. Oogonia divide mitotically a number of times before each cell enters premeiotic S phase. During early meiotic prophase, the chromosomes undergo pairing and genetic recombination, and small amounts of DNA synthesis to ligate exchanged regions of the chromosomes. The primary oocytes arrest at diplotene. The oocyte nucleus may remain in this arrested state for a matter of days, weeks, months, or even years in the case of humans. However, it is in the diplotene stage of meiosis that oocytes are the most synthetically active. During this stage, the oocyte grows and stores all of the RNAs, proteins, and yolk materials needed for early embryonic development. Finally, shortly before ovulation/fertilization, meiosis resumes in the primary oocyte, transcription ceases, and the nucleus divides twice to produce one haploid nucleus and 2 to 3 polar bodies. The haploid oocyte nucleus is then joined by a haploid sperm nucleus and the first somatic diploid cell of the embryo is formed — the zygote. Initally, the zygote's nucleus undergoes numerous rapid and abbreviated cell cycles of S phase and mitosis, in which the primary goal is to rapidly multiply the nuclei and subdivide the cytoplasm in order to reduce the huge cytoplasmic: nuclear ratio of the fertilized egg to that of the typical somatic cell. During this time, only the replication function is stressed. Little or no transcription occurs in these cells. The embryo functions on stored maternal RNAs made during oogenesis either by the oocyte's chromosomes or by those of various adjacent nurse cells or follicle cells. The transcription function appears during mid-to-late blastula as the cell cycle slows down, ending the cleavage stage and adding a G1 and G2 phase to the embryonic cell cycle. Only now do the embryo's genes become active and the embryonic program of differential gene activity begins to unfold for determination and differentiation of the major organ systems and the many specialized cell types of a whole multicellular organism.

These many cellular divisions — mitotic, meiotic, abbreviated mitotic, and then typical mitotic again — require that the chromosomes go through repeated cycles of condensation and decondensation, initiation and cessation of DNA replication, and cessation and resumption of RNA synthesis (transcription). Some of the underlying mechanisms and regulatory factors which are responsible for these cyclic changes are known, while others remain in the realm of educated guesses or mere speculation. Clearly the structure of the nucleus — the nuclear envelope, pores, lamina, matrix, and packaging of the chromatin — are intimately involved in assuring the efficiency and fidelity of the major chromosome functions of transcription and replication.

In the next section of this chapter (II. Oogenesis), we have chosen to focus on specific alterations of the genome which have evolved in order to facilitate the synthesis and storage of RNA and proteins for use by the early embryo. Such alterations may include amplification of the rDNA, formation of extra-chromosomal nucleoli, and elaboration of "lampbrush chromosomes" during diplotene of meiosis. Clearly the interaction of proteins with DNA (and other proteins) in the chromosomes can alter chromosome structure and function. Many nuclear proteins are synthesized during oogenesis. Some are stored in the cytoplasm, and others within the swollen oocyte nucleus (the germinal vesicle). Such proteins will be discussed both within the context of oogenesis and during early embryogenesis, for it is the elucidation of their regulatory functions which is the next horizon for understanding the underlying genetic control of development.

In the third section (Embryogenesis), we examine the interrelationships between the length of the cell cycle and chromosome structure and function, specifically, the enormous replication capacity of the egg, what determines cell cycle length, activation of the embryonic genome, and aspects of the nuclear architecture which may contribute to changes in chromosome function.

II. OOGENESIS

A. Main Stages of Oogenesis

Oogenesis is a period of highly specialized synthetic activity. The chromosomes proceed partially through prophase I of meiosis, then enter into a period of very active RNA transcription. In some organisms (mammals) there is then a quiescent stage of variable duration, followed at some later time by a rapid growth phase in preparation for ovulation. Finally, the fully-grown oocyte is capable of maturation, resumption of meiosis, and ovulation (recent reviews[1-3]). Meiotic prophase I has been divided into several stages, in each of which the activity of the chromosomes differs. It must be stressed that the allocation of each stage is not definitive; one stage merges into the next.

In leptotene, the chromosomes begin to condense from the diffuse interphase chromatin. This is the beginning of meiotic prophase. Chromosomes are visible as fine threads, and may begin to appear beaded at the light microscope level, although most of the chromatin is still diffuse. They are attached by their telomeres to the nuclear membrane. In some organisms, the telomeres are arranged in one region of the nuclear envelope. This formation may facilitate correct association of homologs.

In zygotene, homologous chromosomes closely associate during a process called "synapsis". Synaptonemal complexes form as a proteinaceous ladder between the homologs, beginning at the tips and proceeding toward the center. Synapsis does not rely on DNA sequence homology, but rather on the associated chromosomal proteins (reviewed by Risley[2]). Chromosomes are thought to pair chromomere by chromomere. Synapsis between chromosomes which differ by an inversion may result in an homologously synapsed inversion loop, or an asynaptic segment. Resolution of these structures seems to differ in different organisms. For example, in the mouse the homologous inversion loop is disrupted and later heterologously resynapsed during pachytene.[4] However, in maize the homologously synapsed inversion loops persist through pachytene.[5,6] In *Lilium,* a small amount (0.2 to 0.3%) of the genome is replicated during zygotene.[7,8] The newly synthesized DNA sequences are low copy number sequences interspersed throughout the genome.[9] It is postulated that correct timing of the zygotene DNA synthesis is critical for chromosome synapsis.[10,11]

In pachytene, the chromosomes are fully synapsed, and become maximally condensed with the DNA organized into compact chromomeres along the length of each chromosome. Crossing-over occurs, and chiasmata are formed. There is some DNA synthesis also during pachytene, representing ~0.1% of the genome.[12] Synthesis of pachytene-DNA is mainly of the repair type, and may be involved in crossing-over and chiasma formation.[13] At the end of pachytene, fine lateral loops begin to project from the chromomeres along the chromosomal axis, forming the lampbrush chromosome configuration.

Diplotene is the stage of meiosis in which most mammalian and amphibian oocytes pause. The synaptonemal complexes are disassembled[2,14] and the homologs separate along their length, remaining attached only at chiasmata. The chromosomes retain the lampbrush conformation, and are extremely active in RNA synthesis. The lateral loops contain transcription units with densely packed RNA polymerase molecules and newly synthesized RNA fibrils. In amphibians, towards the end of diplotene, lateral loops of the lampbrush chromosomes retract into the chromomeres, and the chromosomes condense into an aggregate in the center of the germinal vesicle. Mammalian (human, mouse, rat) oocytes enter a resting diplotene stage, termed dictyotene. The oocytes become quiescent, awaiting the signal to resume growth and oocyte maturation (germinal vesicle breakdown and resumption of meiosis). Mouse lampbrush chromosomes uncoil, becoming longer and thinner.

In response to hormonal signals which initiate oocyte maturation, the germinal vesicle membrane breaks down, and the oocyte proceeds through meiotic metaphase I and meiosis II. Matured oocytes (unfertilized eggs) arrest at specific meiotic phases, awaiting fertilization to resume development. The specific time of arrest differs among species.

In the following sections, we will focus on the diplotene stage of oogenesis, when the oocytes are synthetically active. One of the main functions of oogenesis is to stockpile enough cellular components to sustain the embryo through early development. Stored components include ribosomes, mRNA, factors and enzymes responsible for transcription and DNA replication, mitochondria, and yolk. Two major schemes for storage have evolved: (1) the oocyte nucleus is transcriptionally active and synthesizes most or all of its own RNA (e.g., humans, mouse, birds, amphibians); (2) the oocyte nucleus is quiescent and relies on accessory cells, nurse cells, and/or follicle cells to synthesize RNA and proteins for transport to and storage in the oocyte (e.g., *Drosophila* and some other insects). In the former case, the replicated oocyte nucleus has four chromosomal copies of each single copy gene, which are all transcribed during the lampbrush phase of diplotene (see below). In the latter case, the nurse cells and sometimes the follicle cells become polyploid and then supply the oocyte with RNA from many transcriptionally active copies of the maternal genes.

B. NUCLEOLUS ORGANIZING REGIONS, RIBOSOMAL RNA (rRNA) GENE AMPLIFICATION, AND rRNA GENE TRANSCRIPTION IN OOCYTES

The synthesis of the various ribosomal RNAs and ribosomal proteins for assembly and storage of ribosomes in the oocyte is perhaps the greatest challenge for the oocyte. Enough ribosomes must be stored to translate the maternal mRNAs and many embryonic mRNAs during early embryogenesis. The large quantities of rRNA are accumulated in some organisms by activity of the oocyte chromosomal rRNA genes. In other organisms, the oocyte ribosomal RNA genes (rDNA) are amplified during late zygotene to early pachytene into thousands of extrachromosomal copies. These are expressed during oogenesis, while the chromosomal complement of rDNA may or may not be transcriptionally silent. The extrachromosomal copies of rDNA are destroyed shortly after fertilization and do not alter the diploid genome.

During *Xenopus* oogenesis (stages III to V), the 18S, 5.8S, and 28S rRNAs are synthesized from extrachromosomal amplified rRNA gene copies. About 5000 × 2N gene copies are replicated by a rolling circle mechanism, resulting in circles of from one to several hundred tandem copies of the rRNA gene.[15-17] The amplified rDNA then forms extrachromosomal nucleoli, which are arranged around the periphery of the germinal vesicle and actively synthesize rRNA. Electron microscopy shows they have the typical structure normally associated with somatic chromosomal nucleoli, with a granular cortex and fibrillar core.[18] The chromosomal nucleolar organizing regions in *Xenopus* are transcriptionally inactive during oogenesis.[19]

Ribosomal RNA is accumulated in *Acheta* (crickets) by two means. During early oogenesis, the rDNA is amplified[20,21] and exists as circles of rDNA, containing 1 to ~7 gene copies per circle, similar to the situation of *Xenopus*. The fine structural relationship of the amplified rDNA and the chromosomal nucleolus organizing region (NOR) rDNA has been analyzed by Troster et al.[22] During pachytene and early diplotene, the amplified rDNA forms a highly refractile mass. The amplified rDNA is seen as hundreds of transcriptionally active dense micronucleoli, in a hemispherical mass, centered around the "secondary nucleolar component", a highly refractile spherical body, during mid and late diplotene. The chromosomal nucleolus organizing region is visible as a puff-like structure, ~3 × 6 μm. When analyzed by thick section scanning transmission electron microscopy (STEM), the chromosomal nucleolus organizing region consists mainly of transcriptionally active rRNA genes, unlike in *Xenopus*, where the chromosomal rRNA genes are inactive.

In the chicken, although DNA synthesis has been observed by ^{3}H-thymidine incorporation in the oocyte nucleolus at a time typical of rRNA gene amplification,[23] no extrachromosomal amplified nucleoli have been identified.[24]

During human oogenesis, the chromosomal nucleolus organizing regions, which are located on the short arms of some of the acrocentric chromosomes, form a prominent nucleolus. As the oocyte progresses from diplotene to dictyotene stage (during fetal development), the short arms of acrocentric chromosomes associate with the nucleolus.[14] Several acrocentrics can be associated with one nucleolar mass. As the oocyte progresses towards dictyotene stage, blocks of heterochromatin from other chromosomes also become associated with the nucleolus. The nucleolus is transcriptionally very active during diplotene, and slows down somewhat at dictyotene stage.

In mouse, oocyte nucleoli are visible in leptotene and zygotene, but are not synthetically active. The two to three nucleoli which are visible during early prophase later fuse together into one large nucleolus during the dictyotene stage. Transcriptional activity is very high in nucleoli during diplotene and dictyotene stages, slows down when the oocyte follicle becomes fully developed, and is even less in maturing follicles.[25] The nucleolar synthetic activity is achieved by a very few ribosomal gene copies transcribing at maximal rate.

By the end of oogenesis in *Xenopus,* transcription in general ceases. Sims and Bakken[26,27] have demonstrated an inhibitor of transcription specific to rRNA genes which is present in *in vitro* extracts made from mature oocytes and fertilized eggs. This inhibitor may be accumulated gradually during oogenesis, and when concentrations become high enough, can inactivate transcribing rRNA genes. Inactivation of rRNA genes is a gradual process,[26,27] as shown by the electron micrographs in Figures 1 and 2. Active genes show maximal packing of RNA polymerase molecules with attached nascent RNP fibrils, alternating with smooth chromatin of the intergenic regions. The first stage of inactivation is seen as the fibrils become lost from densely packed polymerases through the gene region. Each gene appears to be independently controlled; adjacent genes do not necessarily show inactivation. Individual activation of rRNA genes is observed also in sea urchins,[28] *Drosophila,*[29] and *Oncopeltus.*[30] As the *Xenopus* oocytes mature, elongation of rRNA fibrils is apparently inhibited, and more and more fibrils are lost. Polymerases without RNP fibrils remain attached to the gene, and the intergenic region remains smooth. Subsequently, polymerase density on the gene gradually decreases. Finally, all polymerases are lost, and the genes and intergenic spacer regions adopt the conformation of nucleosomal chromatin.

C. LAMPBRUSH CHROMOSOMES AND TRANSCRIPTION

Synthesis of the accumulated maternal mRNA is accomplished by the lampbrush chromosomes, which transcribe a large array of sequences. Lampbrush chromosomes have been excellently and extensively reviewed.[31,32] We will note here some of the more recent observations and continuing issues.

Lampbrush chromosomes were orginally found in amphibian oocytes, and occur during the diplotene stage of oogenesis. They have been prepared and mapped from oocytes of a variety of organisms, including newts and *Xenopus,*[33-36] chicken,[24] and mouse.[15,118] Lampbrush chromosomes are well-extended chromosomes with large paired lateral loops extending from a chromomeric axis. Sister loops are believed to arise from the closely associated sister chromatids, and are usually symmetrical, although they may sometimes differ in size or transcriptional activity.[32] The chromosomes vary in overall size among different organisms, generally in proportion to the genome size. They are extremely active transcriptionally, as evidenced by light-microscopic ^{3}H-autoradiography and by direct visualization with electron microscopy. Transcription units are very large, often extending around most of the lateral loop.

In amphibians, which have genome sizes from 20 to 100 pg DNA, the lateral loops range in size from 50 to 200 μm. When viewed by scanning electron microscopy, the axis of lampbrush chromosomes in *Triturus* is seen to vary greatly in thickness.[37] Both sister chromatids can sometimes be resolved, entwined around each other. Insertion points of lateral loops are visible on the chromosome axis (Figure 3).

Lampbrush chromosomes have been studied in chicken by Hutchison,[24] where the gen-

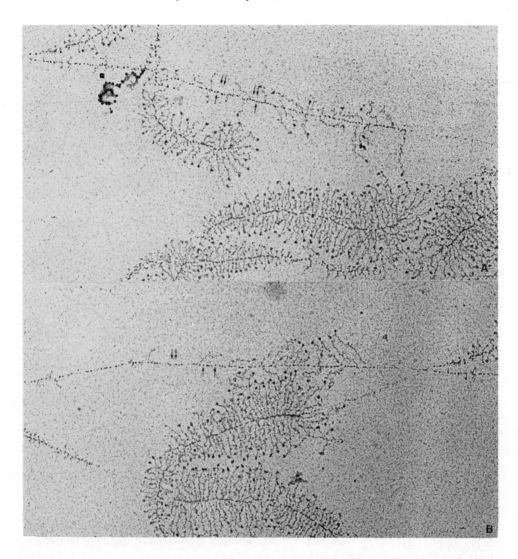

FIGURE 1.   (A) and (B) show several normal, fully active, ribosomal RNA genes. In addition, across the top of each micrograph is a partially loaded rRNA transcription unit reflecting the first events of rRNA gene inactivation in a mature stage VI *Xenopus* oocyte. Note the loss of rRNP fibrils (double arrows) from the gradient, the continued binding of the RNA polymerase I molecules (single arrow), and the fact that in each picture the fully active and partially active genes are present in the same extrachromosomal nucleolus. Each rRNA gene is activated independently and is inactivated independently. As the oocyte matures, a putative inhibitor accumulates in the nucleoplasm and binds to some portion of the active rDNA gene resulting in the progressive loss of nascent rRNA transcripts. (Magnification: A,B = × 26, 411.)

ome size is 1.2 pg. Lampbrush chromosomes are at their maximal extension in chicken oocytes of 1 to 3 mm diameter, in which the germinal vesicle is 200 to 400 μm diameter. The lampbrush chromosomes are similar in structure to amphibian lampbrush chromosomes, although smaller. The bulk of the DNA is in the condensed chromomeres along the chromosome axis. Loops average 10 to 15 μm, or 30 to 45 kbp in size, and contain ~10 to 15% of the total DNA. Transcription units are ~12 to 150 kbp in size. Centromeres are not obvious, and, as in amphibians, the morphology of the loops, spheres, and knobs can be used to identify certain chromosomes.

   Lampbrush chromosomes are found also in diplotene mouse oocytes. In the fetal ovary, oocyte chromosomes show a lampbrush configuration at the light microscope level around

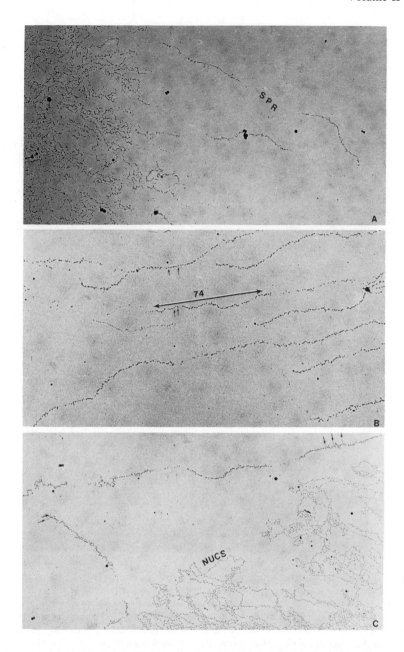

FIGURE 2.   (A), (B), and (C) show a later stage in the gradual inactivation of the ribosomal RNA genes. Note the RNA polymerases remain bound to the coding region while the spacer rDNA (SPR) between transcription units is free of polymerases and nucleosomes. It is not known how much or what protein species remain bound to the spacer DNA, but the fact that it is visible with phosphotungstic acid indicates that some protein is present. (Bar in (B) indicates one rRNA repeat with 74 RNA polymerase molecules: normal range on active genes is 60 to 80. Arrows indicate individual polymerase molecules.) A few short rRNP fibrils are still bound to the gene in (C) (arrows) in the same nucleolus where many genes have lost all fibrils and all polymerase molecules and have returned to the nucleosomal configuration seen in inactive chromatin (NUCS — lower right of picture). Notice that nucleosomes are smaller and stain less intensely than RNA polymerase molecules. (Magnification: A — × 12,600; B — × 18,120; C — × 11,580.)

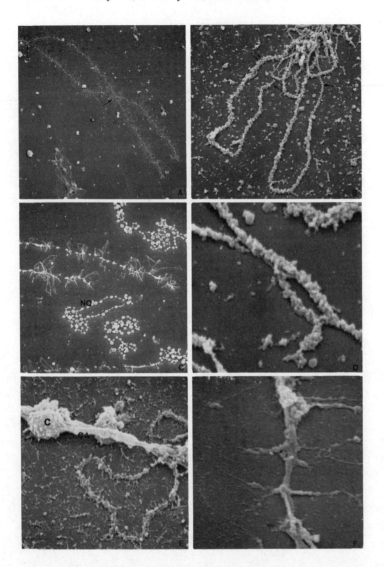

FIGURE 3.   Scanning electron micrographs of lampbrush chromosomes from *Triturus viridescens* diplotene oocytes. (A) A pair of homologous chromosomes joined by two chiasmata (arrows) where genetic recombination occurred during pachytene (× 330). (B) A pair of large lampbrush chromosome loops extending from one homolog. Note the thin end (single arrow) and the thick end (double arrow) of the loop matrix, reflecting the direction of transcription (× 1650). (C) Lampbrush chromosome plus extrachromosomal nucleoli (NO) exhibiting the "necklace" stage of their development (× 660). (D) A higher magnification of portions of several lampbrush loops showing the thick beaded nature of the RNP fibrils when they are isolated at neutral pH in 0.05 *M* KCl, which is slightly hypotonic for the amphibian nuclei and results in some unfolding of the matrix RNPs (× 6500). (E) A portion of one homolog showing a chromomere (C) which is not active in transcription (no lateral loops) and a pair of active lampbrush loops to the right. Note the bipartite nature of the chromosomal axis (CA) showing the two sister chromatids which are tightly paired during the diplotene lampbrush phase (× 3960). (F) This lampbrush chromosome was isolated in 0.025 *M* KCl at pH 7 which has resulted in most of the matrix RNP fibrils unfolding from the DNA so the "loops" are barely visible. The chromosomal axis and chromomeres retain their structure and the two sister chromatids remain wrapped around each other. (Magnification × 6600.)

day 17 to 18, with loops extending 1 μm to either side of the chromosomal axis. The chromosomes become synthetically active during late pachytene and diplotene, as they decondense.[25,38] When Miller spreads are prepared, the loops and chromosomal axis disperse into a disorganized network of chromatin. In comparison with amphibians, there is much less nascent RNA on mouse chromosomes. The transcription units are shorter, and except for 1 to 2 d around the time of birth, the RNA polymerase density is much reduced.[39] Many transcription units are read by only a single or a few polymerases, and the RNP fibrils are very long and folded. Synthetic activity declines once the oocyte becomes surrounded by a single layer of follicle cells. Unlike amphibians, a second phase of RNA synthetic activity occurs in the oocytes of the adult mouse when the oocytes are stimulated to resume growing prior to ovulation, as shown by ³H-uridine incorporation.[40] Miller spreads of these stages show similar low levels of polymerase density on individial transcription units.[41]

Termination of transcription in lampbrush chromosomes appears to be faulty or at least differs from that found in somatic nuclei. One possibility is that transcription does not respond to specific termination signals, but rather reads through a gene and terminates when the transcription complex runs into the base of the loop, or into another transcription unit.[42] *In situ* hybridization to the histone gene transcripts in the newt suggests that transcription of tandemly arrayed histone genes proceeds without terminating at the 3′ end of the gene, through the adjacent highly repeated satellite DNA sequences, up to the next gene cluster.[42] Post-transcriptional processing would then generate mature mRNAs. The observation of portions of untranscribed, condensed DNA between transcription units within the same loop in the chicken lampbrush chromosomes, however, suggests an additional termination mechanism.[24]

Transcription initiation may also be unusual on lampbrush chromosomes. *In situ* hybridization to satellite DNA sequences in the histone gene locus of the American newt shows that transcription seldom begins at the histone gene promoters.[43] Combined with observations that processing at the 5′ end of the RNA is inaccurate in oocytes,[44] it has been proposed that these transcription units do not produce functional (correctly processed and capped) mRNA.[43]

Lampbrush chromosomes are unusual in that they transcribe a vast array of sequences, far greater sequence diversity than is found transcribed by somatic cells. Highly repeated DNA sequences are also transcribed in European newts[45] and in *Xenopus*.[46] Much of the RNA synthesized by lampbrush chromosomes does not persist to embryogenesis. In tube worms, small oocytes have RNA representing 10 to 12% of the complexity of the genome. However, in the large mature oocyte, only 3% of the sequence complexity is represented in the RNA.[47] Similarly, in amphibians, much of the sequence complexity transcribed early in oogenesis is lost by late oogenesis.[48] The fate or function of the excess RNA sequences is as yet unknown.

One unusual activity of RNA synthesized by lampbrush chromosomes is demonstrated by RNA transcripts of a newt satellite DNA sequence, which are produced by read-through transcription from adjacent genes. The satellite RNA transcripts have shown *in vitro* site-specific self-catalyzed cleavage[49] and could possibly be involved in RNA processing.

In amphibian lampbrush chromosomes, transcriptional activity is closely correlated with absence of DNA methylation. Chromosomal rRNA genes are highly methylated, while amplified extrachromosomal rRNA genes are not.[50] 5-Methylcytosine was localized in lampbrush chromosomes of the amphibian *Pleurodeles* using specific antibodies.[51] Methylated DNA was localized by immunofluorescence only on the condensed transcriptionally inactive chromosome axis. A similar reaction to the chromosome axis was observed with immunogold staining and electron microscopy. In addition, occasional labeled globules were observed along a loop. By observing partially spread lampbrush chromosomes, it was concluded that methylated (immunostained) regions of loops could correspond to the occasional occurrence of a nontranscribed portion between two transcription units on a single lateral loop.

Monoclonal antibodies have been prepared against chromosomal proteins of newt lamp-brush chromosomes.[52] Several staining patterns have been observed, staining nucleoli, chromosome axis, and lateral loops. Two of the proteins recognized are found on almost all loops, and may be involved in RNA processing or storage. Two other antigens were specific to a limited number of loops, and may be gene-specific RNA processing or storage proteins.

D. SEQUESTRATION BY THE GERMINAL VESICLE

As the oocyte develops, and goes from egg to embyro, the chromosomes go through several cycles of change in transcriptional activity. It is not unlikely that various associated proteins and/or modifications of those proteins direct these changes. The germinal vesicle plays a role in the correct sequestration of potentially active stored components. The germinal vesicle in *Xenopus* has a larger volume than all nuclei through the early blastula stage. Many proteins are selectively retained in the germinal vesicle during oogenesis and released during germinal vesicle breakdown at oocyte maturation. Some of these proteins migrate back into cleavage-stage nuclei, some at developmentally specified stages, and some in a tissue-specific manner.[53,54] Clark and Masui[55] have shown that transformation of highly compacted mouse sperm nuclei into metaphase chromosomes following fertilization requires elements from the ooctye germinal vesicle, the oocyte cytoplasm, and proteins synthesized during oocyte maturation. Aberrations in the organization of components or timing of events can result in nonproductive fertilization, such as the premature sperm chromosome condensation observed in oocytes arrested at metaphase II.[56]

Protein sorting into the germinal vesicle may be either by selective passage of molecules through nuclear pores, or by nonselective diffusion through pores and selective binding of proteins to elements of the nucleoplasm. At least one case of sorting at the level of the membrane has been demonstrated by Zimmer et al.[57] When nucleoplasmin is microinjected into the cytoplasm of *Xenopus* oocytes, it accumulates up to 30-fold in the germinal vesicle. However, demembranated nuclei actually lose nucleoplasmin content.

In fern oocytes, nuclear protrusions develop which apparently correlate with extrusion of nucleolus-like osmiophilic bodies into the cytoplasm.[58] These bodies consist mostly or entirely of nonhistone protein. This may be a means of temporarily storing nonhistone chromosomal proteins in preparation for sperm chromatin decondensation and activation or for subsequent use by the developing embryo.

Excess snRNP proteins are synthesized and stockpiled in the cytoplasm of fully grown *Xenopus* oocytes. These proteins are normally excluded from the nucleus until endogenous snRNA synthesis is resumed at the midblastula transition. If extra snRNA is microinjected into the oocyte cytoplasm, it will associate with the proteins and migrate into the germinal vesicle, accumulating 30 to 60-fold.[59] The karyophilic signal is only functional when both the RNA and protein are complexed.[60]

Histones are also accumulated and stored during oogenesis by several strategies (reviewed by Poccia[61]). This provides enough histone for the cleaving embryo to reach the stage at which the embryonic genome is activated, ranging from about 4000 nuclear equivalents in *Xenopus* to 2 to 3 nuclear equivalents in mouse. Histone protein or histone mRNA may be accumulated. Histone proteins may be stored in the germinal vesicle and may also be found in lower concentrations in the cytoplasm. In *Xenopus*, histones are found in the germinal vesicle in complex storage particles. In sea urchin unfertilized eggs, mRNA for alpha-histones is localized in the egg pronucleus and perhaps by this is prevented from being translated during oogenesis. By contrast, cleavage-stage histone mRNA is localized in the cytoplasm and is translated during oogenesis and immediately following fertilization. At the time of pronuclear envelope breakdown, the alpha-histone mRNAs are released to the cytoplasm and synthesis of alpha-histones resumes. Mouse oocytes synthesize histone mRNA, which is rapidly degraded following fertilization, but sequester 2 to 3 diploid equivalents of histone H4 (and presumably other histones), which are stored in the germinal vesicle for use immediately following fertilization.

From these examples, it is clear that the germinal vesicle, by virtue of its ability to selectively compartmentalize certain molecules during oogenesis, plays a regulatory role in establishing the metastable state of a fully grown oocyte. Once the germinal vesicle has broken down, the subsequent disorganization and reorganization of the oocyte nucleus and cytoplasm allows for successful fertilization and further development.

III. EMBRYOGENESIS

A. CYTOLOGICAL CHANGES OF EMBRYONIC CHROMOSOMES

It has been known for many years that in some species, for example *Xenopus,* sea urchin, and *Drosophila,* the length of the cell cycle is much shorter during cleavage than it is by mid-to-late blastula stage, and that new RNA synthesis does not begin or does not become needed by the embryo until late blastula to early gastrula stage. This relationship between cell cycle length and the time of embryonic genome activation and its potential significance for regulating both replication and transcription has intrigued investigators for many years. Numerous alterations in cells and their chromosomes occur at the end of cleavage.[62-64] In *Xenopus,* Newport and Kirschner have characterized this "midblastula-transition" (MBT) which occurs after 12 rapid synchronous divisions.[64] They have shown that the cell cycle slows and becomes asynchronous, the cells become motile, and embryonic transcription is initiated. Edgar et al.[65,66] have studied a similar phenomenon in *Drosophila,* which occurs at the time of cellularization of the syncytial blastoderm (cycle 14). In this animal, the slowing of the cell cycle begins at cycle 10, when the nuclei have migrated to the surface of the embryo and are surrounded with cortical cytoplasm. Embryonic transcription is first detectable in cycle 11 or 12, and cell synchrony is lost in cycle 14. These studies have demonstrated that changes in cell cycle length during early embryogenesis have profound implications for patterns of chromosome structure and function.

In this secton we will examine the cytological changes observed in the morphology of embryonic chromosomes at both the light microscope and electron microscope levels of resolution. We will discuss how these morphologies reflect what is known about the chromosome functions of replication and transcription during embryogenesis. In addition, we will attempt to distill what is known about the cause and effect relationships between the changes in the cell cycle and changes in chromosome structure and function. When one considers the underlying genetic control of developmental processes, one must be constantly aware of the temporal aspects of the problem. The embryo is developing from a single diploid cell (the fertilized egg) into a highly complex multicellular organism; thus, there must be differential activation of specific genes at different times in different cells. It is well established that all the cells resulting from division of the initial zygote have the same genome. Even in highly mosaic embryos the nuclei in cleavage stage and, in some cases, even blastula embryos can be shown to still be totipotent, when transplanted into an appropriate enucleated early embryo (ascidian, snails, *Drosophila,* etc., see Reference 67 for a superb review). It has been proven that various cytoplasmic determinants are laid down in the developing oocyte by the maternal genome and the totipotent zygotic nuclei, which become segregated into the cytoplasm during cleavage, will follow those determined paths of development (e.g., polar lobe determinants for snail muscle cells, germ cell determinants in the vegetal, or posterior pole of the frog or *Drosophila* embryo). Therefore, those cytoplasmic substances must exert an effect on the chromosomes in the totipotent nuclei. However, when and how that occurs is unknown. Thus, we must look at changes in the early embryo's chromosomes which may alter their receptivity to such determinative influences.

B. CLEAVAGE STAGE CELL CYCLE AND REPLICATION

In normal mitotically active somatic cells, the length of S phase is species-specific and to some degree tissue specific. Most actively growing plant and animal cells complete a full cell cycle every 10 to 20 h. For an average mammalian cell, S phase for DNA synthesis lasts about 7 h out of a 16-h cycle. In contrast, S phase during embryonic cleavage may be as short as 4 min in *Drosophila*, less than 20 min in *Xenopus*, or as much as 4 h in the mouse.[68] How is this possible? In most cases it appears to be the result of replication being initiated simultaneously at many more sites along the DNA in cleavage nuclei. Electron microscopy and light microscope fiber autoradiography have demonstrated much shorter distances between origins of replication in early embryos than are found in somatic cells. In *Drosophila*, the distance between origins in cleavage nuclei is 2 to 4 μm of chromatin (6 to 12 kb of DNA) and 10 to 20 μm in cultured somatic ells.[69-72] In tissue culture cells from Chinese hamster or rat, the origin to origin distance is 50 to 120 μm.[73] In cultured *Xenopus* and *Triturus* cells, the origin to origin distance is 60 to 100 μm, but in early blastula stage *Triturus* embryos, it is only 6 to 10 μm.[74] There may be alterations in replication enzyme efficiency from species to species, but it is believed that the primary contributor to the fast pace of DNA synthesis during cleavage is the number of replication origins *and* the ready availability of DNA polymerase and other replication factors which are necessary both for initiation and elongation of replication forks. Transplantation of nuclei from differentiated tissues and from other species into *Xenopus* eggs presented a graphic demonstration of the enormous replication capacity of the egg.[75] Not only did these transplanted nuclei cease RNA transcription (similar to cleavage nuclei), but they became swollen with nuclear proteins migrating in from the egg cytoplasm and began synthesizing DNA within 90 min after transfer. The rapid replication for the transplanted nuclei was in sharp contrast to their normal 7 to 10 h S phase. A similar increase in nuclear volume is observed during the cleavage cell cycle both in *Xenopus* embryos[76] and in *Drosophila* embryos.[77-79] After the MBT in *Xenopus*, the interphase nuclear diameter is about 12.5 μm, whereas in early cleavage nuclei it swells to 16.7 μm. Similarly in *Drosophila*, the nuclear diameter swells to 11 μm during cleavage interphase, while it is only 5 to 7 μm after the cellular blastoderm stage. The increased size of the nucleus is correlated both with the rapid influx of particular nuclear proteins and also with the decondensation of the telophase chromosomes, which makes them accessible to the replication machinery. Foe and Alberts[79] observed the largest increase in interphase nuclear volume in the earlier (1 to 9) cleavage cycles of *Drosophila* embryos. These values decreased in cycles 10 to 14. These authors suggested that nuclear volume may be directly proportional to the instantaneous rate of DNA synthesis. Alternatively, it could be determined by the degree of chromosome condensation at telophase. The nuclear membranes reform around the telophase chromosomes,[76] and thus the length of the chromosomes may well affect the total surface area of the reformed interphase nucleus. For example, the degree of condensation is less in cleavage stage *Drosophila* metaphase chromosomes than is seen in the meiotic divisions of the oocyte genome[118] and also is less than is seen later in larval neuroblasts.[119] Less condensation of the chromosomes through the cleavage cell cycles agrees with the observations that no cytologically condensed heterochromatin is formed until cellular blastoderm formation at cycle 14[80,81] (and Schultz, Ashton, and Mahowald as cited in Reference 82).

C. DETERMINATION OF THE LENGTH OF THE CELL CYCLE

It is thought that the length of S is determined by the amount of DNA (genome size) and the number of replication origins used in that cell type. Thus, *Drosophila*, with a relatively small genome size, 1.4×10^8 bp of DNA, and no cytokinesis during its syncytial cleavage divisions, has a very fast cleavage cycle (9 min for the first 10 cycles lengthening to 10′ = cycle 10, 12′ = cycle 11, 16′ = cycle 12, 25′ = cycle 13, >80′ = cycle 14 at 22°C). *Xenopus*, with a large genome size, 3.1×10^9 bp, and a large pool of stored histones and DNA replication proteins, can replicate all of its DNA in less than 20 min per 30 to 35 min cleavage cycle, lengthening to one or more hours after the MBT. In addition, there is considerable evidence that the nuclear:cytoplasmic ratio in cleaving embryos influences the length of interphase. However, recent studies have called into question the role of nuclear components in determining the timing of the cell cycle. Hara et al.[83] enucleated fertilized *Xenopus* eggs and observed that they underwent periodic surface contraction waves (SCWs) that were synchronous with the actual cleavages in control embryos. Thus, nuclear syntheses and divisions can be separated from an inherent cell cycle clock, which itself may regulate those nuclear activities. However, when the enucleation is performed on *oocytes*, no SCWs are observed after oocyte activation.[84] Maturation or M phase promoting activity (MPF) and mitotic kinase oscillations do occur and by themselves can induce simple repeated cell cycles of alternating M and S phases.[85-87] Thus, something released by the oocyte nucleus into the egg cytoplasm during maturation appears to be required for the SCWs in the embryos, but the oscillations of MPF and mitotic kinases are independent of the mechanism underlying the SCWs. What does regulate the timing of the MBT which was postulated to result from the decreased cytoplasmic:nuclear ratio? Several pieces of data point to the lengthening of interphase being the result of a delay in the onset of metaphase, possibly because some component which is required for stimulating/activating MPF becomes titrated by the increasing number of nuclei.[85] The resolution of these confusing data will require further investigation. Cyert and Kirschner[87] have established an *in vitro* system, which regulates MPF activity and thus should facilitate this endeavor.

D. EFFECT OF THE SHORTENED CELL CYCLE ON THE POTENTIAL FOR RNA TRANSCRIPTION

Autoradiographic and/or radioisotope incorporation studies demonstrated that RNA synthesis was detectable in the ovulated oocyte and continued in increasing amounts in early *Xenopus* embryos, although ribosomal RNA (rRNA) synthesis was not detected until gastrulation.[88] A similar pattern of RNA synthesis was observed in early sea urchin embryos[89,90] and in *Drosophila* at preblastoderm stages.[91] More recent studies using improved methodologies (better permeabilization of eggs and embryos, microinjection techniques, electron microscopy for direct visualization of stage-specific transcription, and identification of mitochondrial vs. nuclear transcription) have resulted in more accurate timing of the actual onset of nuclear non-ribosomal (RNA polymerase II and III) and ribosomal RNA (pol I) transcription.

McKnight and Miller[29,72] analyzed transcription in *Drosophila* embryos by electron microscopy. At syncytial blastoderm (cycles 11 to 13), they observed a low frequency of short (1 to 2 μm) transcription units which were fully packed with RNA polymerase molecules, but were clearly non-ribosomal in their morphology. Active ribosomal RNA transcription units were first observed in the first 5 to 10 min of cellular blastoderm (cycle 14) which correlated well with the first appearance of nucleoli in these embryos.[92] In addition, a second class of non-ribosomal transcription units appeared at this stage which were $4 \times$ as long (8 μm) and frequently had a much lower density of RNA polymerases and attached RNP fibrils. These authors suggested three interesting relationships between the length of the cell cycle and the *Drosophila* transcription patterns: (1) that the large increase in length of the non-ribosomal transcription units (seen after the cell cycle had slowed in cycle 14) perhaps reflected the fact that, given known transcription rates in other systems, there would not be sufficient time in earlier cell cycles to transcribe such a long "gene"; (2) that the nucleoli and the individual, active ribosomal RNA transcription units appeared before the onset of S phase in cycle 14, heralding the addition of a G1 phase and more time for transcription; and (3) if replication and transcription were incompatible on the same length of chromatin, this new pattern of early transcription might well alter the pattern, accessibility, and number of replication origins, thereby lengthening S phase.

Similar observations were made in sea urchin embryos.[28,93] Much longer transcription units were visualized at gastrulation than were seen during cleavage and blastula stage embryos. This was correlated with the lengthening cell cycle observed after midblastula stage. These authors suggested that the longer gastrula genes could be a new class of genes, activated for embryonic determination and further development, or they might actually be embryonic genes whose transcription was initiated and then terminated prematurely during earlier stages by the short cell cycle. The shorter length of the transcription units during cleavage was calculated to be appropriate for the length of time allowed for transcription during these shorter cell cycles. It was not possible to determine or compare the identity of these non-ribosomal genes at the different stages of embryogenesis.

The importance of the relationship between the cell cycle length at cleavage stage and the potential for RNA transcription has been examined in depth by Newport and Kirschner[63,64] and Kimelman et al.[94] in *Xenopus* embryos and by Edgar et al.[65,66] in *Drosophila* embryos. Newport and Kirschner hypothesized that the fertilized egg has a titratable factor which interacts with and blocks transcription from the nuclear DNA. They showed that plasmid DNAs microinjected into fertilized eggs were prematurely transcribed, suggesting that when this factor is sufficiently titrated out, then generalized transcription can begin along with the other aspects of the midblastula-transition. Edgar et al. have reported elegant experiments with *Drosophila* which show that the nuclear-cytoplasmic ratio of the cleavage stage embryo has a definite effect on the length of the cell cycle. Haploid embryos compensate for their reduced amount of DNA by going through one additional round of nuclear division before the onset of the MBT. Reducing the amount of cytoplasm per nucleus by ligation of the *Drosophila* embryo can cause an early MBT. Similar relationships have been demonstrated in starfish embryos,[95,96] in the Japanese newt,[62] and in sea urchins.[97]

Edgar et al.[65,66] then went on to show that the duration of the mitotic cycles is slowed even before cellular blastoderm formation, i.e., the cycles begin slowing down from cycles 10 through 14. Transcription of some genes has been detected as early as cycle 10,[98,99] but maximal transcription is attained in cycle 14.[100] The transcription patterns are not a direct result of the alteration in nuclear-cytoplasmic ratio, but rather a response to the gradually lengthening cell cycle, specifically, the lengthening of interphase. When very short (10 min) pulses of ^{32}P-labeled RNA are analyzed from precisely timed embryos, the initiation of synthesis of rRNA, tRNAs, 5S RNA, snRNAs, polyA$^+$ mRNAs, and histone mRNAs are detectable during cycle 11 or 12. Histone genes are transcribed during the S phase in conjunction with DNA replication. The other RNAs are transcribed only in a putative G2 period, which is presumed to be added on to the length of S phase. Interestingly, rRNA synthesis seems to increase in a cumulative fashion from one cycle to the next, i.e., the rate of rRNA synthesis 20 min into cycle 14 was significantly greater than that measured after 20 min into cycle 13. This suggests to us that those ribosomal genes which were activated in cycle 13 retained their "activated state" through the next round of replication, and perhaps additional rRNA genes were then activated during cycle 14. This would be a very interesting system in which to look for the activating factors and coincident alterations in chromatin structure during and following gene activation.

E. THE CURRENT VIEW REGARDING THE MBT AND THE ACTIVATION OF THE EMBRYO'S GENES

Fertilization activates a series of molecular and morphogenetic events, which are initially governed by stored maternal gene products. Then, cleavage ensues — a series of rapid cell cycles involving S phase and mitosis which may permit histone mRNA transcription during S phase but not other gene transcription. Toward the end of cleavage, the cell cycles start to slow down, approaching the normal rate for somatic cells. The lengthening of the cell cycles appears to be regulated in some as yet unelucidated fashion by the decreasing cytoplasmic:nuclear ratio of the embryo. The mechanism for activation of embryonic transcription also has not been elaborated. Low levels of transcription occur before the MBT (in *Xenopus*) or cycle 14 (in *Drosophila*). Premature slowing or blocking of cell cycle progression turns on transcription prematurely. Blocking transcription does not affect the onset of cell motility or the changes in cell cycle length associated with the MBT or cycle 14 (in *Drosophila*). Blocking protein synthesis and/or blocking DNA replication during cleavage both result in the premature activation of transcription of RNA pol II and III genes (from cycle 10 onward in *Drosophila*).

These data, in conjunction with our studies on transcription extracts from fertilized eggs, cleavage, blastula, and gastrula stage *Xenopus* embryos,[26,27] indicate that all the transcriptional machinery is present, but something is blocking the assembly of active transcription complexes on the chromatin. In the case of the ribosomal RNA genes, which are activated 4 h after the MBT in *Xenopus* (and 1 to 2 cycles after other genes in *Drosophila*), there appears to be a specific inhibitor. This was shown first by microinjecting rDNA plasmids into *Xenopus* embryos[101] and observing that they cannot be prematurely activated. They await activation of the embryo's rRNA genes. This is in contrast to the premature transcription of tRNA (pol III) genes microinjected by Newport and Kirschner. Thus, the ribosomal RNA genes appear to have additional requirements for activation during embryogenesis. Similar results are achieved with our embryo extracts. There is an inhibitor present in the fertilized egg extracts, which can be diluted out in a test tube, and is naturally degraded or diluted out in extracts from gastrulae and neurulae. Whether or not new factors must be synthesized by the pol II genes to activate the embryonic ribosomal genes is being tested.

F. ACTIVATION OF EMBRYONIC GENES
1. Chromatin

Significant changes in the chromatin surrounding particular genes have been documented in a number of systems (see Reference 102 for an excellent review). Such changes as demethylation, appearance of nuclease hypersensitive sites within 1 kb upstream of the gene, torsional stress on the domain surrounding a gene, and the acquisition of specific protein factors have all been implicated in the activation and maintenance of a gene in a transcriptionally activated or committed state. Numerous examples have shown that *in vitro* assembly of transcriptionally active genes (and/or hypersensitive sites on those genes) requires that specific factors must be present during chromatin assembly in order to form an active gene.[102] Thus, it is likely that some embryonic gene activation occurs during DNA replication. If such protein factors are not present until, e.g., cycle 10 in *Drosophila*, transcription cannot be activated. Nuclear migration from the interior yolk to the cortical cytoplasm is completed during cycle 10. One hypothesis is that "activation proteins" are being made at the time and/or they have accumulated to the required concentration to confer activational competence on all portions of the genome. Zalokar[77] showed that the pole cells (which are the forerunners of the germ cells) do not become transcriptionally active until several cycles after cycle 14. Even though their nuclei also migrate into the cortical cytoplasm, they become cellularized prematurely, at cycle 10. Their cell membranes may serve to protect or isolate the pole cell nuclei from activation factors which are putatively synthesized in the cortical cytoplasm of the syncytial blastoderm.

2. Nuclear proteins

Numerous examples of the synthesis, storage and/or transport of nuclear proteins in embryos have appeared recently. When the germinal vesicle breaks down, late in oogenesis, proteins are released to the cytoplasm including DNA replication enzymes, actin, and nucleoplasmin, which is crucial for assembly of histones into nucleosomes. A large store of both histones and nucleoplasmin are present, e.g., in the *Xenopus* egg, in readiness for the rapid rounds of DNA replication. In oocytes of many species, additional histones will have to be made during cleavage, to meet the demands of the vast amounts of newly replicated DNA. In a number of species, different classes of histones are produced as development proceeds. Early and late histones are found in *Xenopus, Drosophila,* sea urchins, echiurid worms, clams, and snails.[61] The most complex transitions in histone types occur in sea urchins. Variants of H1 histone range from embryonic to cleavage to blastula type and late forms of postblastula forms.

Although the nearly universal distribution of histones and their high degree of amino acid conservation argue that they must play a critical role in the cell's physiology, in large measure their functions or modulations of chromatin structure and activity remain obscure. Early type H1 histone accumulates in some embryonic cells which cease division early in sea urchin development.[103] Perhaps these histones play some role in programming the cell's DNA or preventing it from being altered by the activities of surrounding cells. Histone variants in mammalian cells have been classified as replication-dependent, partially replication-dependent, replication-independent, minor, and tissue-specific.[104] This list suggests several characteristics of these proteins: their synthesis may or may not be tightly linked to DNA replication; they may be tissue specific such as the H5 variant in erythrocytes; certain variants may be found abundantly in differentiated tissues which have ceased cell division; histone types change through several distinct periods of early embryogenesis. Thus, histone remodeling, substitution of one DNA binding protein for another, is assumed to have major effects on the functioning of that chromatin. It is suggested that the different histone variants may serve different functions and that some might bind to restricted portions of the genome, thereby conferring a specificity of function on histones that has yet to be verified.

Other oocyte nuclear proteins may be released to the egg cytoplasm and then differentially reaccumulated in the embryonic nuclei. Dreyer[105] has described two classes of such proteins; one which is accumulated in *Xenopus* pronuclei and cleavage nuclei, and another class, which is sequestered in the cytoplasm until blastula or gastrula stage, at which point each protein species re-enters the nucleus according to its own timetable. Dreyer tested the possibility that different translocation efficiencies could account for the stage-specific differences in accumulation by altering the length of the cell cycles, but certain proteins were still excluded from the nucleus. Thus, it appears that some proteins may await a factor for transport, such as an RNA molecule, or be modified to reveal a karyophilic domain, which then targets them to the nucleus. Small nuclear RNA proteins are synthesized in the oocyte and remain in the *Xenopus* embryo cytoplasm until snRNAs are synthesized at the 12th cleavage.[60,106] The combined snRNPs are then transported to the nucleus. Similar phenomena have been documented in mouse embryos, but the time frame is moved forward greatly. Embryonic genes are activated in the 2 cell stage. Maternal mRNAs disappear shortly thereafter. Newly synthesized proteins in the first 3 cleavages do not show any male vs. female differences, but the new embryonic proteins drastically alter the gel patterns of nuclear proteins as compared with those from the 1 cell stage, in which only maternal proteins are present. Presumably, these embryonic nuclear proteins are remodeling the embryonic chromatin, although gene-specific examples are as yet lacking.

G. NUCLEAR PROTEINS, NUCLEAR ARCHITECTURE, AND SUBSEQUENT CHANGES IN GENE EXPRESSION

In the discussions above we have referred to changes in nuclear volume, changes in degree of chromosome condensation, changes in the programming or packaging of chromatin, and alterations in the kinds of nuclear proteins sequestered by or excluded from the nucleus of the oocyte and the embryonic cells. These dynamic aspects of nuclear structure, through these cycles of meiotic and mitotic divisions, have until recently been inaccessible to experimental investigation. Recently, a number of *in vitro* assembly systems have been developed which should increase our understanding of these processes in the near future (see Reference 107 for an excellent review). The study of cleavage stage embryos, with their simplified cell cycle, may help to unravel the mysterious, rapid reformation of the nuclear envelope from components surrounding the chromosomes during division. Montag et al.[76] have analyzed the reassembly and massive fusion of the nuclear membrane components using pre-MBT *Xenopus* embryos prepared for electron microscopy. Thus, the membrane components surrounding each chromosome contribute enough material to assemble the unusually large surface area of the cleavage stage interphase nuclear envelope. Beneath this layer is the nuclear lamina, a network polymerized from four different lamin proteins. The lamins are thought to be important for stabilizing the nuclear envelope and for serving as attachment sites for chromosomes (chromatin) to the nuclear envelope. In this regard, it may be significant that in *Xenopus* the nuclear lamina of the germinal vesicle in mature oocytes and in all cleavage nuclei up to the MBT is made of only one species of lamin (L III). At MBT, a second type of lamin (L I) is synthesized and added to the lamina, and at gastrulation, a third, L II, is newly synthesized and added to the lamina. In adult nuclei, L I and L II predominate, whereas the nuclei of terminally differentiated cells, such as neurons and heart muscle cells, contain all three lamins. What role these different proteins may play in regulating nuclear structure and function is yet to be determined.[108-110]

Nucleoplasmin is the most abundant "nuclear protein" in *Xenopus* oocytes and is also found plentifully in early embryos. It was originally described as the nucleosome assembly factor[111] and has been proposed also as a factor that might disassemble nucleosomes to permit transcription of active genes.[112] Nucleoplasmin isolated from eggs is larger and assembles nucleosomes *in vitro* much more efficiently than the species found in oocytes.[113] Such modified forms could theoretically facilitate the reprogramming of embryonic DNA in conjunction with the stage-specific histone variants described in different species of embryos. Most of the stored oocyte histones are complexed with nucleoplasmin and two other karyophilic proteins, N1 and N2. How these stored protein complexes interact with the embryonic chromatin in cleavage is not known, but N1 and N2 have nuclear targeting signals in their amino acid sequence. They may therefore have some function in supplying the "right" histones to the nucleus at the right time during early development. N1 and N2 are also found in somatic cells.[114]

Several examples have been given in previous sections in which nuclear proteins have been shown to be released from the germinal vesicle and then reaccumulated at different stage-specific times during early embryogenesis.[60,105] Other types of studies also indicate that specific proteins enter the embryonic nuclei at stage-specific times to activate or initiate specific gene expression. Maxson et al.[115] have demonstrated the differential stimulation of sea urchin early vs. late histone genes in *Xenopus* oocytes. They injected cloned sea urchin histone genes with and without a gastrula-stage nuclear extract made from sea urchin embryos. A factor from the extract bound specifically to a 105 bp region surrounding the initiation site of the late histone genes and stimulated that gene preferentially over early histone genes. In another study, an enhancer sequence has been identified in *Xenopus* which is responsible for activating transcription of its adjacent sequence precisely at the MBT.[116] This 74 bp sequence is located 700 bp upstream of the promoter. Examples of stage-specific gene expression have been documented also in *Drosophila,* when cloned genes are microinjected into early embryos.[117] Of the injected DNA, 20% enters the embryonic nuclei and presumably becomes subject to the same stage-specific and tissue-specific regulatory factors as the endogenous genes. The actual mechanism by which these protein factors and their target sequences are regulated is currently under investigation.

With the availability of cloned gene sequences and our increasing capability to compare their transcription and regulation *in vitro* and *in vivo,* many of the aspects of regulating chromosome structure and function during oogenesis and early development are now within reach of experimental solution.

REFERENCES

1. **Austin, C. R. and Short, R. F.,** *Reproduction in Mammals. 1. Germ Cells and Fertilization,* Cambridge Unviersity Press, New York, 1982.
2. **Risley, M. S.,** The organization of meiotic chromosomes and synaptonemal complexes, in *Chromosome Structure and Function,* Risley, M. S., Ed., Van Nostrand Reinhold, New York, 1986, 126.
3. **Browder, L. W., Ed.,** *Developmental Biology: Oogenesis,* Vol. I, Plenum Press, New York, 1985.
4. **Moses, M. J., Poorman, P. A., Roderick, T. H., and Davisson, M. T.,** Synaptonemal complex analysis of mouse chromosomal rearrangements. IV. Synapsis and synaptic adjustment in two paracentric inversion, *Chromosoma,* 84, 457, 1982.
5. **Maguire, M. P.,** A search for the synaptic adjustment phenomenon in maize, *Chromosoma,* 81, 717, 1981.
6. **Anderson, L. K., Stack, S. M., and Sherman, J. D.,** Spreading synaptonemal complexes from *Zea Mays.* I. No synaptic adjustment of inversion loops during pachytene, *Chromosoma,* 96, 295, 1988.
7. **Hotta, Y. and Stern, H.,** Analysis of DNA synthesis during meiotic prophase in *Lilium, J. Mol. Biol.,* 55, 337, 1971.
8. **Ito, M. and Hotta, Y.,** Radioautography of incorporated ³H-thymidine and its metabolism during meitoic prophase in microsporocytes of *Lilium, Chromosoma,* 43, 391, 1973.
9. **Hotta, Y. and Stern, H.,** Zygotene and pachytene-labeled sequences in the mieotic organization of chromosomes, in *The Eukaryotic Chromosome,* Peacock, N. S. and Brock, R. D., Eds., Australian National University Press, Canberra, 1975, 283.
10. **Takegami, M. H. and Ito, M.,** Effect of hydroxyurea on mitotic and meiotic divisions in explanted lily microsporocytes, *Cell Struct. Funct.,* 7, 29, 1982.
11. **Hotta, Y., Tabata, S., and Stern, H.,** Replication and nicking of zygotene DNA sequences, *Chromosoma,* 90, 243, 1984.
12. **Howell, S. H. and Stern, H.,** The appearance of DNA breakage and repair activities in the synchronous meiotic cycle of *Lilium, J. Mol. Biol.,* 55, 357, 1971.
13. **Stern. H.,** Chromosome organization and DNA metabolism in meiotic cells, in *Chromosomes Today,* Vol. 7, Bennett, M. D., Bobrow, M., and Hewitt, G., Eds., Unwin-Hyman, Winchester, MA, 1981, 94.
14. **Vagner-Capodano, A. M., Hartung, M., and Stahl, A.,** Nucleolus, nucleolar chromosomes, and nucleolus-associated chromatin from early diplotene to dictyotene in the human oocyte, *Hum. Genet.,* 75, 140, 1987.
15. **Hourcade, D., Dressler, D., and Wolfson, J.,** The nucleolus and the rolling circle, *Cold Spring Harbor Symp. Quant. Biol.,* 38, 537, 1973.
16. **Bird, A. P., Rochaix, J.-D., and Bakken, A. H.,** The mechanism of gene amplification in *Xenopus laevis* oocytes, in *Molecular Cytogenetics,* Hamkalo, B. A., and Papaconstantinou, J., Eds., Plenum Press, New York, 1973, 49.
17. **Rochaix, J.-D., Bird, A., and Bakken, A.,** Ribosomal RNA gene amplification by rolling circles, *J. Mol. Biol.,* 87, 473, 1974.
18. **Miller, O. L.,** Fine structure of lampbrush chromosomes, *Natl. Cancer Inst. Monogr.,* 18, 79, 1965.
19. **Morgan, G. T., Macgregor, H. D., and Colman, A.,** Multiple ribosomal gene sites revealed by *in situ* hybridization of *Xenopus* rDNA to *Triturus* lampbrush chromosomes, *Chromosoma,* 80, 309, 1980.
20. **Cave, M. D.,** Morphological manifestation of ribosomal DNA amplification during insect oogenesis, in *Insect Ultrastructure,* King, R. D. and Akai, H., Eds., Plenum Press, New York, 1982, 86.
21. **Trendelenburg, M. F., Franke, W. W., and Scheer, U.,** Frequencies of circular units of nucleolar DNA in oocytes of two insects, *Acheta domesticus* and *Dytiscus marginalis,* and change of nucleolar morphology during oogenesis, *Differentiation,* 7, 133, 1977.
22. **Troster, H., Spring, H., Meissner, B., Schultz, P., Oudet, P., and Trendelenburg, M. F.,** Structural organization of an active, chromosomal nucleolar organizer region (NOR) identified by light microscopy, and subsequent TEM and STEM electron microscopy, *Chromosoma,* 91, 151, 1985.
23. **Wylie, C. C.,** Nuclear morphology and nucleolar DNA synthesis during meiotic prophase in oocytes of the chick *(Gallus domesticus), Cell Differen.,* 1, 325, 1972.

24. **Hutchison, N.,** Lampbrush chromosomes of the chicken, *Gallus domesticus, J. Cell Biol.,* 105, 1493, 1988.
25. **Bakken, A. H. and McClanahan, M.,** Patterns of RNA synthesis in early meiotic prophase oocytes from fetal mouse ovaries, *Chromosoma,* 67, 21, 1978.
26. **Sims, S. H. and Bakken, A. H.,** Developmental regulation of ribosomal genes in *Xenopus laevis, J. Cell Biol.,* 103, 499a, 1986.
27. **Sims, S. H. and Bakken, A. H.,** Ribsomal gene regulation in *Xenopus* oocytes and early embryos, in preparation.
28. **Busby, S. and Bakken, A. H.,** Quantitative electron microscopic analysis of transcription in sea urchin embryos, *Chromosoma,* 71, 249, 1979.
29. **McKnight, S. L. and Miller, O. L.,** Ultrastructural patterns of RNA synthesis during early embryogenesis of *Drosophila melanogaster, Cell,* 8, 305, 1976.
30. **Foe, V. E., Wilkinson, L. E., and Laird, C. D.,** Comparative organization of active transcription units in *Oncopeltus fasciatus, Cell,* 9, 131, 1976.
31. **Callan, H. G.,** *Lampbrush Chromosomes,* Springer-Verlag, Berlin, 1986.
32. **Macgregor, H. C.,** Lampbrush chromosomes, *J. Cell Sci.,* 88, 7, 1987.
33. **Callan, H. G.,** The nature of lampbrush chromosomes, *Int. Rev. Cytol.,* 15, 1, 1963.
34. **Callan, H. G. and Lloyd, L.,** Lampbrush chromosomes of crested newts *Triturus cristatus* (Laurenti), *Philos. Trans. R. Soc. London. Ser. B.,* 243, 135, 1960.
35. **Gall, J. G.,** Lampbrush chromosomes from oocyte nuclei of the newt, *J. Morphol.,* 94, 283, 1954.
36. **Callan, H. G., Gall, J. G., and Berg, C. A.,** The lampbrush chromosomes of *Xenopus laevis:* preparation, identification, and distribution of 5S DNA sequences, *Chromosoma,* 95, 236, 1987.
37. **Bakken, A. H. and Graves, B.,** Visualization of the teritary structure of lampbrush chromosomes with the scanning electron microscope, *J. Cell Biol.,* 67, 17a, 1975.
38. **Bakken, A. H.,** Lampbrush chromosomes in mouse oocytes, *J. Cell Biol.,* 70, 144a, 1976.
39. **Miller, O. L. and Bakken, A. H.,** Morphological studies of transcription, *Acta Endocrinol.* (Suppl.), 168, 155, 1972.
40. **Bachvarova, R.,** Synthesis, turnover, and stability of heterogeneous RNA in growing mouse oocytes, *Dev. Biol.,* 86, 384, 1981.
41. **Bachvarova, R., Burns, J. P., Spiegelman, I., Choy, J., and Chaganti, R. S. K.,** Morphology and transcriptional activity of mouse oocyte chromosomes, *Chromosoma,* 86, 181, 1982.
42. **Diaz, M. O. and Gall, J. G.,** Giant readthrough transcription units at the histone loci on lampbrush chromosomes of the newt *Notophthalmus, Chromosoma,* 92, 243, 1985.
43. **Bromley, S. E. and Gall, J. G.,** Transcription of the histone loci on lampbrush chromosomes of the newt *Notophthalmus viridescens, Chromosoma,* 95, 396, 1987.
44. **Georgiev, O., Mous, J., and Birnstiel, M. L.,** Processing and nucleocytoplasmic transport of histone gene transcripts, *Nucleic Acids Res.,* 12, 8539, 1984.
45. **Macgregor, H. C.,** *In situ* hybridization of highly repetitive DNA to chromosomes of *Triturus cristatus, Chromosoma,* 71, 57, 1979.
46. **Jamrich, M., Warrior, R., Steele, R., and Gall, J. G.,** Transcription of repetitive sequences on *Xenopus* lampbrush chromosomes, *Proc. Natl. Acad. Sci., U.S.A.,* 80, 3364, 1983.
47. **Lee, Y. R. and Whiteley, A. H.,** Gene transcription during oogenesis of *Schizobranchia insignis,* a tubiculous polychaete, *Fortshcr. Zool.,* 29, 167, 1984.
48. **Roshbash, M. and Ford, P. J.,** Polyadenylic acid-containing RNA in *Xenopus laevis* oocytes, *J. Mol. Biol.,* 85, 87, 1974.
49. **Epstein, L. M. and Gall, J. G.,** Self-cleaving transcripts of satellite DNA from the newt, *Cell,* 45, 535, 1987.
50. **Dawid, I. B., Brown, D. D., and Reeder, R. H.,** Composition and structure of chromosomal and amplified ribosomal DNAs of *Xenopus laevis, J. Mol. Biol.,* 51, 341, 1970.
51. **Angelier, N., Bonnanfant-Jais, M. L., Moreau, N., Gounon, P., and Lavaud, Z.,** DNA methylation and RNA transcription activity in amphibian lampbrush chromosomes, *Chromosoma,* 94, 169, 1986.
52. **Roth, M. B. and Gall, J. G.,** Monoclonal antibodies that recognize transcription unit proteins on newt lampbrush chromosomes, *J. Cell Biol.,* 105, 1047, 1987.
53. **Dreyer, C. and Hausen, P.,** Two-dimensional gel analysis of the fate of oocyte nuclear proteins in the development of *Xenopus laevis, Dev. Biol.,* 100, 412, 1983.
54. **Dequin, R., Saumweber, H., and Sedat, J. W.,** Proteins shifting from the cytoplasm into the nuclei during early embryogenesis of *Drosophila melanogaster, Dev. Biol.,* 104, 37, 1984.
55. **Clarke, H. J. and Masui, Y.,** Dose dependent relationship between oocyte cytoplasmic volume and transformation of sperm nuclei to metaphase chromosomes, *J. Cell Biol.,* 104, 831, 1987.
56. **Schmiady, H., Sperling, K., Kentenich, H., and Stauber, M.,** Prematurely condensed human sperm chromosomes after *in vitro* fertilization, *Hum. Genet.,* 74, 441, 1986.

57. **Zimmer, F. J., Dreyer, C., and Hausen, P.,** The function of the nuclear envelope in nuclear protein accumulation, *J. Cell Biol.,* 106, 1435, 1988.
58. **Bell, P. R. and Pennell, R. I.,** The origin and composition of nucleolus-like inclusions in the cytoplasm of fern egg cells, *J. Cell Sci.,* 87, 283, 1987.
59. **Zeller, R. Nyffenegger, T., and DeRobertis, E. M.,** Nucleocytoplasmic distribution of snRNPs and stockpiled snRNA binding proteins during oogenesis and early development of *Xenopus laevis, Cell,* 32, 425, 1983.
60. **Mattaj, I. W. and DeRobertis, E. M.,** Nuclear segregation of U2 snRNA requires binding of specific snRNP proteins, *Cell,* 40, 111, 1985.
61. **Poccia, D.,** Remodeling of nucleoproteins during gametogenesis, fertilization and early development, *Int. Rev. Cytol.,* 105, 1, 1986.
62. **Kobayakawa, Y. and Kubota, H.,** Temporal pattern of cleavage and the onset of gastrulation in amphibian embryos developed from eggs with the reduced cytoplasm, *J. Embryol. Exp. Morphol.,* 62, 83, 1981.
63. **Newport, J. and Kirschner, M.,** A major developmental transition in early *Xenopus* embryos. I. Characterization and timing of cellular changes at the midblastula stage, *Cell,* 30, 675, 1982.
64. **Newport, J. and Kirschner, M.,** A major developmental transition in early *Xenopus* embryos. II. Control of the onset of transcription, *Cell,* 30, 687, 1982.
65. **Edgar, B. A., Kiehle, C. P., and Schubiger, G.,** Cell cycle control by the nucleo-cytoplasmic ratio in early *Drosophila* development, *Cell,* 44, 365, 1986.
66. **Edgar, B. A. and Schubiger, G.,** Parameters controlling transcriptional activation during early *Drosophila* development, *Cell,* 44, 871, 1986.
67. **Davidson, E. H.,** *Gene Activity in Development,* Academic Press, Orlando, FL, 1986.
68. **Bolton, V. N., Oades, P. J., and Johnson, M. H.,** The relationship between cleavage, DNA replication, and gene expression in the mouse 2-cell embryo, *J. Embryol. Exp. Morphol.,* 79, 139, 1984.
69. **Wolstenholme, D. R.,** Replicating DNA molecules from eggs of *Drosophila melanogaster, Chromosoma,* 43, 1, 1973.
70. **Blumenthal, A. B., Kriegstein, H. J., and Hogness, D. S.,** The units of DNA replication in *Drosophila melanogaster* chromosomes, *Cold Spring Harbor Symp. Quant. Biol.,* 38, 205, 1973.
71. **Zakian, V. A.,** Electron microscopic analysis of DNA replication in main band and satellite DNAs of *Drosophila virilis, J. Mol. Biol.,* 108, 305, 1976.
72. **McKnight, S. L. and Miller, O. L.,** Electron microscopic analysis of chromatin replication in the cellular blastoderm *Drosophila melanogaster* embryo, *Cell,* 12, 795, 1977.
73. **Huberman, J. A. and Riggs, A. D.,** On the mechanism of DNA replication in mammalian chromosomes, *J. Mol. Biol.,* 32, 327, 1968.
74. **Callan, H. G.,** DNA replication in the chromosomes of eukaryotes, *Cold Spring Harbor Symp. Quant. Biol.,* 38, 195, 1973.
75. **Gurdon, J. B. and Woodland, H. R.,** The cytoplasmic control of nuclear activity in animal development, *Biol. Rev.,* 43, 233, 1968.
76. **Montag, M., Spring, H., and Trendelenberg, M. F.,** Structural analysis of the mitotic cycle in pre-gastrula *Xenopus* embryos, *Chromosoma,* 96, 187, 1988.
77. **Zalokar, M.,** Autoradiographic study of protein and RNA formation during early development of *Drosophila* eggs, *Dev. Biol.,* 49, 425, 1976.
78. **Foe, V. E. and Alberts, B. M.,** Studies of nuclear and cytoplasmic behaviour during the five mitotic cycles that precede gastrulation in *Drosophila* embryogenesis, *J. Cell Sci.,* 61, 31, 1983.
79. **Foe, V. E. and Alberts, B. M.,** Reversible chromosome condensation induced in *Drosophila* embryos by anoxia: visualization of interphase nuclear organization, *J. Cell Biol.,* 100, 1623, 1985.
80. **Fristrom, J. W.,** The developmental biology of *Drosophila, Annu. Rev. Genet.,* 4, 325, 1970.
81. **Chen, P. S.,** *Biochemical Aspects of Insect Development,* Karger, Basel, 1971.
82. **Mahowald, A. P.,** Polar granules of *Drosophila.* II. Ultrastructural changes during early embryogenesis, *J. Exp. Zool.,* 167, 237, 1968.
83. **Hara, K., Tydeman, P., and Kirschner, M.,** A cytoplasmic clock with the same period as the division cycle in *Xenopus* eggs, *Proc. Natl. Acad. Sci., U.S.A.,* 77, 462, 1980.
84. **Oshumi, K., Shinagawa, A., and Datagiri, C.,** Periodic changes in the rigidity of activated eggs depend on germinal vesicle materials, *Dev. Biol.,* 118, 467, 1986.
85. **Dabauvalle, M. C., Doree, M., Bravo, R., and Karsenti, E.,** Role of nuclear material in the early cell cycle of *Xenopus* embryos, *Cell,* 52, 525, 1988.
86. **Newport, J. and Kirschner, M.,** Regulation of the cell cycle during early *Xenopus* development, *Cell,* 37, 731, 1984.
87. **Cyert, M. S. and Kirschner, M.,** Regulation of MPF activity *in vitro, Cell,* 53, 185, 1988.
88. **Brown, D. D. and Littna, E.,** RNA synthesis during the development of *Xenopus laevis,* the South African clawed toad, *J. Mol. Biol.,* 8, 669, 1964.

89. **Emerson, C. P. and Humphreys, T.**, Regulation of DNA-like RNA and the apparent activation of ribosomal RNA synthesis in sea urchin embryos: quantitative measurements of newly synthesized RNA, *Dev. Biol.*, 23, 86, 1970.

90. **Emerson, C. P. and Humphreys, T.**, Ribosomal RNA synthesis and the multiple atypical nucleoli in cleaving embryos, *Science*, 171, 898, 1971.

91. **Fausto-Sterling, A., Zheutlin, L. M., and Brown, P. R.**, Rates of RNA synthesis during early embryogenesis in *Drosophila melanogaster*, *Dev. Biol.*, 40, 78, 1974.

92. **Mahowald, A. P.**, Ultrastructural differentiations during formation of the blastoderm in the *Drosophila melanogaster* embryo, *Dev. Biol.*, 8, 186, 1963.

93. **Busby, S. and Bakken, A. H.**, Transcription in developing sea urchins: electron microscope analysis of cleavage, gastrula and prism stages, *Chromosoma*, 79, 85, 1980.

94. **Kimelman, D., Kirschner, M., and Scherson, T.**, The events of the midblastula transition in *Xenopus* are regulated by changes in the cell cycle, *Cell*, 48, 399, 1987.

95. **Mita, I.**, Studies on the factors affecting the timing of early morphogenetic events during starfish embryogenesis, *J. Exp. Zool.*, 225, 293, 1983.

96. **Mita, I. and Obata, C.**, Timing of morphogenetic events in tetraploid starfish embryos, *J. Exp. Zool.*, 229, 215, 1984.

97. **Langelan, R.**, *A Study of Unequal Cleavage, Micromere Determination and Spicule Differentiation in the Sea Urchin Embryo*, Ph.D. dissertation, University of Washington, Seattle, 1985.

98. **Anderson, K. V. and Lengyel, J. A.**, Changing rates of histone mRNA synthesis and turnover in *Drosophila* embryos, *Cell*, 21, 171, 1980.

99. **Weir, M. P. and Kornberg, T.**, Patterns of engrailed and fushi tarazu transcripts reveal novel intermediate stages in *Drosophila* segmentation, *Nature (London)*, 318, 433, 1985.

100. **Anderson, K. V. and Lengyel, J. A.**, Rates of synthesis of major classes of RNA in *Drosophila* embryos, *Dev. Biol.*, 70, 217, 1979.

101. **Busby, S. and Reeder, R. H.**, Fate of amplified nucleoli in *Xenopus laevis* embryos, *Dev. Biol.*, 91, 458, 1982.

102. **Sheffery, M.**, Chromatin structure, gene expression, and differentiation, in *Chromosome Structure and Function*, Risley, M. S., Ed., Van Nostrand Reinhold, New York, 1986, 39.

103. **Pehrson, J. R. and Cohen. L. H.**, Distribution of histone H1-alpha among cells of the sea urchin embryo, *Dev. Biol.*, 111, 530, 1985.

104. **Zweidler, A.**, in *Histone Genes*, Stein, G. S., Stein, J. L., and Marzluff, W. F., Eds., John Wiley & Sons, New York, 1984, 339.

105. **Dreyer, C.**, Differential accumulation of oocyte nuclear proteins by embryonic nuclei of *Xenopus*, *Development*, 101, 829, 1987.

106. **Fritz, A., Parisot, R., Newmeyer, D., and DeRobertis, E. M.**, Small nuclear U-RNPs in *Xenopus laevis* development: uncoupled accumulation of the protein and RNA components, *J. Mol. Biol.*, 178, 273, 1984.

107. **Newport, J. and Forbes, D. J.**, The nucleus: structure, function and dynamics, *Annu. Rev. Biochem.*, 56, 535, 1987.

108. **Krohne, G. and Benavente, R.**, The nuclear lamins: a multigene family of proteins in evolution and differentiation, *Exp. Cell Res.*, 162, 1, 1986.

109. **Benavente, R., Krohne, G., and Franke, W.**, Cell type-specific expression of nuclear lamina proteins during development of *Xenopus laevis*, *Cell*, 41, 177, 1985.

110. **Stick, R. and Hausen, P.**, Immunological analysis of nuclear lamina proteins, *Chromosoma*, 80, 219, 1980.

111. **Laskey, R. A., Honda, B. M., Mills, A. D., and Finch, J. T.**, Nucleosomes are assembled by an acidic protein which binds histones and transfers them to DNA, *Nature (London)*, 275, 416, 1980.

112. **Laskey, R. A. and Earnshaw, W. C.**, Nucleosome assembly, *Nature (London)*, 286, 763, 1980.

113. **Sealy, L., Cotten, M., and Chalkley, R.**, *Xenopus* nucleoplasmin: egg vs. oocyte, *Biochemistry*, 25, 3064, 1986.

114. **Kleinschmidt, J. A., Dingwall, C., Maier, G., and Franke, W. W.**, Molecular characterization of a karyophilic, histone-binding protein: cDNA cloning, amino acid sequence and expression of nuclear protein N1/N2 of *Xenopus laevis*, *EMBO J.*, 5, 3547, 1985.

115. **Maxson, R., Ito, M., Balcells, S., Thayer, M., French, M., Lee, F., and Etkin, L.**, Differential stimulation of sea urchin early and late H2B histone gene expression by a gastrula nuclear extract after injection into *Xenopus laevis* oocytes, *Mol. Cell Biol.*, 8, 1236, 1988.

116. **Krieg, P. A. and Melton, D.**, An enhancer responsible for activating transcription at the midblastula transition in *Xenopus* development, *Proc. Natl. Acad. Sci., U.S.A.*, 84, 2331, 1987.

117. **Steller, H. and Pirrotta, V.**, Regulated expression of genes injected into early *Drosophila* embryos, *EMBO J.*, 3, 165, 1984.

118. **Bakken, A. H.**, unpublished, 1970.

119. **Wakimoto, B.**, personal communication, 1988.

Chapter 3

CHROMATIN ORGANIZATION IN SPERM

Michael S. Risley

TABLE OF CONTENTS

I. INTRODUCTION

Spermatogenesis is the cellular differentiation pathway responsible for the production of the male gametes. Sperm development occurs in three principal phases. The first is a mitotic phase of stem cell proliferation, renewal, and commitment to differentiation. A meiotic phase follows during which genomic recombination and reduction to haploidy occur. The final phase is spermiogenesis in which haploid round spermatids proceed through complex morphogenetic steps to acquire the final unique sperm shape. Although the details of morphogenesis are species specific, key features include development of a motile apparatus (flagellum), formation of the acrosome, a Golgi-derived organelle whose contents are important for penetration of egg investments, and development of membrane specializations for fusion of sperm and egg. Spermiogenesis in most species (except crab) is also characterized by transcriptional inactivation and condensation of DNA into a highly compact mass. The last process is the focus of this chapter. Discussions of chromosome and chromatin structure in meiotic cells are available in recent reviews.[1,2]

The extreme condensation of sperm DNA is a poorly understood process for which there are a number of assumptions regarding both purpose and mechanism. For example, it is often assumed that DNA condensation is important for the formation of sperm nuclei with hydrodynamically efficient shapes. Condensation will certainly reduce volume and drag, but nuclear shapes in sperm from certain organisms (e.g., rat) appear to be poor designs for swimming. Indeed, it can be argued that for many species the shaping of sperm nuclei is directed by processes independent of the DNA condensation process.[3]

Another assumption is that condensation of DNA in sperm is important for protection of the genome from environmental mutagens. The protection afforded by DNA condensation appears to be relatively weak, however, since sperm and late spermatids are sensitive to chemically induced mutations[4,5] and DNA damage.[6] Indeed, protamines, the basic proteins which replace histones in sperm, are themselves targets of alkylating agents and thus may enhance the sensitivity of sperm to certain types of mutagenesis.[7] It seems that protection of the sperm genome would be accomplished most readily through the presence of efficient DNA repair mechanism, the principal cellular mechanism for protection from mutagenesis. Yet, DNA repair mechanisms are disabled in late spermatids and sperm.[8,9] Damaged sperm DNA is repaired only after fertilization.[10]

A third assumption regarding the purpose of genomic compaction in sperm is that this process produces a male gamete with a genetic blank slate on which the egg can inscribe the molecular information essential for the direction of embryonic development. By this mechanism, the egg would gain immediate control of the male genome, enhancing the prospects for a coordinated regulation of expression of the embryonic genome. This concept is supported by the observed low levels of efficiency of normal development in embryos containing transplanted nuclei from late embryos or adults, a problem often ascribed to asynchrony between the transplanted nucleus and the egg cytoplasm.[11]

The phenomenon of chromosome imprinting suggests that the blank slate concept may be an over-simplification. Chromosome imprinting, a term coined by Crouse[12] in 1960, refers to the differential behavior of homologous chromosomes according to their germline origin.[13,14] Thus, in the coccid insects, the entire paternal chromosome set is either eliminated or becomes heterochromatic.[15] The paternal X is selectively eliminated in Sciarids and preferentially inactivated in all cells of marsupials.[16] In eutherian females, the paternal X chromosome is preferentially inactivated in extraembryonic tissues, while random inactivation occurs in the embryonic tissues.[17] Nuclear transplantation studies have shown that the maternal and paternal pronuclei are not equivalent in mouse embryos; both pronuclei are required for normal embryonic development to term and they each possess complementary information.[18,19] The paternal genome is important for formation of extraembryonic tissues, while the maternal genome contributes predominantly to the embryonic tissues. Evidently, the passage of chromosomes through the male or female germ-line results in an imprinting of the genome which is essential for normal development and is sufficiently stable to survive the activation and epigenetic programming of the embryonic genome.[20]

Genomic imprinting may be accomplished by chemical or structural modifications of DNA in developing germ cells. Since selective chromosome elimination or inactivation is commonly directed to the paternal chromosome, it is reasonable to suggest that DNA modifications during spermatogenesis designate the paternal chromosomes as targets of the embryonic inactivation process. On the other hand, it has been argued that the maternal chromosomes are imprinted during oogenesis to protect them from inactivation or elimination.[13] A more complex mechanism may be operating; imprinting of paternal chromosomes during spermatogenesis could block subsequent embryonic modifications that confer resistance to inactivation. Maternal chromosomes would thus be selectively protected from inactivation during early embryogenesis, but not as a consequence of imprinting during oogenesis.

An understanding of chromosome imprinting and the role of spermatogenesis in this process will require more detailed information about the timing of imprinting as well as the molecular mechanism of imprinting and chromosome inactivation. Investigations of chromatin organization in developing sperm should shed light on the potential role of DNA condensation and modification in paternal chromosome imprinting. Attention is being focused on the roles of DNA methylation and DNA-binding proteins during gametogenesis and embryogenesis. The remainder of this chapter will discuss our current understanding of sperm chromatin structure particularly as it may relate to chromosome imprinting.

II. CHROMATIN STRUCTURE IN SPERMATIDS AND SPERM

A. SPERM HISTONES AND PROTAMINES

The structural subunit of chromatin, the nucleosome, is fundamental to the condensation of DNA in the nuclei of all eukaryotic somatic cells. These disk-shaped structures (110 × 57 Å) are formed by the association of approximately 200 bp of DNA with a single histone H1 molecule and a histone octamer consisting of 2 molecules each of histones H2A, H2B, H3, and H4.[21,22] Each nucleosome consists of a chromatosome, containing all of the histones plus 165 bp of DNA, and about 35 bp of linker DNA.[23] Within the chromatosome, DNA is wrapped around the outside of the histone octamer in a two-turn left-handed toroidal supercoil with histone H1 bound to the DNA at the entry and exit points of the coil. The linker DNA that connects adjacent chromatosomes can vary in length among somatic cell types with a range of 0 to 60 bp.[24,25] The linker length may also vary within a single cell type.[26] Although the functional significance has not been demonstrated, length variation in linker DNA represents a means of varying the higher order packing of nucleosomes in chromatin.

The composition of nucleosomes also varies as a consequence of the existence of non-allelic variants of the major histone types. With the exception of histone H4, each of the histone types may consist of 2 or more sequence variants with the exact composition varying phylogenetically[27,28] and during development and differentiation within a species.[29] For example, there are at least 5 H1 histones in the frog *Xenopus laevis* that vary among adult tissues,[30] but the core histones in *Xenopus* appear to be constant and without variants during embryogenesis.[31] In contrast, the sea urchin *S. purpuratus* contains variants of H1, H2A, and H2B which are synthesized in a precise sequence during early development.[29] Chickens[32] and mammals[28] contain sequence variants of all the major histone types except H4. The variants in both chickens[33] and mouse[28] change in composition during postnatal development. Histone H3 variants switch during myogenesis in chicken[34] and, along with H2A variants, also switch during neuronal differentiation in the rat brain.[35] As with linker length variation, the physiological significance of histone variant switching has been not demonstrated; however, it is reasonable to expect that each variant may have a specific influence on higher-order nucleosome packing and nucleosome stability.

During spermatogenesis in most species the histone composition of germ cell chromatin is altered by the partial or total replacement of somatic type histones by sperm-specific histones, protamines, or other basic proteins. Detailed discussions of the cytochemical and biochemical properties of the sperm basic proteins are available in several reviews.[29,36-39] In contrast to the relative conservation of somatic cell histones, the composition of chromatin basic proteins in sperm is highly divergent. Bloch[36,37] identified five major categories of sperm protein composition largely based on cytochemical data; the categories are: (1) sperm containing low molecular weight, arginine-rich protamines (e.g., trout, salmon, rooster), (2) sperm with protamines rich in arginine and cysteine (e.g., eutherian mammals), (3) sperm proteins with arginine and lysine contents intermediate to those of histones and protamines (e.g., molluscs, *Xenopus laevis*), (4) sperm with histones (e.g., goldfish, sea urchins, *Rana*), and (5) sperm whose nuclei lack detectable basic proteins. This final category has been found only among crustaceans which have immotile sperm with uncondensed DNA.

Biochemical analyses of sperm proteins from different species in Bloch's four categories have demonstrated substantial heterogeneity of the proteins in each grouping. Thus far, only sperm from the goldfish *Carassius auratus*[40] and the carp *Ctenopharyngodon idella*[41] have been found to contain a somatic-type histone complement. Sperm from other members of this group contain either sperm-specific H1 histones (e.g., *Rana pipiens*,[42] *Holuthuria tubulosa*[43]), sperm specific H1 and H2B histones (e.g., sea urchins[44]), or somatic-type histones and a family of high molecular weight basic proteins (e.g., winter flounder[45]). The heterogeneity of sperm basic proteins in category 4 is further exemplified by the species specificity of sea urchin histone H2B variants.[29]

Within Bloch's group 3, the intermediate protein category, extensive heterogeneity has been demonstrated in the sperm proteins of the single genus *Xenopus*.[46] Electrophoretic analysis of sperm proteins from 12 different *Xenopus* species and six different subspecies resolved 17 distinct basic protein complements (Figure 1). There was sufficient diversity in the sperm proteins to permit reliable identification of each species. Notably, two of the species (*Xenopus tropicalis*, *Xenopus* sp. n. [Zaire]) in this genus were shown to belong to Bloch's category 4 since they contained sperm-specific histone variants but lacked intermediate-type protamines. Diversity can also extend to the sperm proteins of individuals in a single subspecies (*Xenopus laevis laevis*) due to the occurrence (see Figure 2) of two variants of the protamine SP2 that appear to segregate allelically.[47]

'Bloch's categories 1 and 2 consist of species whose sperm contain either the typical monoprotamines about 30 amino acids long or the cysteine-rich protamines of about 50 amino acids, respectively. Although there is some sequence divergence in the protamines of each category, the protein diversity within category 1 or 2 is significantly less than that in categories 3 and 4. In fact, mammalian protamines in category 2 contain a central 25 amino acid arginine-rich domain which shows strong sequence homology to fish protamines. Nevertheless, an interesting example of protamine diversity in mammals has been described. All mammalian sperm examined thus far contain protamine P1. Another protamine (P2), characterized by a high histidine content, has been found in sperm from mice[48] and humans.[49-51] Yet, mRNA transcripts homologous to mouse P2 have been found on testis polysomes from rat and hamster which lack the P2 protamine in mature sperm.[52] The mechanism responsible for determining the final P1/P2 ratio in sperm is not clear; however, synthesis of P1 and P2 differ in that P2 is first synthesized as a precursor 43 amino acids longer than mature P2 (63 amino acids).[53]

It will be important to determine the extent to which differential processing and stability of protamine precursors contribute to sperm protein diversity in other genera or phyla. Such

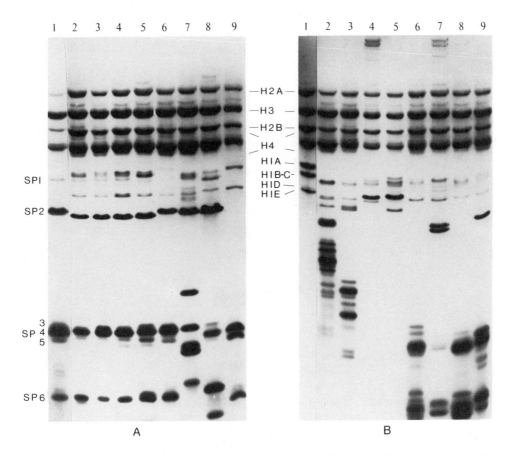

FIGURE 1.   Basic chromatin proteins isolated from testes or sperm and late spermatids and electrophoresed on 12% acrylamide, 2.5 M urea, 6 mM Triton X-100, 5% acetic acid (AUT) slab gels. A. *X. laevis laevis* sperm (1), *X. laevis petersi* sperm (2), *X. laevis victorianus* sperm (3), *X. laevis bunyoniensis* testes (4), *X. laevis sudanensis* testes (5), *X. laevis sp. n* (Malaxi) sperm (6), *X. clivii* sperm (7), *X. sp. n* (Ethiopia) sperm (8), *X. fraseri* sperm; B. *X. laevis* erythrocytes (1), *X. borealis* sperm (2), *X. muelleri* sperm (3), *X. tropicalis* testes (4), *X. sp. n.* (Zaire) sperm (5), *X. ruwenzoriensis* sperm (6), *X. vestitus* testes (7), *X. amieti* sperm (8), *X. wittei* sperm (9). H, histones; Sp, sperm proteins. (From Mann, M., Risley, M. S., Eckhardt, R. A., and Kasinsky, H. E., *J. Exp. Zool.*, 222, 173, 1982. With permission.)

information would provide insights into genetic and evolutionary aspects of protamine function in sperm. The genus *Xenopus* is particularly attractive for such studies. Since many of the species in this genus appear to have evolved through polyploidization of an ancestral genome, the members of this genus have a similar genetic background.[54,55] Recent studies of protamine synthesis and stability in *Xenopus laevis laevis* spermatids have provided preliminary evidence for precursors for both protamines SP2 and SP6 and a high rate of turnover for SP1.[56] Differences in precursor processing and protamine stability among *Xenopus* species may be key determinants of sperm protein diversity in this genus. In addition, there may be multiple protamine genes shared by all *Xenopus*, but each species may express only a select combination of protamine genes. Hybridization of protamine cDNAs from one *Xenopus* species to genomic DNAs from other *Xenopus* species could provide a better understanding of the origins of protamine diversity.

B. SPERM NUCLEOSOMES AND NUCLEOPROTAMINES

The composition of sperm basic proteins has a primary role in determining the organization of sperm DNA. As may be expected, the diversity in these proteins is correlated with a diversity in sperm chromatin structure. At the nucleosomal level of chromatin or-

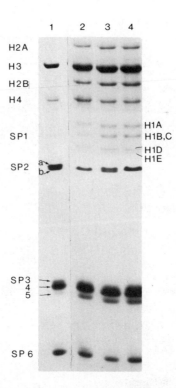

FIGURE 2.   Electrophoresis on AUT slab gels of X. *laevis laevis*
chromatin basic proteins isolated from the combined sperm from
12 frogs (lane 1) or the testes of three separate frogs (lanes 2 to
4). H, histones; Sp, sperm proteins. (From Risley, M. S. and
Eckhardt, R. A., *J. Exp. Zool.*, 242, 373, 1987. With permission.)

ganization, the weight of evidence suggests that these structural subunits are present primarily
in sperm that contain a full complement of core histones.[29,38] Nuclease digest studies have
revealed cutting frequencies typical of repeating nucleosomal subunits in bulk sperm chromatin from sea urchins,[57-59] starfish,[60] clams,[61,62] sea cucumber,[63] goldfish,[40] carp,[64] and
winter flounder,[65] but not in sperm chromatin from trout,[65] rabbit,[67] rooster,[67] or ram[68] which
contain protamines predominantly. Even *Xenopus laevis* sperm chromatin, which contains
histones H3 and H4 as well as 6 protamines,[30] is not cleaved by nucleases into nucleosome-
length DNA fragments.[67]

Electron microscopy of spread spermatid chromatin from cricket,[69,70] mouse,[71] and ram[68]
has provided further evidence that nucleosomes are lost as histones are replaced during
spermiogenesis. Each of these studies has demonstrated a transformation of beaded chromatin
fibers to smooth chromatin fibers in late spermatids. An analysis of DNA topology in *Rana
catesbeiana*, *Xenopus laevis*, and *Bufo fowleri* sperm nucleoids also suggested that nucleosomes are lost as histones are replaced.[72] Only *Rana* sperm contained DNA which was
negatively superhelical after extraction with 2 *M* NaCl; *Xenopus* and *Bufo* sperm (which
contain protamines) both contained relaxed DNA after 2 *M* NaCl extraction. The negative
superhelicity was correlated only with the occurrence of a full complement of core histones
and nucleosomes. In *Xenopus*, the transition from negatively supercoiled DNA to relaxed
DNA occurred during late spermatid stages. Further evidence that histone replacement during
spermatogenesis is accompanied by loss of nucleosomes has been obtained from X-ray
diffraction analysis of sperm nuclei from fish,[73] molluscs,[74] and ram[68] and from considerations
of the chromatin packing density required for condensation of DNA into the final volume
of the mouse sperm nucleus.[75]

Sperm which contain only histones have most of their DNA organized into nucleosomes; however, the nucleosomes in such sperm often have a unique DNA repeat length. Sperm from goldfish[40] and carp[41] appear to contain somatic-type histones and typical nucleosome repeat lengths of 205 and 200 bp, respectively. Starfish *(A. japonica)*[60] and sea cucumber *(Holothuria tubulosa)*[63] sperm contain large H1 variants and nucleosomes with 224 and 227 bp of DNA, respectively. In the sea urchins, which contain sperm-specific histones H1 and H2B, nucleosome repeat lengths vary from 231 bp in *S. intermedium*[59] to 241 bp in *A. lixula*[57] and 250 bp in *Strongylocentrotus purpuratus*.[58] The nucleosomal repeat length in the sperm from the sea urchin *Lutechinus variegatus* is unlike that in other sea urchins since it is only 201 bp, 10 bp less than that in gastrula stage embryos.[76] In winter flounder sperm, which contain somatic-type histones and high molecular weight basic proteins,[45] the nucleosome repeat length increases from 195 bp in prespermatids to 222 bp in sperm.[65]

With the exception of the winter flounder, increased nucleosomal repeat lengths in sperm of most species correlate with the occurrence of large sperm-specific H1 histones. Thus, although the sea urchins all contain large sperm-specific H2B histones, the relatively short nucleosomal repeat length found in *L. variegatus* sperm is best explained by the presence of a relatively small H1 in the sperm of this species.[76] Detailed studies of sea urchin sperm nucleosomes have demonstrated the presence of a typical chromatosome length of 164 bp, suggesting that the increased repeat length is due almost entirely to an increased length of linker DNA which is presumably protected by the large H1 variant in sea urchin sperm.[77] The H2B variant in sea urchin sperm may contribute to the resistance of the nucleosomes to nucleolytic trimming to a typical core particle size.[85] In winter flounder sperm, the increased nucleosomal repeat length appears to be due to binding of the linker DNA by the high molecular weight basic proteins.[65]

The structure of nucleoprotamine is significantly different from that in nucleosomes. Essentially complete neutralization of DNA negative charge by basic amino acids in fish and mammalian protamines results in a tight packing of DNA. X-ray diffraction of fish sperm nuclei indicates that nucleoprotamine is nearly crystalline.[38,73] Protamines appear to bind and stabilize DNA in an extended B form and the nucleoprotamine fibers condense into parallel bundles with hexagonal packing. Sperm of the mollusc *Mytilus* also contain parallel bundles of condensed nucleoprotamine but the order is not as regular as that in fish.[74] Mammalian nucleoprotamine is also less regular than that in fish,[38,78] but it is highly stabilized by disulfide bonds which form between neighboring protamines.[79]

A major obstacle encountered in structural studies of chromatin in sperm which contain protamines is the relative insolubility of nucleoprotamine at physiological ionic strength. The highly condensed mass of nucleoprotamine is usually resistant to gentle methods of spreading for electron microscopy and may also resist digestion by nucleases. Failure to detect nucleosomes in sperm of some species could therefore arise from the inaccessibility of probes to protected chromatin regions. For example, certain molluscs produce sperm which contain both histones and protamines. Nuclease digestion of condensed sperm nuclei from the mollusc *Mytilus edulis* generates a polydisperse cutting pattern suggesting that nucleosomes are absent.[74] If *Mytilus* sperm nuclei are swollen hypotonically prior to nuclease digestion, however, 20 to 25% of the DNA is cut by nucleases in a repeating fragment-length pattern.[61] These observations suggest that *Mytilus* sperm DNA may be organized into both nucleosomes and nucleoprotamine structures. A recent nuclease digest analysis of sperm chromatin from the surf clam *Spisula solidissima* has also provided evidence for the coexistence of nucleosomal and nucleoprotamine structures in sperm.[62] The occurrence of small amounts of tightly bound (resistant to 2 M NaCl, 6 M urea extraction) core histones in trout[80] and ram[81] sperm suggests the possibility of a dual nucleosomal/nucleoprotamine organization in these sperm also. Even trace amounts of core histones in sperm chromatin could establish discrete nucleosomal domains which are functionally important in early postfertilization events.

Direct evidence for a dual nucleohistone/nucleoprotamine organization in mammalian sperm has been obtained for sperm from one species, *Homo sapiens*. Human sperm, unlike sperm from most mammals, retain about 15% of the usual haploid histone complement and also contain a heterogeneous mixture of protamines and intermediate basic proteins.[82,83] Electron microscopy of spread human sperm chromatin has resolved globular structures which correspond in size to nucleosomes.[84-86] Gatewood et al.[87] have recently demonstrated that the nucleohistone and nucleoprotamine portions of human sperm chromatin form discrete structures which are at least partly sequence specific. DNA was selectively solubilized from sperm nucleohistone by *Bam*H1 nuclease digestion of chromatin stripped of histones by extraction with 0.65 *M* NaCl. DNA was also purified from the insoluble nucleoprotamine fraction and restricted with *Bam*H1. Size-selected, single-copy DNA fragments were then cloned. Two clones from nucleohistone DNA preferentially hybridized to total nucleohistone DNA versus nucleoprotamine DNA, while two nucleoprotamine DNA clones preferentially hybridized to total nucleoprotamine DNA. This study did not examine the organization of the nucleohistone chromatin, but the low salt concentration required for histone extraction suggests that typical nucleosome structure is altered and may resemble that in transcriptionally active *Physarum* rDNA.[87]

Although the functional significance is yet to be demonstrated, these studies show that the differential distribution of histones and protamines in human sperm establishes a definitive, sequence-specific biochemical and structural imprint on the paternal genome. Similar experiments using sperm from other species that contain both histones and protamines should demonstrate the generality of sequence-specific chromatin structures in sperm. It will also be important to determine if the chromatin in sperm which contain only histones exhibit any variegation in chromatin structure that may serve as a means of imprinting the paternal genome. Histone genes in sperm of the sea urchin *Paracentrotus lividus* are organized into nucleosomes which are more resistant than bulk chromatin to micrococcal nuclease digestion, suggesting that structural variegation is also present in sperm with histones.[88]

Sequence-specific chromatin structures have also been identified in rooster sperm.[89] DNase-I-hypersensitive sites relatively specific to sperm have been located in the genes for thymidine kinase, β^A-globin, and the inactive ev-1 retrovirus.[89] These hypersensitive sites do not correspond to the sites which are associated with either the active or inactive genes in somatic cells. Thus, the sperm hypersensitive sites may not be directly involved in transcriptional regulation and may result from the torsional strain associated with the extreme condensation of sperm DNA. Nevertheless, the sperm hypersensitive sites are sequence-specific structures which may serve as signals important in the discrimination of paternal and maternal genes in the early embryo.

C. NUCLEOSOME MODIFICATIONS IN DEVELOPING SPERMATIDS

The organization of DNA in sperm nuclei is the result of sequential changes in chromatin composition and structure in developing spermatids. Insights into the mechanisms important to the DNA condensation process could be obtained from analyses of spermatid chromatin biochemistry and structure, but relatively few species have been employed for such studies due to experimental difficulties in isolating and culturing large numbers of spermatids from successive developmental stages. Nevertheless, some important observations on spermatid chromatin transitions have been obtained from studies with sea urchin, trout, winter flounder, *Xenopus*, rodents, and ram.

In the sea urchin, *S. purpuratus*, the "sperm-specific" histones SpH1 and SpH2B are found, surprisingly, in chromatin from early stages (spermatogonia-spermatocytes) of spermatogenesis, well before the DNA condensation stages in spermatids.[90] These histones appear to be synthesized early in germ cell development and are then conserved. This contrasts with the typical pattern of sperm-specific basic protein accumulation in spermatid stages and it will be important to determine if the "sperm-specific" histones identified in sperm from other species are also synthesized in early germ cell stages.

The SpH1 and SpH2B N-terminal arms are maintained in a phosphorylated state during sea urchin spermatogenesis until late stages of DNA condensation in spermatids. At that time, dephosphorylation occurs concomitant with increases in nucleosome linker length, resistance to micrococcal nuclease digestion, and thermal stability.[91] Dephosphorylated, the N-terminal arms of SpH1 and SpH2B may be capable of tight binding and cross-linking of linker DNA, thus facilitating the terminal, tight condensation of DNA. It should be noted, however, that other mechanisms are also important since condensation begins prior to de-phosphorylation of SpH1 and SpH2B.[90]

The winter flounder is an example of a species which adds high molecular weight basic proteins (HMWBP) to spermatid chromatin without displacing most histones.[45] In addition, synthesis of histones (H3 and H4) may continue into spermatid stages in this species.[92] The histones appear similar to those in somatic cells and all are retained in similar proportions in sperm. During the transition from early germ cell stages to spermatid stages, the histones H2A and H4 become extensively modified by phosphorylation.[45,93] The HMWBP are also extensively phosphorylated as they are added to chromatin, then at late spermatid stages the histones and HMWBP are dephosphorylated.[93] Dephosphorylation is accompanied by an increase in the nucleosome linker length from 195 to 222 bp.[65]

In *Xenopus laevis,* histones are partially replaced by 6 intermediate-type protamines termed SPs.[30,47,94] The histones in spermatogenic cells are similar to those in somatic cells and histone synthesis occurs in both mid-spermatid stages and premeiotic stages.[94] SP synthesis and accumulation begin in spermatids with developing acrosomal vacuoles but are most intense in early elongate spermatids with condensed acrosomes, condensing DNA, and round nuclei. SP1 appears early in spermatids, is synthesized throughout spermiogenesis, and remains at relatively constant levels throughout spermatid development. SP2—SP5 accumulate continuously from mid to late spermatid stages, and SP6 accumulates predominantly in elongate spermatids. Only SP3—SP5 appear to be highly phosphorylated. Histones are extensively modified by phosphorylation (H2A and H4) and acetylation (H3 and H4) in mid through elongate spermatid stages, and the modifications are independent of histone synthesis.[94] Eventually, all of histone H1 is replaced and 90% of H2A and H2B are removed, but only 50% of H3 and H4 are lost. Although nuclease digest studies suggest that nucleosomes are not present in *Xenopus* sperm,[67] it will be important to rule out the possibility that a small fraction of chromatin contains nucleosomes.

In trout, the majority of histones are replaced by three arginine-rich protamines and nucleosomal structures are removed.[66] The histones in spermatogenic cells are similar to those in somatic cells, and, in contrast to *Xenopus*[94] and winter flounder,[92] trout histones are synthesized only in prespermatid stages.[95,96] Histones are modified by both phosphorylation[95-97] and acetylation,[96,98,99] but histone phosphorylation is restricted to early spermatogenic stages where it is coupled to the synthesis and binding of histones to DNA.[95-97] This also contrasts with the pattern of histone phosphorylation in winter flounder [45,93] and *Xenopus.*[94] On the other hand, histone acetylation is similar to that seen in *Xenopus* spermatids in that it is intense and coordinated with the accumulation of protamines on spermatid DNA.[98,99] The protamines are also highly phosphorylated after synthesis and dephosphorylated at late spermatid stages of DNA condensation.[95,100] The sequence of histone replacement in trout appears opposite from that in *Xenopus* since H4 is lost first followed by H2A, H2B and H3, then H1.[101,102]

The nucleosome to nucleoprotamine transitions in organisms with cysteine-rich protamines (e.g., cricket, dogfish, eutherian mammals) appear more complex than those in other organisms. During spermatogenesis in crickets, two testis-specific histones, TH1 and TH2, accumulate in spermatid chromatin just prior to the time that protamines begin to accumulate.[103,104] The basic protein composition then undergoes several successive changes in which somatic-type histones are lost first, followed by replacement of TH1 and TH2 by a complex array of protamines, before the mature composition of 2 major and 2 minor protamines is assumed in the spermatheca.[104] In dogfish spermatids, histones are also gradually replaced by intermediate proteins (S1, S2) and these are subsequently replaced by protamines (Z1 to Z4).[105]

In eutherian mammals, three successive changes occur in the basic protein composition of chromatin.[29,106,107] First, testis-specific variants of histones H1, H2A, H2B, and H3 are added to chromatin in spermatogonial or spermatocyte stages and are conserved until late spermatid stages since histone synthesis does not occur postmeiotically.[108,109] In late spermatids, histones are replaced by low molecular weight transitional proteins (TPs). Mice contain four TPs which are synthesized in step 12 spermatids when protamine synthesis begins.[48] Rat TPs appear in steps 13 to 15 (TP, TP2, TP4) and steps 16 to 19 (TP3) spermatids.[110,111] Histones are also replaced by a complex set of TPs in ram spermatids.[112] TPs of elongate spermatids are eventually replaced by protamines which undergo a cycle of phosphorylation and dephosphorylation during their assembly into nucleoprotamine.[113] Final maturation of the mammalian cysteine-rich nucleoprotamine occurs with the formation of cysteinyl disulfide cross-links between protamine molecules during sperm transit through the epididymis.[79]

The mechanism which govern the replacement of nucleosomal chromatin by nucleoprotamine are not understood. Observations described above suggest that certain details are species specific and may be determined primarily by the types of protamines involved in nucleoprotamine formation. Thus, trout and *Xenopus* sperm contain different protamines and the order and extent of histone replacement differ in spermatids from these species. The arginine-rich trout protamines may be more effective than the intermediate-type protamines of *Xenopus* in displacing histones from DNA. Such differences have been demonstrated between fish and rooster (galline) protamines; at physiological concentrations *in vitro*, galline displaces nucleosomes from DNA[117] while fish protamines displace only histone H1.[117-119]

Mechanisms which render nucleosomes susceptible to disruption and replacement by nucleoprotamine may be more common in diverse organisms. The observation that fish protamines do not displace core histones from DNA *in vitro* suggests that *in vivo* histone-histone or histone-DNA interactions must be destabilized to facilitate protamine binding and histone replacement.[118,119] Destabilization may be achieved through enzymatic modification of the histones.

Most attention has been given to histone acetylation as a primary mechanism for nucleosome destabilization. All species examined to date exhibit extensive histone acetylation (particularly of H4) in spermatids replacing histones with either protamines (e.g., trout,[98,99] *Xenopus*,[94] rooster[117,120]) or TPs (e.g., rat[114,115]). In contrast, histone acetylation levels are very low in spermatids of the winter flounder[45] or carp[121] which do not replace their histones. Early attempts to show that chemically acetylated histones are more readily displaced from DNA by protamines were unsuccessful,[122,123] prompting the suggestion that other factors, such as histone proteolysis, are important.[123] These early studies employed sheared chromatin, however, and the possibility of histone rearrangement in such preparations is good. A more recent study employed isolated nucleosomes with different levels of acetylation and found that *in vitro* protamine displacement of histones is facilitated by acetylation.[117]

Bode et al.[124] have isolated nucleosome monomers with varied levels of acetylated histones from lymphoid cells and demonstrated that acetylation results in more open conformational states when there are at least 10 acetylated residues per nucleosome. A hyperacetylated oligonucleosome fraction has been isolated from trout spermatids by a mild nuclease release of 10 to 15% of chromatin from nuclei and a subsequent salt fractionation.[121] The hyperacetylated fraction of chromatin was soluble in 0.1 *M* NaCl, devoid of protamine, and exhibited a relaxed structure with reduced capability to form higher-order structures. The lengths of chromatin exhibiting hyperacetylation contained 10 to 50 nucleosomes but most fragments were <20 nucleosomes. These observations indicate that hyperacetylation occurs in chromatin domains; it precedes protamine deposition and it alters chromatin structure.

Structural chromatin domains have also been noted by electron microscopy of spread spermatid chromatin from cricket,[70] mouse,[71] and ram.[68] At least three structural states were noted in these studies: a completely nucleosomal state in round spermatids, a completely nonnucleosomal chromatin consisting of thin (30 to 48 Å), smooth filaments in late elongate spermatids, and a transitional state consisting of both nucleosomal fibers and smooth fibers in early elongate spermatids. In crickets,[70] the transitional state occurs prior to addition of protamines to DNA, while the transition in mammalian spermatids corresponds with the period of histone replacement by TPs.[68,71] Importantly, the transitional chromatin was seen in individual spermatid nuclei.[68,70] Also, in ram spermatids, the smooth chromatin first appears in the anterior portion of the nucleus and later in the posterior portion, a pattern which mimics the topology of histone replacement.[68]

The observations discussed above suggest that early stages of the nucleohistone to nucleoprotamine transition occur in discrete domains. The thin, nucleosome-free fibers seen in elongate spermatids by electron microscopy may correspond to hyperacetylated domains. The smooth thin fibers may be generated artifactually by the selective mechanical disruption of unstable, hyperacetylated nucleosomes. On the other hand, the thin fibers may exist *in vivo* and consist of hyperacetylated nucleosomes which have unfolded to form half-nucleosomes. The histones may also be partially or totally replaced by TPs in the thin fibers, but this does not appear to be the case in cricket spermatids where the thin fibers occur prior to the stages of histone replacement.[70] Knowledge of the protein composition of thin fibers could help define their relationship to histone hyperacetylation.

It will be important to determine if the early chromatin modifications which facilitate the conversion of nucleosomal domains to nucleoprotamine are ordered with respect to DNA sequence and the organization of transcriptional units. That this may occur is suggested by the demonstration that certain single copy DNA sequences are found selectively in either nucleohistone or nucleoprotamine of human sperm.[87] These sequences have not been characterized and their functional significance is unknown. Presumably the nucleohistone sequences reside in domains which are unavailable for binding at the time of protamine or TP synthesis. The nucleohistone sequences reside in destabilized domains since the histones are extractable at relatively low salt concentrations and both H3 and H4 are extensively acetylated.[87] This suggests that nucleosome modification may occur in a defined order with the sperm nucleohistone sequences being modified late. Alternatively, nucleosome acetylation and destabilization may be relatively random with respect to DNA sequence, but certain sequences may be protected from protamine binding. It would be informative to compare the DNA sequences present in hyperacetylated chromatin isolated from different spermatid stages. Cloned DNA sequences isolated from human sperm nucleohistone or nucleoprotamine could be used as probes to study the packaging of these sequences throughout spermatid development.

D. HIGHER-ORDER STRUCTURE

In somatic cells, filaments of nucleosomes organize into 30 nm "thick" fibers.[21,125] Structural models for the thick fiber include a solenoid with 6 to 8 nucleosomes/turn,[126,127] a solenoid consisting of a double-start, twisted ribbon,[128,129] and superbeads or clustered arrays of nucleosomes.[130,131] At present, the solenoid is the favored model.[21,125] The 30 nm fibers are folded into numerous 30 to 100 kbp loops in both interphase[132,133] and metaphase[134,135] chromosomes. These loops are topologically independent domains due to nonhistone constraints at the base of each loop.[132,133]

Like somatic cells, sperm that contain nucleosomes also contain thick chromatin fibers, but the fibers are tightly packed in the nucleus. Electron microscopy of chromatin released from nucleosomal sperm suggests that the "thick" fibers may be formed from linear arrays of superbeads. Thus, superbeads have been observed in dispersed sperm chromatin from the sea urchins *Paracentrotus lividus*[136,137] and *Strongylocentrotus purpuratus,*[138] the mussel, *Mytilus galloprovincialis,*[139] and the sea cucumber, *Holothuria tubulosa.*[140] In sea urchin sperm, superbeads are 40[138] to 48[137] nm in diameter corresponding to a cluster of approximately 48 nucleosomes.[137] *Holothuria* superbeads are about 30 nm in diameter[139] and *Mytilus* superbeads are 21 to 25 nm.[139] The large size of the sea urchin sperm superbeads may result from the greater ability of the germ-line histones SPH1 and SPH2B to form internucleosomal cross-links.[91,137]

The higher-order organization of nucleoprotamine fibers is not well defined due to its extreme condensation and uncertainties regarding structural artifacts that may be generated by harsh decondensing and dispersal methods. Electron microscopy of sectioned spermatids from diverse species has revealed significant diversity in the pattern of chromatin fiber condensation which probably reflects an underlying biochemical diversity.[141] Nevertheless, three major patterns of condensation have been described: the fibrous, lamellar, and granular patterns.[141-143] In the fibrous pattern, such as that seen in the cricket *Acheta domesticus,* there is a progressive increase in fiber thickness and decrease in number of fibrous structures as chromatin condenses.[144] Lamellae or sheets of interconnected fibers are intermediates in the lamellar pattern of condensation. This pattern is common in insects like the grasshopper, *Melanoplus femor-rubrum.*[145] Granules of increasing diameter and decreasing number characterize the granular pattern seen in the fowl, *Gallus domesticus,*[146] and the frog, *Xenopus laevis.*[143]

Ultrastructural analyses of condensing chromatin in spermatids suggest that the formation of nucleoprotamine results from a lateral association of multiple thin filaments into fibers of increasing thickness rather than an increasing thickening of individual fibers.[141] Thus, smooth, thin (30 to 60 Å) filaments replace nucleosomal filaments in spermatids of crickets,[69,70] mouse,[71] ram,[68,147,148] and humans,[147] and the thin filaments associate stepwise into fibers of increasing thickness.

A somewhat different model has been proposed for rat sperm nucleoprotamine which has been resolved by freeze fracture into stacked lamellar plates 13 to 26 nm thick and 7.2 μm wide.[150] Koehler et al.[150] suggest that the fibers which form the lamellae are comprised of thin nucleoprotamine filaments coiled through two orders of supercoiling into 13 to 26 nm thick fibers which associate laterally into lamellae. This is consistent with linear dichroism measurements of chromatin released from equine sperm which suggest that the DNA helix is perpendicular to the longitudinal axis of the chromatin fiber.[151] The supercoiling model appears to be inconsistent with X-ray diffraction studies, however, which have failed to resolve regular higher-order supercoils of nucleoprotamine fibers.[38,73,74,78]

Although sperm chromatin is considered most often as a homogeneous, nearly crystalline array of fibers,[38,73,82] there is morphological evidence for structural domains in nucleoprotamine. For example, chromatin fibers from decondensed sperm nuclei often appear "knobby." Gusse and Chevaillier have argued that nucleoprotamine in trout, dogfish, bull, and human sperm is organized into fibers with repeating globular subunits.[84] Although inconsistent with X-ray diffraction data,[68,73,78] repeating structural subunits of nucleoprotamine are consistent with the particulate model proposed by Benzett-Jones and Ottensmeyer.[153] Larger knobby structures about 200 to 300 Å thick have been noted in chromatin of rat,[150] ram,[68,147] and human[85,86,149] sperm. Further evidence comes from the observation that chromatin decondensation produces fibers of discrete size classes, suggesting preferred, relatively stable fiber associations.[68,86] Random lateral associations of fibers should generate a structure which decondenses into a continuous range of fiber sizes.

Structural units in nucleoprotamine may arise as a consequence of an ordered packaging of chromatin domains in spermatid nuclei. Kierszenbaum and Tres[69] have resolved structural packaging units in spreads of chromatin from cricket spermatids. These units are 23 nm in diameter and 330 nm in length and, in sections, appear to be oriented perpendicular to the nuclear envelope. Each packaging unit consists of several laterally associated loops of thin (30 Å), smooth chromatin fibers. At late spermatid stages, the packaging units associate laterally into thick bundles with paracrystalline features. Tsanev and Avramova have noted the presence of similar packaging units in trout sperm chromatin partially decondensed with 0.5 M NaCl.[152]

Further evidence for structural domains in sperm chromatin has been obtained from ultrastructural analyses of decondensed sperm nuclei from trout,[152] mussels,[140] and ram.[154] Sperm from each of these organisms contain clusters of DNA fibers attached radially to granular or knobby structures 30 to 70 nm in size. In trout sperm, the clusters appear when packaging units are disrupted by extraction with 0.5 M NaCl and 5 M urea,[152] suggesting that the packaging unit consists of a condensed cluster of DNA fibers. A similar structure was proposed for the packaging unit in cricket spermatids.[69] The DNA clusters in trout sperm are resistant to disruption by 2 M NaCl, 2% SDS, or extraction with urea and reducing agents.[152] Tightly bound nonhistones of 60 to 80 kDa may be recovered from the detergent extracted chromatin only after digestion with DNase I.

The organization of sperm chromatin into structural domains has also been inferred from a study of DNA topology in *Rana catesbeiana, Bufo fowleri,* and *Xenopus laevis* sperm nuclei extracted with 2 M NaCl and Triton X-100.[72] Decondensed sperm nuclei (nucleoids) from each species consist of DNA-protein networks which, although swollen, retain shapes similar to those of control nuclei (Figures 3 and 4). DNA in the sperm nucleoids is organized into supercoiling domains, since 100 μg/ml ethidium bromide condenses the nucleoids while UV illumination of the ethidium stained nucleoids causes decondensation. Mild nicking of nucleoid DNA by DNase I also relaxes the nucleoid DNA. *Xenopus* (Figure 5) and *Bufo* (Figure 6b) sperm nucleoid DNA condenses gradually when stained with increasing concentrations (1 to 100 μg/ml) of ethidium bromide. In contrast, *Rana* sperm nucleoid DNA displays a biphasic transition, first expanding as ethidium bromide is raised to 8 μg/ml and subsequently condensing at ethidium concentrations above 8 μg/ml (Figure 6a). The biphasic transition is typical of negatively supercoiled DNA in nucleoids from *Xenopus* erythrocytes, spermatocytes, and round spermatids,[72] and somatic cell nucleoids from diverse sources.[132,133] The absence of a biphasic transition in *Xenopus* and *Bufo* nucleoid DNA suggests that negative supercoils in nucleosomes are removed from DNA as histones are replaced by protamines during spermatogenesis.

Xenopus nucleoids were unaffected by RNase or dithiothreitol, but were fragmented when digested with DNase I or proteases. Extraction of nucleoids in 2 M NaCl with 4 M urea relaxed ethidium-stained nucleoid DNA, but did not disperse the DNA. Electron microscopy of *Xenopus* nucleoids resolved a network of knobby (20 to 30 nm) fibers with little evidence for DNA loops (Figure 7). Many of the fibers were arranged parallel to the longitudinal axis of the nucleoid. Morphological evidence for a sperm nuclear matrix was not obtained.

A similar study of hamster sperm nucleoids has led to the conclusion that mammalian sperm also contain DNA organized into looped or supercoiling domains.[155] Similar to DNA in *Xenopus* and *Bufo* sperm nucleoids, DNA in hamster sperm nucleoids lacks negative superhelicity.[190] The DNA in the hamster sperm nucleoids appears to be anchored as loops to a nonhistone protein matrix which maintains its structural integrity during DNase I digestion.

The observations described above suggest that sperm DNA is knotted periodically to form independent domains of supercoiling. Nonhistone proteins appear to be responsible for this organization since histones and protamines are extracted by 2 M NaCl and supercoiled

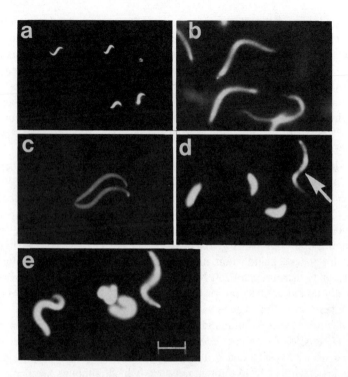

FIGURE 3.   Fluorescence photomicrographs of *Xenopus laevis* sperm nuclei (a) and nucleoids (b to e) stained with ethidium bromide at 2.5 (a), 0.5 (b), 6 (c), and 100 μg/ml (d, e). The same microscopic field is shown in d and e after a 4s exposure to UV for photography (d) and a subsequent 15 s exposure to UV prior to a second photographic exposure of 4 s (e). The arrow in d indicates a relaxed nucleoid. Bar in e represents 25 μm, and all panels are at the same magnification. (From Risley, M. S., Einheber, S., and Bumcrot, D. A., *Chromosoma*, 94, 217, 1986. With permisison.)

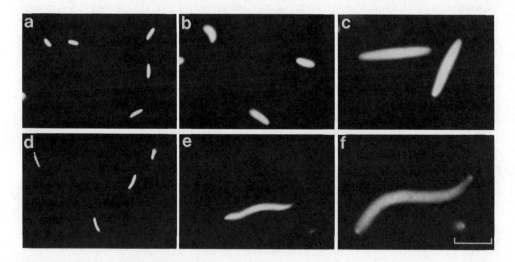

FIGURE 4.   Fluorescence photomicrographs of *Rana catesbeiana* sperm nuclei (a) and nucleoids (b, c), and *Bufo fowleri* sperm nuclei (d) and nucleoids (e, f). Ethidium bromide concentrations were 4 (a, d) and 100 μg/ml (b, c, e, f). UV exposures were 2 s (b, e) or 17 s (c, f). Bar in f represents 50 μm, and all panels are the same magnification. (From Risley, M. S., Einheber, S., and Bumcrot, D. A., *Chromosoma*, 94, 217, 1986. With permission).

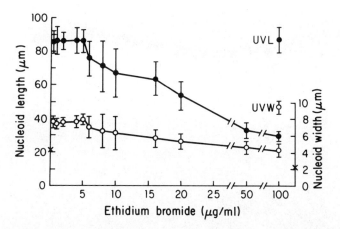

FIGURE 5.   Lengths (-●-) and widths (-○-) of *Xenopus laevis* sperm nucleoids at different ethidium bromide concentrations. Nucleoid width was measured in the center where nucleoids are cylindrical. Nucleoid length was measured in sections by moving the ocular micrometer along the length of each nucleoid. Nucleoids were protected from UV damage by 2% mercaptoethanol. The length and width of sperm nuclei are indicated by x on the appropriate axis. UVL and UVW, length and width of nucleoids stained with 100 µg/ml ethidium bromide and relaxed by 30 s UV exposure (in the absence of mercaptoethanol). (From Risley, M. S., Einheber, S., and Bumcrot, D. A., *Chromosoma,* 94, 217, 1986. With permission.)

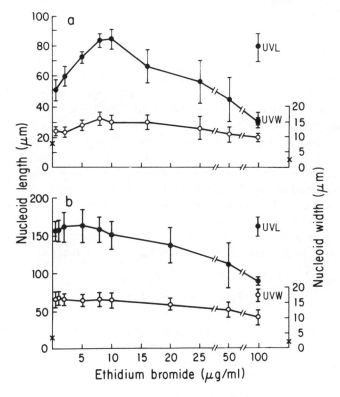

FIGURE 6.   Lengths (-●-) and widths (-○-) of sperm nucleoids from *Rana catesbeiana* (a) and *Bufo fowleri* (b). See Figure 5 for description of methods and symbols. (From Risley, R. S., Einheber, S., and Bumcrot, D. A., *Chromosoma,* 94, 217, 1986. With permission.)

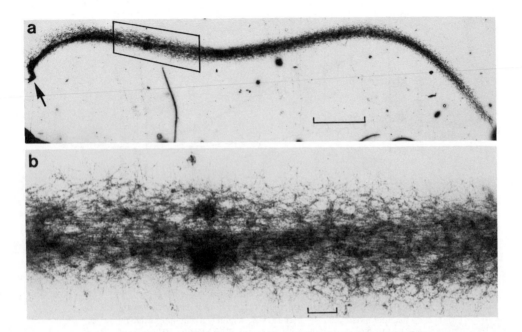

FIGURE 7. Electron micrographs of a *Xenopus laevis* sperm nucleoid centrifuged onto poly-D-lysine treated Formvar coated grids and stained with 0.5% uranyl acetate. (a) The arrow indicates the posterior tip of the nucleoid, while the box shows the region displayed at higher magnification in b. Bars represent 10 μm (a) and 1 μm (b). (From Risley, M. S., Einheber, S., and Bumcrot, D. A., *Chromosoma*, 94, 217, 1986. With permission.)

domains become relaxed by protease digestion.[72,152,154] The nonhistones present in *Xenopus* sperm nucleoids are a complex mixture. They do not appear to form a nuclear matrix since a residual structural matrix could not be isolated from *Xenopus* nucleoids digested with DNase I. The tight binding nonhistones recovered from ram sperm DNA after high salt and detergent extractions appear to be glycoproteins with sequence homologies to each other but not to lamins.[156] Recent evidence suggests that the tight binding proteins on ram and rat sperm DNA are homologous to those on somatic cell DNA and may be attached covalently to DNA.[157]

The sperm chromatin domains and nonhistones have not been sufficiently characterized to establish their functional significance. On the one hand, the domains may be important for the orderly packaging of nucleoprotamine fibers. Attachment of DNA fibers to nonhistones may align the fibers and facilitate their lateral association into thicker bundles. On the other hand, attachment of nonhistones to DNA may be important for imprinting specific regions of the paternal genome. Further information will be required to evaluate these possibilities.

III. DNA METHYLATION

DNA in many eukaryotes is modified by methylation at the 5-position of cytosine residues in specific CpG dinucleotides. This postreplicative modification occurs in 2 to 5% of cytosines and in tissue-specific patterns which suggest a role for methylation in gene expression.[158-161] Generally, low levels of methylation in specific sequences correlate with transcriptional activity while high levels are common in inactive genes. Indeed, site-specific methylation of promoters *in vitro* leads to inactivation of the flanking genes when the DNA in transferred to cells and assayed for transcriptional activity.[162-164] DNA methylation patterns are propagated during replication by a maintenance methylase that recognizes hemimethylated sites.[165,166] Thus, DNA methylation appears to be capable of both regulation of gene expression and propagation through multiple cell divisions. These properties suggest a potential role for methylation in genomic imprinting and the epigenetic control of cell differentiation.[167-169] If methylation is important for imprinting, DNA methylation patterns should differ in the male and female gametes.

Total 5-methylcytosine (5mC) is generally higher in somatic cells than sperm,[170-172] but sperm DNA is hypermethylated relative to DNA from oocytes, 8-cell embryos, and blastocysts.[171] Analysis of 5mC in specific sequences has shown that the methylation patterns in sperm are gene specific. Thus, bovine[173] and mouse[174,175] satellite sequences are undermethylated in sperm relative to somatic cells, but interspersed repetitive sequences are hypermethylated in sperm.[171,174] Hypermethylation of endogenous structural genes is also common in sperm. For example, genes for chicken ovalbumin,[89,176] globin,[89,177] thymidine kinase,[89] and the ev-3 retrovirus[89] are heavily methylated in sperm. Rabbit β-globin genes,[178] human globin genes,[179] *Xenopus* rDNA,[180] and mouse unique sequence DNA[181] are also highly methylated in sperm. An endogenous structural gene that is not hypermethylated in sperm is TPG-3, a trout protamine gene.[182] This gene is methylated at only one of 5 CpG sites in testes, nucleoprotamine, liver, and erythrocytes. Failure of this gene to be heavily methylated in spermatogenesis may be related to its specific expression in the male germ line. Alternatively, TPG-3 expression may not be regulated by methylation.

Detailed mapping by Groudine and Conklin[89] of site-specific methylations have shown that although most CpG sites in structural genes are hypermethylated in sperm, specific sites are protected from methylation. In the constitutively expressed chicken thymidine kinase gene and the ev-3 retroviral gene, protected sites occur in DNase-I-hypersensitive regions which are potentially important for the expression of these genes. In contrast, the tissue-specific globin and ovalbumin genes are fully methylated in sperm. These results suggest that control regions in constitutive genes expressed during spermatogenesis and in early embryos may be selectively undermethylated to maintain signals essential for expression, even as hypersensitivity in the control regions is abolished during chromatin condensation in sperm.[89] The spermatogenic stage at which methylation occurs in roosters was found to be between the spermatogonial and primary spermatocyte stages,[89] possibly during premeiotic S-phase.

The observations discussed above demonstrate that a *de novo* methylation process is operative in early spermatogenic stages and it is both gene specific and site specific within a gene. Methylation of sperm DNA is therefore a potential mechanism for regulating gene expression in the early embryo. Recent studies have employed transgenes to determine if methylation varies with parental sex, since transgenes can be maintained in a hemizygous state and their parental history can be controlled.[183]

The transgenic constructs that have been studied for heritable methylation patterns were SV CAT IgH (CAT = chloramphenicol acetyl transferase),[184] RSV Ig c-*myc*,[185] quail troponin I genes,[186] and the gene for a hepatitis B antigen.[187] With the exception of one troponin I transgenic line,[185] all transgenes were relatively undermethylated in somatic cells if transmitted paternally, but hypermethylated if transmitted maternally. In males that received the transgene maternally, transgenes in testes were undermethylated relative to those in somatic cells. In all but one case (hepatitis B antigen[187]), the methylation state was reversed by passage of the transgene through the germ line of the opposite sex. These results are consistent with the hypothesis that site-specific DNA methylation patterns are heritable, dependent on parental sex, and may differ in sperm and eggs.

The transgene studies have generated considerable excitement, but also some questions. For example, transgenes appear different from many endogenous structural genes, in that they are usually hypomethylated in testes or somatic cells when inherited paternally. Is this an effect of the male pronucleus (site of transgene injection) on DNA methylation? Does the site of transgene integration influence methylation during spermatogenesis? An insertion site dependence of germ line methylation is suggested by the observation that troponin I genes could be either hypomethylated or hypermethylated when transmitted paternally.[185] Nonrandom insertion of transgenes into imprinted domains in the male pronucleus may lead to their protection from methylation. Thus, the transgenes may not be subject to the *de novo* methylation processes in spermatogenesis[89] or the early embryo.[171,188,189] Paternally inherited transgenes would, therefore, remain hypomethylated unless they were passed through the maternal germ line. Clearly, this explanation suggests that the transgenes are preferentially integrating into domains which are imprinted during gametogenesis or in the fertilized egg.

Transgenes passed through the maternal germ line are subject to extensive methylation, but methylation of transgenes in eggs has not been directly assessed.[184-187] Transgenes passed through oogenesis may indeed be hypomethylated, consistent with the global undermethylation of egg DNA.[171] This would require, however, that an imprint which protects the transgene from methylation during spermatogenesis be removed during oogenesis. The transgene would then be subject to embryonic methylation in the subsequent generation leading to hypermethylation in somatic cells. This hypothesis can be tested by direct analysis of the 5mC content of transgenes in eggs.

The studies discussed in this section suggest that sperm development is accompanied by widespread, but selective, DNA methylation which may be important for regulation of genomic expression in the early embryo or for processes specific to spermatogenesis. DNA methylation in early spermatogenic cells may be involved in establishing meiotic or post meiotic patterns of gene expression.[89] Patterns of methylation established in early germ cells may also direct the packaging of DNA in sperm by influencing DNA-protein interactions.

IV. CONCLUSION

Proposed models of sperm chromatin structure have focused entirely on structures generated by protamine-DNA and protamine-protamine interactions.[38,73,78] Experimental evidence summarized here, however, suggests that sperm chromatin is biochemically and structurally complex and that current structural models are incomplete. In the sperm of certain species, like *Homo sapiens,* chromatin consists of both nucleohistone and nucleoprotamine organized in a partly sequence-dependent manner. There is also a sequence specificity to the distribution of DNase-I-hypersensitive sites in sperm chromatin. DNA in the sperm of several species is bound to nonbasic proteins as well as histones or protamines. Certain nonbasic proteins assemble sperm DNA into independent structural domains and some nonbasic proteins appear to be covalently linked to DNA. Finally, the methylation of CpG dinucleotides is site-specific in sperm and methylation patterns are sperm-specific. The organization of sperm DNA is thus considerably more complex and sequence specific than expected for a *tabula rasa* or blank slate.

REFERENCES

1. **Risley, M. S.,** The organization of meiotic chromosomes and synaptonemal complexes, in *Chromosome Structure and Function,* Risley, M. S., Ed., Van Nostrand Reinhold, New York, 1986, 126.
2. **Moens, P. B.,** *Meiosis,* Academic Press, New York, 1987.
3. **Risley, M. S., Eckhardt, R. A., Mann, M., and Kasinsky, H. E.,** Determinants of sperm nuclear shaping in the genus *Xenopus, Chromosoma,* 84, 557, 1982.
4. **Hemsworth, B. N. and Wardhaugh, A. A.,** Embryopathies due to spermatozoal impairment in *Xenopus laevis, Mutat. Res.,* 51, 45, 1978.

5. **Lyon, M. F.,** Sensitivity of various germ-cell stages to environmental mutagens, *Mutat. Res.*, 87, 323, 1981.

6. **Sega, G. A., Sluder, A. E., McCoy, L. S., Owens, J. G., and Generoso, E. E.,** The use of alkaline elution procedures to measure DNA damage in spermiogenic stages of mice exposed to methyl methane-sulfonate, *Mutat. Res.*, 159, 55, 1986.

7. **Sega, G. A. and Owen, J. G.,** Ethylation of DNA and protamine by ethyl methanesulfonate in the germ cells of male mice and the relevancy of these molecular targets to the induction of dominant lethals, *Mutat. Res.*, 52, 87, 1978.

8. **Kofman-Alfaro, S. and Chandley, A. C.,** Radiation-initiated DNA synthesis in spermatogenic cells of the mouse, *Exp. Cell Res.*, 69, 33, 1971.

9. **Sega, G. A.,** Unscheduled DNA synthesis (DNA repair) in the germ cells of male mice. Its role in the study of mammalian mutagenesis, *Genetics*, 92, 49, 1979.

10. **Generoso, W. M., Cain, K. T., Krishna, M., and Huff, S. W.,** Genetic lesions induced by chemicals in spermatozoa and spermatids of mice are repaired in the egg, *Proc. Natl. Acad. Sci. U.S.A.*, 76, 435, 1979.

11. **Debaradino, M. A., Hoffner, N. J., and Etkin, L. D.,** Activation of dormant genes in specialized cells, *Science*, 224, 946, 1983.

12. **Crouse, H. V.,** The controlling element in sex chromosome behavior in *Sciara*, *Genetics*, 45, 1429, 1960.

13. **Lyon, M. F. and Rastan, S.,** Parental source of chromosome imprinting and its relevance for X chromosome inactivation, *Differentiation*, 26, 63, 1984.

14. **Surani, M. A. H., Barton, S. C., and Norris, M. L.,** Experimental reconstruction of mouse eggs and embryos: an analysis of mammalian development, *Biol. Reprod.*, 36, 1, 1987.

15. **Chandra, H. A. and Brown, S. W.,** Chromosome imprinting and the mammalian X chromosome, *Nature (London)*, 253, 165, 1975.

16. **Cooper, D. W., Vandeberg, J. L., Sharman, G. B., and Poole, W. E.,** Phosphoglycerate kinase polymorphism provides further evidence for paternal X inactivation, *Nature (London)*, 230, 155, 1971.

17. **Lock, L. F. and Martin, G. R.,** Dosage compensation in mammals: X chromosome inactivation, in *Chromosome Structure and Function*, Risley, M. S., Ed., Van Nostrand Reinhold, New York, 1986, 197.

18. **McGrath, J. and Solter, D.,** Completion of mouse embryogenesis requires both the maternal and paternal genomes, *Cell*, 37, 179, 1984.

19. **Surani, M. A. H., Barton, S. C., and Norris, M. L.,** Development of reconstituted mouse eggs suggests imprinting of the genome during gametogenesis, *Nature (London)*, 308, 548, 1984.

20. **Surani, M. A. H., Barton, S. C., and Norris, M. L.,** Nuclear transplantation in the mouse: heritable differences between parental genomes after activation of the embryonic genome, *Cell*, 45, 127, 1986.

21. **McGhee, J.,** The structure of interphase chromatin, in *Chromosome Structure and Function*, Risley, M. S., Ed., Van Nostrand-Reinhold, New York, 1986, 2.

22. **Pederson, D. S., Thoma, F., and Simpson, R. T.,** Core particle, fiber and transcriptionally active chromatin structure, *Annu. Rev. Cell Biol.*, 2, 117, 1986.

23. **Simpson, R. T.,** Structure of the chromatosome, a chromatin particle containing 160 base pairs of DNA and all the histones, *Biochemistry*, 17, 5524, 1978.

24. **Morris, N. R.,** A comparison of the structure of chicken erythrocyte and chicken liver chromatin, *Cell*, 9, 627, 1976.

25. **Lohr, D., Corden, J., Tatchell, K., Kovacic, R. T., and Van Holde, K. E.** Comparative subunit structure of HeLa, yeast, and chicken erythrocyte chromatin, *Proc. Natl. Acad. Sci. U.S.A.*, 74, 79, 1977.

26. **Prunell, A. and Kornberg, R. D.,** Variable center to center distance of nucleosomes in chromatin, *J. Mol. Biol.*, 154, 515, 1982.

27. **Zweidler, A.,** Nonallelic histone variants in development and differentiation, *Dev. Biochem.*, 15, 47, 1980.

28. **Zweidler, A.,** Core histone variants of the mouse: primary structure and differential expression, in *Histone Genes*, Stein, G. S., Stein, J. L., and Marzluff, W. F., Eds., John Wiley & Sons, New York, 1984, 339.

29. **Poccia, D.,** Remodeling of nucleoproteins during gametogenesis, fertilization and early development, *Int. Rev. Cytol.*, 105, 1, 1986.

30. **Risley, M. S. and Eckhardt, R. A.,** H1 histone variants in *Xenopus laevis*, *Dev. Biol.*, 84, 79, 1981.

31. **Perry, M., Thomsen, G. H., and Roeder, R. G.,** Major transitions in histone gene expression do not occur during development in *Xenopus laevis*, *Dev. Biol.*, 116, 532, 1986.

32. **Urban, M. K., Franklin, S. G., and Zweidler, A.,** Isolation and characterization of histone variants in chicken erythrocytes, *Biochemistry*, 18, 3952, 1979.

33. **Urban, M. K. and Zweidler, A.,** Changes in nucleosomal core histone variants during chicken development and maturation, *Dev. Biol.*, 95, 421, 1983.

34. **Wunsch, A. M. and Lough, J.,** Modulation of histone H3 variant synthesis during the myoblast-myotube transition of chicken myogenesis, *Dev. Biol.*, 119, 94, 1987.

35. **Piña, B. and Suau, P.,** Changes in histones H2A and H3 variant composition in differentiating and mature rat brain cortical neurons, *Dev. Biol.*, 123, 51, 1987.

36. **Bloch, D. P.,** A catalog of sperm histones, *Genetics* (Suppl.), 61, 93, 1969.
37. **Bloch, D. P.,** Histones of sperm, in *Handbook of Genetics,* Vol. 5, King, R. C., Ed., Plenum Press, New York, 1976, 139.
38. **Subirana, J. A.,** Nuclear proteins in spermatozoa and their interactions with DNA, in *The Sperm Cell,* Andre, J., Ed., Martinus Nijhoff, The Hague, 1983, 197.
39. **Kasinsky, H. E.,** Specificity and distribution of sperm basic proteins, in *Histones and Other Basic Nuclear Proteins,* Hnilica, L., Stein, G., and Stein, J., Eds., CRC Press, Boca Raton, FL, 1989.
40. **Muñoz-Guerra, S., Azorin, M., Casas, M. T., Marcel, X., Maristancy, M. A., Roca, J., and Subirana, J. A.,** Structural organization of sperm chromatin from the fish *Carassius auratus, Exp. Cell Res.,* 137, 47, 1982.
41. **Kadura, S. N., Khrapunov, S. N., Chabanny, V. N., and Berdyshev, G. D.,** Changes in chromatin basic proteins during male gametogenesis of grass carp, *Comp. Biochem. Physiol.,* 74B, 343, 1983.
42. **Alder, D. and Gorovsky, M. A.,** Electrophoretic analysis of liver and testis histones of the frog *Rana pipiens, J. Cell Biol.,* 64, 389, 1975.
43. **Phelan, J. J., Subirana, J. A., and Cole, R. D.,** An unusual group of lysine-rich histones from gonads of a sea cucumber, *Holothuria tubulosa, Eur. J. Biochem.,* 31, 63, 1972.
44. **Easton, D. and Chalkley, R.,** High-resolution electrophoretic analysis of the histones from embryos and sperm of *Arbacia punctulata, Exp. Cell Res.,* 72, 502, 1972.
45. **Kennedy, B. P. and Davies, P. L.,** Acid-soluble nuclear proteins of the testis during spermatogenesis in the winter flounder, *J. Biol. Chem.,* 255, 2533, 1980.
46. **Mann, M., Risley, M. S., Eckhardt, R. A., and Kasinsky, H. E.,** Characterization of spermatid/sperm basic chromosomal proteins in the genus *Xenopus* (Anura, pipidae), *J. Exp. Zool.,* 222, 173, 1982.
47. **Risley, M. S. and Eckhardt, R. A.,** Protamine polymorphism in *Xenopus laevis laevis, J. Exp. Zool.,* 242, 373, 1987.
48. **Balhorn, R., Weston, S., Thomas, C., and Wyrobek, A. J.,** DNA packaging in mouse spermatids. Synthesis of protamine variants and four transition proteins, *Exp. Cell Res.,* 150, 298, 1984.
49. **Puwaravutipanich, T. and Panyim, S.,** The nuclear basic proteins of human testis and ejaculated spermatozoa, *Exp. Cell Res.,* 90, 153, 1975.
50. **Ammer, H., Henschen, A., and Lee, C.,** Isolation and amino acid sequence analysis of human sperm protamines P1 and P2. Occurrence of two forms of protamine P2, *Biol. Chem. Hoppe Seyler,* 367, 515, 1986.
51. **McKay, D. J., Renaux, B. S., and Dixon, G. H.,** Human sperm protamines. Amino acid sequences of two forms of protamine P2, *Eur. J. Biochem.,* 156, 5, 1986.
52. **Bower, P. A., Yelick, P. C., and Hecht, N. B.,** Both P1 and P2 protamine genes are expressed in mouse, hamster, and rat, *Biol. Reprod.,* 37, 479, 1987.
53. **Yelick, P. C., Balhorn, R., Johnson, P. A., Corzett, M., Mazrimas, J. A., Kleene, K. C., and Hecht, N. B.,** Mouse protamine 2 is synthesized as a precursor whereas mouse protamine 1 is not, *Mol. Cell. Biol.,* 7, 2173, 1987.
54. **Tymowska, J. and Fischberg, M.,** Chromosome complements of the genus *Xenopus, Chromosoma,* 44, 335, 1973.
55. **Bisbee, C. A., Baker, M. A., Wilson, A. C., Hedji-Azimi, I., and Fischberg, M.,** Albumin phylogeny for clawed frogs *(Xenopus), Science,* 195, 785, 1977.
56. **Risley, M. S.,** unpublished observations.
57. **Spadafora, C., Bellard, M., Compton, J. L., and Chambon, P.,** The DNA repeat length in chromatins from sea urchin sperm and gastrula cells are markedly different, *FEBS Lett.,* 69, 281, 1976.
58. **Keichline, L. D. and Wasserman, P. M.,** Structure of chromatin in sea urchin embryos, sperm, and adult somatic cells, *Biochemistry,* 18, 214, 1979.
59. **Arceci, R. J. and Gross, P. R.,** Sea urchin chromatin structure as probed by pancreatic DNase I: evidence for a novel cutting periodicity, *Dev. Biol.,* 80, 210, 1980.
60. **Zalenskaya, I. A., Pospelov, V. A., Zalensky, A. O., and Vorobev, V. I.,** Nucleosomal structure of sea urchin and starfish sperm chromatin. Histone H2B is possibly involved in determining the length of linker DNA, *Nucleic Acids Res.,* 9, 473, 1981.
61. **Zalensky, A. O. and Avramova, Z. V.,** Nucleosomal organization of a part of chromatin in mollusc sperm nuclei with a mixed basic protein composition, *Mol. Biol. Rep.,* 10, 69, 1984.
62. **Ausio, J. and Van Holde, K.,** A dual chromatin organization in the sperm of the bivalve mollusc *Spisula solidissima, Eur. J. Biochem.,* 165, 363, 1987.
63. **Cornudella, L. and Rocha, E.,** Nucleosome organization during germ cell development in the sea cucumber *Holothuria tubulosa, Biochemistry,* 18, 3724, 1979.
64. **Kadura, S. N., Khrapunov, S. N., Chadanny, V. N., and Berdyshev, G. D.,** Chromatin structure during male gametogenesis of grass carp, *Comp. Biochem. Physiol.,* 74B, 819, 1983.
65. **Kennedy, B. P. and Davies, P. L.,** Chromatin reorganization during spermatogenis in the winter flounder, *J. Biol. Chem.,* 257, 11160, 1982.

66. **Honda, B. M., Baillie, D. L., and Candido, E. P. M.,** The subunit structure of chromatin: characteristics of nucleohistone and nucleoprotamine from developing trout testis, *FEBS Lett.,* 48, 156, 1974.
67. **Young, R. J. and Sweeney, K.,** The structural organization of sperm chromatin, *Gamete Res.,* 2, 265, 1979.
68. **Loir, M., Bouvier, D., Fornells, M., Lanneau, M., and Subirana, J. A.,** Interactions of nuclear proteins with DNA, during sperm differentiation in the ram, *Chromosoma,* 92, 304, 1985.
69. **Kierszenbaum, A. L. and Tres, L. L.,** The packaging unit: a basic structural feature for the condensation of late cricket spermatid nuclei, *J. Cell Sci.,* 33, 265, 1978.
70. **McMaster-Kaye, R. and Kaye, J. S.,** Organization of chromatin during spermiogenesis: beaded fibers, partly beaded fibers, and loss of nucleosomal structure, *Chromosoma,* 77, 41, 1980.
71. **Kierszenbaum, A. L. and Tres, L. L.,** Structural and transcriptional features of the mouse spermatid genome, *J. Cell Biol.,* 65, 258, 1975.
72. **Risley, M. S., Einheber, S., and Bumcrot, D. A.,** Changes in DNA topology during spermatogenesis, *Chromosoma,* 94, 217, 1986.
73. **Suau, P. and Subirana, J. A.,** X-ray diffraction studies of nucleoprotamine structure, *J. Mol. Biol.,* 117, 909, 1977.
74. **Ausio, J. and Subirana, J. A.,** Nuclear proteins and the organization of chromatin in spermatozoa of *Mytilus edulis, Exp. Cell Res.,* 141, 39, 1982.
75. **Pogany, G. C., Corzett, M., Weston, S., and Balhorn, R.,** DNA and protein content of mouse sperm, *Exp. Cell Res.,* 136, 127, 1981.
76. **Rowland, R. D. and Rill, R. L.,** Atypical changes in chromatin structure during development in the sea urchin, *Lytechinus variegatus, Biochim. Biophys. Acta,* 908, 169, 1987.
77. **Puigdomenech, P., Romero, M. C., Allan, J., Sautiere, P., Giancotti, V., and Crane-Robinson, C.,** The chromatin of sea urchin sperm, *Biochim. Biophys. Acta,* 908, 70, 1987.
78. **Balhorn, R.,** A model for the structure of chromatin in mammalian sperm, *J. Cell Biol.,* 93, 298, 1982.
79. **Calvin, H. I. and Bedford, J. M.,** Formation of disulfide bonds in the nucleus and accessory structures of mammalian spermatozoa during maturation in the epididymis, *J. Reprod. Fertil.,* 13, 65, 1971.
80. **Avramova, Z., Uschewa, A., Stephanova, E., and Tsanev, R.,** Trout sperm chromatin. I. Biochemical and immunological study of the protein composition, *Eur. J. Cell Biol.,* 31, 137, 1983.
81. **Uschewa, A., Avramova, Z., and Tsanev, R.,** Tightly bound somatic histones in mature ram sperm nuclei, *FEBS Lett.,* 138, 50, 1982.
82. **Tanphaichitr, N., Sobhon, P., Talupphet, N., and Chalermisarachai, P.,** Basic nuclear proteins in testicular cells and ejaculated spermatozoa in man, *Exp. Cell Res.,* 117, 347, 1978.
83. **Gusse, M., Sautiere, P., Belaiche, D., Martinage, A., Roux, C., Dadoune, J.-P., and Chevaillier, P.,** Purification and characterization of nuclear basic proteins of human sperm, *Biochim. Biophys. Acta,* 884, 124, 1986.
84. **Gusse, M. and Chevaillier, P.,** Electron microscope evidence for the presence of globular structures in different sperm chromatins, *J. Cell Biol.,* 87, 280, 1980.
85. **Wagner, T. E. and Yun, J. S.,** Human sperm chromatin has a nucleosomal structure, *Arch. Androl.,* 7, 251, 1981.
86. **Sobhon, P., Tanphaichitr, N., Chutatape, C., Vongpayabal, P., and Panuwatsuk, W.,** Electron microscopic and biochemical analyses of the organization of human sperm chromatin decondensed with sarkosyl and dithiothreitol, *J. Exp. Zool.,* 223, 277, 1982.
87. **Gatewood, J. M., Cook, G. R., Balhorn, R., Bradbury, E. M., and Schmid, C. W.,** Sequence-specific packaging of DNA in human sperm chromatin, *Science,* 236, 962, 1987.
88. **Spinelli, G., Albanese, I., Anello, L., Ciaccio, M., and DiLegro, I.,** Chromatin structure of histone genes in sea urchin sperms and embryos, *Nucleic Acids Res.,* 10, 1977, 1982.
89. **Groudine, M. and Conklin, K. F.,** Chromatin structure and de novo methylation of sperm DNA: implications for activation of the paternal genome, *Science,* 228, 1061, 1985.
90. **Poccia, D. L., Simpson, M. V., and Green, G. R.,** Transitions in histone variants during sea urchin spermatogenesis, *Dev. Biol.,* 121, 445, 1987.
91. **Green, G. R. and Poccia, D. L.,** Interaction of sperm histone variants and linker DNA during spermiogenesis in the sea urchin, *Biochemistry,* in press.
92. **Kennedy, B. P., Crim, L. W., and Davies, P. L.,** Expression of histone and tubulin genes during spermatogenesis, *Exp. Cell Res.,* 158, 445, 1985.
93. **Kennedy, B. P. and Davies, P. L.,** Phosphorylation of a group of high molecular weight basic nuclear proteins during spermatogenesis in the winter flounder, *J. Biol. Chem.,* 256, 9254, 1981.
94. **Risley, M. S.,** Basic Chromatin Protein Changes During Spermatogenesis in *Xenopus laevis,* Ph.D. thesis, The City University of New York, 1977.
95. **Louie, A. J. and Dixon, G. H.,** Trout testis cells. II. Synthesis and phosphorylation of histones and protamines in different cell types, *J. Biol. Chem.,* 247, 5498, 1972.

96. **Louie, A. J. and Dixon, G. H.**, Synthesis, acetylation, and phosphorylation of histone IV and its binding to DNA during spermatogenesis in trout, *J. Biol. Chem.*, 69, 1975, 1972.
97. **Louie, A. J., Sung, M. T., and Dixon, G. H.**, Modification of histones during spermatogenesis in trout. III. Levels of phosphohistone species and kinetics of phosphorylation of histone IIb1, *J. Biol. Chem.*, 248, 3335, 1973.
98. **Candido, E. P. M. and Dixon, G. H.**, Trout testis cells. III. Acetylation of histones in different cell types from developing trout testis, *J. Biol. Chem.*, 247, 5506, 1972.
99. **Christensen, M. E. and Dixon, G. H.**, Hyperacetylation of histone H4 correlates with the terminal, transcriptionally inactive stages of spermatogenesis in rainbow trout, *Dev. Biol.*, 93, 404, 1982.
100. **Louie, A. J. and Dixon, G. H.**, Kinetics of enzymatic modification of the protamines and a proposal for their binding to chromatin, *J. Biol. Chem.*, 247, 7962, 1972.
101. **Marushige, K. and Dixon, G. H.**, Developmental changes in chromosomal composition and template activity during spermatogenesis in trout testis, *Dev. Biol.*, 19, 397, 1969.
102. **Marushige, K. and Dixon, G. H.**, Transformation of trout testis chromatin, *J. Biol. Chem.*, 246, 5799, 1971.
103. **Kaye, J. S. and McMaster-Kaye, R.**, Histones of spermatogenous cells in the house cricket, *Chromosoma*, 96, 397, 1974.
104. **McMaster-Kaye, R. and Kaye, J. S.**, Basic protein changes during the final stages of sperm maturation in the house cricket, *Exp. Cell Res.*, 97, 378, 1976.
105. **Gusse, M. and Chevaillier, P.**, Microelectrophoretic analysis of basis protein changes during spermiogenesis in the dogfish *Scylliorhinus caniculus* (L.), *Exp. Cell Res.*, 136, 391, 1981.
106. **Meistrich, M. L., Trostle, P. K., and Brock, W. A.**, Association of nucleoprotein transitions with chromatin changes during rat spermatogenesis, in *Bioregulators of Reproduction*, Jagiello, G. and Vogel, H., Eds., Academic Press, New York, 1981, 151.
107. **Grimes, S. R., Jr.**, Nuclear proteins in spermatogenesis, *Comp. Biochem. Physiol.*, 83B, 495, 1986.
108. **Bhatnagar, Y. M., Romrell, L. J., and Bellve, A. R.**, Biosynthesis of specific histones during meiotic prophase of mouse spermatogenesis, *Biol. Reprod.*, 32, 599, 1985.
109. **Meistrich, M. L., Bucci, L. R., Trostle-Weige, P. K., and Brock, W. A.**, Histone variants in rat spermatogonia and primary spermatocytes, *Dev. Biol.*, 112, 230, 1985.
110. **Platz, R. D., Grimes, S. R., Jr., Meistrich, M. L., and Hnilica, L.**, Changes in nuclear proteins of rat testis cells separated by velocity sedimentation, *J. Biol. Chem.*, 250, 5791, 1975.
111. **Grimes, S. R., Jr., Chae, C. B., and Irvin, J. L.**, Nuclear protein transitions in rat testis spermatids, *Exp. Cell Res.*, 110, 31, 1977.
112. **Loir, M. and Lanneau, M.**, Transformation of ram spermatid chromatin, *Exp. Cell Res.*, 115, 231, 1978.
113. **Marushige, Y. and Marushige, K.**, Transformation of sperm histone during formation and maturation of rat spermatozoa, *J. Biol. Chem.*, 250, 39, 1975.
114. **Grimes, S. R., Jr. and Henderson, N.**, Hyperacetylation of histone H4 in rat testis spermatids, *Exp. Cell Res.*, 152, 91, 1984.
115. **Grimes, S. R., Jr. and Henderson, N.**, Acetylation of histones during spermatogenesis in the rat, *Arch. Biochem. Biophys.*, 221, 108, 1983.
116. **Tanphaichitr, N., Sobhon, P., Teluppeth, N., and Chalermisarachai, P.**, Basic nuclear proteins in testicular cells and ejaculated spermatozoa in man, *Exp. Cell Res.*, 117, 347, 1978.
117. **Oliva, R. and Mezquita, C.**, Marked differences in the ability of distinct protamines to disassemble nucleosomal core particles *in vitro*, *Biochemistry*, 25, 6508, 1986.
118. **Wong, T. K. and Marushige, K.**, Modification of histone binding in calf thymus chromatin by protamine, *Biochemistry*, 14, 122, 1975.
119. **Bode, J., Willmitzer, L., and Opatz, K.**, On the competition between protamines and histones: studies directed towards the understanding of spermiogenesis, *Eur. J. Biochem.*, 72, 393, 1977.
120. **Oliva, R. and Mezquita, C.**, Histone H4 hyperacetylation and rapid turnover of its acetyl groups in transcriptionally inactive rooster testis spermatids, *Nucleic Acids Res.*, 10, 8049, 1982.
121. **Christensen, M. E., Rattner, J. B., and Dixon, G. H.**, Hyperacetylation of histone H4 promotes chromatin decondensation prior to histone replacement by protamines during spermatogenesis in rainbow trout, *Nucleic Acids Res.*, 12, 4575, 1984.
122. **Wong, T. K. and Marushige, K.**, Modification of histone binding in calf thymus chromatin and in the chromatin-protamine complex by acetic anhydride, *Biochemistry*, 15, 2041, 1976.
123. **Marushige, K., Marushige, Y., and Wong, T. K.**, Complete displacement of somatic histones during transformation of spermatid chromatin: a model experiment, *Biochemistry*, 15, 2047, 1976.
124. **Bode, J., Gomez-Lira, M. M., and Schroter, H.**, Nucleosomal particles open as the histone core becomes acetylated, *Eur. J. Biochem.*, 130, 437, 1983.
125. **Butler, P. J. G.**, The organization of the chromatin fiber, in *Chromosomes and Chromatin*, Vol. 1, Adolph, K. W., Ed., CRC Press, Boca Raton, FL, 1988, 57.

126. **Finch, J. T. and Klug, A.,** Solenoidal model for superstructure of chromatin, *Proc. Natl. Acad. Sci. U.S.A.,* 73, 1897, 1976.

127. **Thoma, F., Koller, T., and Klug, A.,** Involvement of histone H1 in the organization of the nucleosome and of the salt dependent superstructures of chromatin, *J. Cell Biol.,* 83, 403, 1979.

128. **Worcel, A., Strogatz, S., and Riley, D.,** Structure of chromatin and the linking number of DNA, *Proc. Natl. Acad. Sci. U.S.A.,* 78, 1461, 1981.

129. **Woodcock, C. L. F., Frado, L. L. Y., and Rattner, J. B.,** The higher-order structure of chromatin: evidence for a helical ribbon arrangement, *J. Cell Biol.,* 99, 42, 1984.

130. **Hozier, J., Renz, M., and Nehls, P.,** The chromosome fiber: evidence for an ordered superstructure of nucleosomes, *Chromosoma,* 62, 301, 1977.

131. **Zentgraf, H., Muller, U., and Franke, W. W.,** Reversible *in vitro* packing of nucleosomal filaments into globular supra-nucleosomal units in chromatin of whole chick erythrocyte nuclei, *Eur. J. Cell Biol.,* 23, 171, 1980.

132. **Cook, P. R. and Brazell, I. A.,** Conformational constraints in nuclear DNA, *J. Cell Sci.,* 22, 287, 1976.

133. **Benyajati, C. and Worcel, A.,** Isolation, characterization, and structure of the folded interphase genome of *Drosophila melanogaster, Cell,* 9, 393, 1976.

134. **Paulson, J. R. and Laemmli, U. K.,** The structure of histone-depleted metaphase chromosomes, *Cell,* 12, 817, 1977.

135. **Marsden, M. P. F. and Laemmli, U. K.,** Metaphase chromosome structure: evidence for a radial loop model, *Cell,* 17, 849, 1979.

136. **Zentgraf, H., Müller, U., and Franke, W. W.,** Supranucleosomal organization of sea urchin sperm chromatin in regularly arranged 40 to 50 nm large granular subunits, *Eur. J. Cell Biol.,* 20, 254, 1980.

137. **Zentgraf, H. and Franke, W. W.,** Differences of supranucleosomal organization in different kinds of chromatin: cell type-specific globular subunits containing different numbers of nucleosomes, *J. Cell Biol.,* 99, 272, 1984.

138. **Aboukarsh, N. and Kunkle, M.,** Ultrastructural organization of heterochromatin within sea urchin sperm nuclei, *Gamete Res.,* 12, 55, 1985.

139. **Avramova, Z., Zalensky, A., and Tsanev, R.,** Biochemical and ultrastructural study of the sperm chromatin from *Mytilus galloprovincialis, Exp. Cell Res.,* 152, 231, 1984.

140. **Subirana, J. A., Muñoz-Guerra, S., Martinez, A. B., Perez-Grau, L., Marcet, X., and Fita, I.,** The subunitstructure of chromatin fibers, *Chromosoma,* 83, 455, 1981.

141. **Kaye, J. S.,** The ultrastructure of chromatin in nuclei of interphase cells and in spermatids, in *Handbook of Molecular Cytology,* Lima-de Faria, A., Ed., North Holland, Amsterdam, 1969, 361.

142. **Walker, M. H.,** Studies on the arrangement of nucleoprotein in elongate sperm heads, *Chromosoma,* 34, 340, 1971.

143. **Walker, M. H. and MacGregor, H. C.,** Chromosomes in elongate sperm heads, in *Chromosomes Today,* Vol. 5, John Wiley & Sons, New York, 1974, 13.

144. **Kaye, J. S. and McMaster-Kaye, R.,** The fine structure and chemical composition of nuclei during spermiogenesis in the house cricket. I. Initial stages of differentiation and the loss of non-histone protein, *J. Cell Biol.,* 31, 159, 1966.

145. **Gall, J. G. and Bjork, L. B.,** The spermatid nucleus in two species of grasshopper, *J. Biophys. Biochem. Cytol.,* 4, 479, 1958.

146. **McIntosh, J. R. and Porter, K. R.,** Microtubules in the spermatids of the domestic fowl, *J. Cell Biol.,* 35, 153, 1967.

147. **Loir, M. and Courtens, J. L.,** Nuclear reorganization in ram spermatids, *J. Ultrastruct. Res.,* 67, 309, 1979.

148. **Loir, M. and Lanneau, M.,** Structural function of the basic nuclear proteins in ram spermatids, *J. Ultrastruct. Res.,* 86, 262, 1984.

149. **Evenson, D. P., Darzynkiewicz, Z., and Melamed, M. R.,** Comparison of human and mouse sperm chromatin structure by flow cytometry, *Chromosoma,* 78, 225, 1980.

150. **Koehler, J. K., Wurschmidt, U., and Larsen, M. P.,** Nuclear and chromatin structure in rat spermatozoa, *Gamete Res.,* 8, 357, 1983.

151. **Sipski, M. L. and Wagner, T. E.,** The total structure and organization of chromosomal fibers in eutherian sperm nuclei, *Biol. Reprod.,* 16, 428, 1977.

152. **Tsanev, R. and Avramova, Z.,** Trout sperm chromatin. II. Ultrastructural aspects after salt dissociation of proteins, *Eur. J. Cell Biol.,* 31, 143, 1983.

153. **Bazett-Jones, D. P. and Ottensmeyer, F. P.,** A model for the structure of nucleoprotamine, *J. Ultrastruct. Res.,* 67, 255, 1979.

154. **Tsanev, R. and Avramova, Z.,** Nonprotamine nucleoprotein ultrastructures in mature ram sperm nuclei, *Eur. J. Cell Biol.,* 24, 139, 1981.

155. **Ward, W. S. and Coffey, D. S.,** DNA in hamster sperm nuclear matrix is organized in loop domains, *J. Cell Biol.,* 103, 51a, 1986.

156. **Avramova, Z. and Tasheva, B.,** Tightly bound nonprotamine proteins from ram sperm nuclei studied by one- and two-dimensional peptide mapping, *Mol. Cell. Biochem.,* 74, 67, 1987.
157. **Avramova, Z. and Tsanev, R.,** Stable DNA-protein complexes in eukaryotic chromatin, *J. Mol. Biol.,* 196, 437, 1987.
158. **Ehrlich, M. and Wang, R. Y. H.,** 5-Methylcytosine in eukaryotic DNA, *Science,* 212, 1350, 1981.
159. **Cooper, D. N.,** Eukaryotic DNA methylation, *Human Genet.,* 64, 315, 1983.
160. **Doerfler, W.,** DNA methylation and gene activity, *Annu. Rev. Biochem.,* 52, 93, 1983.
161. **Bird, R.,** CpG-rich islands and the function of DNA methylation, *Nature (London),* 321, 209, 1986.
162. **Busslinger, M., Hurst, J., and Flavell, R. A.,** DNA methylation and the regulation of globin gene expression, *Cell,* 34, 197, 1983.
163. **Langner, K.-D., Vardimon, L., Renz, D., and Doerfler, W.,** DNA methylations of three 5′-C-C-G-G-3′ sites in the promoter and 5′ region inactivated the E2a gene of adenovirus type 2, *Proc. Natl. Acad. Sci. U.S.A.,* 81, 2950, 1984.
164. **Keshet, I., Yisraeli, J., and Cedar, H.,** Effect of regional DNA methylation on gene expression, *Proc. Natl. Acad. Sci. U.S.A.,* 82, 2560, 1985.
165. **Wigler, M. H.,** The inheritance of methylation patterns in vertebrates, *Cell,* 24, 285, 1981.
166. **Gruenbaum, Y., Cedar, H., and Razin, A.,** Substrate and sequence specificity of a eukaryotic DNA methylase, *Nature (London),* 295, 620, 1982.
167. **Holiday, R. and Pugh, J. E.,** DNA modification mechanisms and gene activity during development, *Science,* 187, 226, 1975.
168. **Razin, A. and Riggs, A. D.,** DNA methylation and gene function, *Science,* 210, 604, 1980.
169. **Holiday, R.,** The inheritance of epigenetic defects, *Science,* 238, 163, 1987.
170. **Singer, J., Roberts - Ems, J., Luthardt, F. W., and Riggs, A. D.,** Methylation of DNA in mouse early embryos, teratocarcinoma cells, and adult tissues of mouse and rabbit, *Nucleic Acids Res.,* 7, 2369, 1979.
171. **Monk, M., Boubelik, M., and Lehnert, S.,** Temporal and regional changes in DNA methylation in the embryonic, extraembryonic and germ cell lineages during mouse embryo development, *Development,* 99, 371, 1987.
172. **Rocamora, N. and Mezquita, C.,** Hypomethylation of DNA in meiotic and postmeiotic rooster testis cells, *FEBS Lett.,* 177, 81, 1984.
173. **Sturm, K. S. and Taylor, J. H.,** Distribution of 5-methyl-cytosine in DNA of somatic and germ line cells from bovine tissues, *Nucleic Acids Res.,* 9, 4537, 1981.
174. **Sanford, J., Forrester, W., Chapman, V., Chandley, A., and Hastie, N.,** Methylation patterns of repetitive DNA sequences in germ cells of *Mus musculus, Nucleic Acids Res.,* 12, 2823, 1984.
175. **Ponzetto-Zimmerman, C. and Wolgemuth, D. J.,** Methylation of satellite sequences in mouse spermatogenic and somatic DNAs, *Nucleic Acids Res.,* 12, 2807, 1984.
176. **Mandel, J. L. and Chambon, P.,** DNA methylation; organ specific variations in the methylation pattern within and around ovalbumin and other chicken genes, *Nucleic Acids Res.,* 7, 2081, 1979.
177. **Haigh, L. S., Owens, B. B., Hellewell, S., and Ingram, V. M.,** DNA methylation in chicken α-globin gene expression, *Proc. Natl. Acad. Sci. U.S.A.,* 79, 5332, 1982.
178. **Shen, C. J. and Maniatis, T.,** Tissue specific DNA methylation in a cluster of rabbit β-like globin genes, *Proc. Natl. Acad. Sci., U.S.A.,* 77, 6634, 1980.
179. **van der Ploeg, L. H. T. and Flavell, R. A.,** DNA methylation in the human γδβ-globin locus in erythroid and nonerythroid tissues, *Cell,* 19, 947, 1980.
180. **Bird, A., Taggart, M., and MacLeod, D.,** Loss of rDNA methylation accompanies the onset of ribosomal gene activity in early development of *X. laevis, Cell,* 26, 381, 1981.
181. **Rahe, B., Erickson, R. P., and Quinto, M.,** Methylation of unique sequence DNA during spermatogenesis in mice, *Nucleic Acids Res.,* 11, 7947, 1983.
182. **Delcuve, G. P. and Davie, J. R.,** DNA methylation pattern and restriction endonuclease accessibility in chromatin of a germ line specific gene, the rainbow trout protamine gene, *Nucleic Acids Res.,* 15, 3385, 1987.
183. **Surani, M. A., Reik, W., and Allen, N. D.,** Transgenes as molecular probes for genomic imprinting, *Trends Genet.,* 4, 59, 1988.
184. **Reik, W., Collick, A., Norris, M. L., Barton, S. C., and Surani, M. A.,** Genomic imprinting determines methylation of parental alleles in transgenic mice, *Nature (London),* 328, 248, 1987.
185. **Sapienza, C., Peterson, A. C., Rossant, J., and Balling, R.,** Degree of methylation of transgenes is dependent on gamete of origin, *Nature (London),* 328, 251, 1987.
186. **Swain, J. L., Stewart, T. A., and Leder, P.,** Parental legacy determines methylation and expression of an autosomal transgene: a molecular mechanism for parental imprinting, *Cell,* 50, 719, 1987.
187. **Hadchouel, M., Farza, H., Simon, D., Tioelais, P., and Pourcel, C.,** Maternal inhibition of hepatitis B surface antigen gene expression in transgenic mice correlates with *de novo* methylation, *Nature (London),* 329, 454, 1987.

188. **Rossant, J., Sanford, J. P., Chapman, V. M., and Andrews, G. K.,** Undermethylation of structural gene sequences in extraembryonic lineages of the mouse, *Dev. Biol.,* 117, 567, 1986.
189. **Jahner, D., Stuhlmann, H., Stewart, C. L., Harbers, K., Lohler, J., Simon, J., and Jaenisch, R.,** *De novo* methylation and expression of retroviral genomes during mouse embryogenesis, *Nature (London),* 298, 623, 1982.
190. **Ward, W. S. and Coffey, D. S.,** personal communication, 1987.

Section II. Eukaryotic Chromosomes: Special, Plant, and Organelle

INTRODUCTION

An overview of the most significant aspects of DNA and protein packaging in chromosomes cannot be restricted to the primary chromosomes of animal and bacterial cells. The chapters of this section are therefore devoted to special chromosomes (polytene chromosomes, yeast) and chromosomes of higher plants and organelles such as chloroplasts. These chromosomes invite study because of the role of plants in the ecology of the earth and in human nutrition. In addition, some, particularly yeast and polytene chromosomes, serve as model systems for molecular biology studies of gene activity and chromosome organization. Progress in understanding these systems has advanced greatly in recent years due to the application of the techniques of modern biological research that were developed with bacterial and animal cells.

Polytene chromosomes, particularly those of *Drosophila melanogaster,* have been the subject of many investigations. In salivary glands of *Drosophila,* polytene chromosomes consist of about 1000 DNA molecules aligned side-by-side to produce a massive and highly specialized chromosome. Interest in polytene chromosomes stems from the importance of *Drosophila* in genetics. A precise pattern of chromosome banding and the development of puffs at actively transcribed loci have facilitated detailed studies.

Yeast, along with ciliated protozoa and slime molds, are lower eukaryotes that have provided valuable information concerning the molecular biology of eukaryotes. The role of DNA sequences in the replication and maintenance of yeast chromosomes is an example of a problem that has been fruitfully investigated. Studies have been concerned with replication of the primary yeast chromosomes, which are linear molecules, and circular plasmids known as 2 micron DNA. Yeast centromeric DNA, found at the attachment point of spindle microtubules, prevents chromosome loss through cycles of replication. Protozoan telomere DNA, found at the ends of linear chromosomes, consists of unusual DNA sequences that act as protective chromosome caps.

Plant cells, being involved in photosynthesis, are distinguished by the presence of chloroplasts as DNA-containing organelles. The contribution of chloroplasts to total cellular DNA synthesis is substantial. Steady progress is being made in understanding the interrelationship between DNA replication in plant cell nuclei and chloroplasts. The involvement of chloroplast gene products is becoming clearer as the transcription of individual genes is examined. Plant chromatin structure is another research problem that is yielding exciting results with the application of molecular biology techniques. A good correlation has been found between the openness of genes, as detected by restriction endonuclease digestion, and the active transcriptional state. Even with these modern approaches, the value of experiments in classical plant genetics remains undiminished. Plants are unsurpassed for certain studies such as the manipulation of whole sets of chromosomes.

Chapter 4

STRUCTURE AND FUNCTION OF POLYTENE CHROMOSOMES

Milan Jamrich

TABLE OF CONTENTS

I. INTRODUCTION

In 1881, more than a century ago, Balbiani[1] discovered unusual nuclear structures in the salivary glands of *Chironomus* larvae. Their identity remained unknown until 1912 when Rambousek[2] concluded that these "banded ribbons" were giant chromosomes. Nevertheless, more than 20 years went by before this view became generally accepted. The experiments of Heitz and Bauer,[3] Painter,[4,5] King and Beams,[6] Koltzoff,[7] Bauer,[8] Bridges,[9] and Caspersson[10] showed conclusively that these banded structures indeed represent giant interphase chromosomes. Over the next 50 years these polytene chromosomes developed into an extremely useful tool for the analysis of chromosome structure, DNA organization, and gene expression. In many ways this was due to the fortunate fact that the favorite organism of geneticists — *Drosophila* — proved to have excellent polytene chromosomes in their salivary glands. This review will attempt to highlight the most important findings of the last 100 years of research on polytene chromosomes and point out the questions yet to be answered. For summaries of the literature of a given historical period, the interested reader is advised to consult the articles of Alverdes,[11] Tanzer,[12] Bauer,[8] Mets,[13] Alfert,[14] and Beermann.[15]

Polytene chromosomes are giant chromosomes usually found in specialized, terminally differentiated cells. The DNA of these chromosomes is replicated to a high degree, in some cases thousands of DNA strands remain aligned side by side without separation of daughter chromatids. In contrast to metaphase chromosomes, the polytene chromosomes are transcriptionally active interphase chromosomes. They can be found in protists, plants, as well as in lower and higher animals.[16] The degree of polyteny varies from organism to organism and from tissue to tissue. Different degrees of polyteny can also be found within the same tissue. The degree of polyteny is reflected in the size of the nuclei since larger space is needed to accommodate the additional amount of DNA. This property is dramatically visualized in ethidium-bromide-stained *Drosophila melanogaster* salivary glands (Figure 1). The size difference between diploid cells of the imaginal disks and polytene cells of the excretory duct and secretory cells is obvious. The final level of polyteny attained in the euchromatic regions of polytene chromosomes of secretory cells is 1024, yet significantly higher levels of polyteny are found in *Chironomus* and *Rhyncosciara*, where the degree of polyteny can be as much as 16,384 and 32,768, respectively.[15] The resulting chromosomes can be hundreds of μm long and up to 25 μm in diameter. The nuclei can reach 150 μm in diameter. It has been determined that the position of the chromosomes within the nucleus is not random. Chromosomes are preferentially located on the periphery of the nucleus leaving the center relatively free. Certain chromosomal sites, such as the chromocenter and intercalary heterochromatin, have an attachment to the nuclear membrane.[17,18] The reason for this attachment is not understood, but it appears that it is not related to transcriptional processes since attachment sites are not altered when drastic changes in transcription occur.[19,20]

II. BANDS AND INTERBANDS

The most intriguing feature of polytene chromosomes is their banding pattern. Figure 2 shows a preparation of *Drosophila melanogaster* polytene chromosomes from a single salivary gland cell. The darkly stained regions, the bands (better visible in Figure 3), alternate with regions of lesser staining called the interbands. The question has arisen: what is the molecular basis of this banding pattern? Early on it was established that at least some DNA molecules run along the entire length of the chromosome uninterrupted.[21,22] Nevertheless, when the DNA content of bands and interbands was compared, it was found that 75 to 95% of the DNA was in the bands.[23-25] Each band has on the average 30 kb of DNA and each interband correspondingly less. However, not all bands are equivalent and range in DNA content from 5 to 100 kb depending on their thickness.

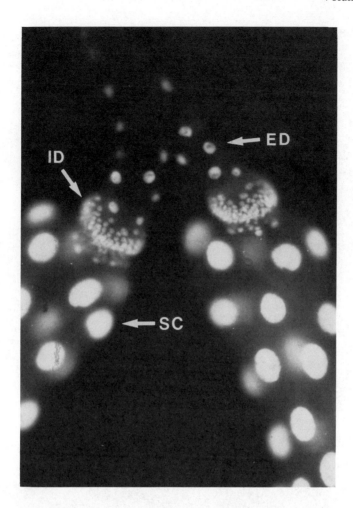

FIGURE 1.   Salivary glands of *Drosophila melanogaster* stained with ethidium bromide. ID — imaginal discs, SC — secretory cells, ED — excretory duct. Note the size difference between the diploid nuclei of the ID and polytene nuclei of ED and SC.

Two models were offered to explain the discrepancies in DNA content between bands and interbands. The first model proposed differential folding of the basic chromatin fiber.[26] According to this model, the DNA would be more condensed in the bands than in the interbands. The second model, based on DNA content obtained for bands and interbands by dry-mass determination, proposed differential replication of the DNA.[25,27,28] This model suggested that the DNA in bands was replicated to a higher degree than in the interbands.

In recent years, these two models have been critically tested. Using restriction enzyme analysis combined with Southern blotting analysis,[29] DNA fragments covering many bands and interbands could be tested for their degree of polytenization. Spierer and Spierer[30] and Spierer et al.[31] analyzed 315 kb of *Drosophila* DNA spanning 13 adjacent bands and interbands. They found that all of the 84 restriction fragments in this region, regardless of whether they were located in the band or interband, had the same degree of polyteny. Lifschytz[32] came to the same conclusions by analyzing randomly selected sequences which had been mapped to various positions on the polytene chromosomes. Based on these results, it seems likely that the model proposing higher order of DNA folding in the bands is the correct one. The picture emerging from studies done so far is that in the euchromatic regions of polytene

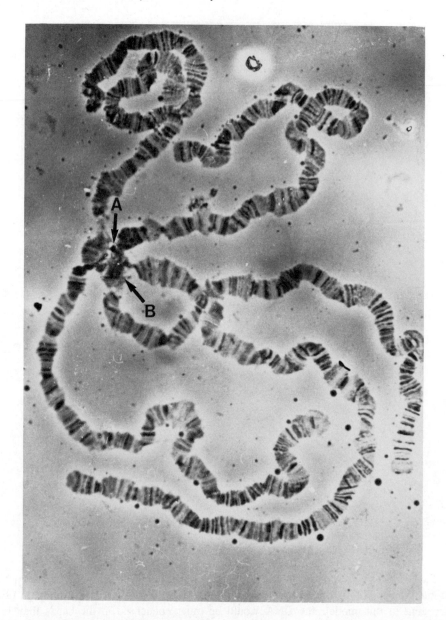

FIGURE 2.   Polytene chromosomes of *Drosophila melanogaster*. A — α-heterochromatin;
B — β-heterochromatin.

chromosomes the double stranded DNA complexed with histone proteins forms a 10 nm
fiber. This fiber coils into a 25 nm thick chromatin fiber, which is present in an extended
form in the interbands of polytene chromosomes and is highly coiled in the bands.[33] It is
not quite certain how this DNA fiber is organized within the band. Data from Mortin and
Sedat,[34] based on optical sections of chromosomes, suggest that this condensed chromatin
forms a toroidal structure. However, in mechanically sectioned chromosomes such structure
was not observed.[8]

Although the differences in band-interband morphology are not based on differential
replication, this does not mean that all of the DNA is replicated to the same degree. Early
investigators noticed that some DNA sequences are heavily underreplicated.[8,35-37] Heterochromatic satellite DNA, ribosomal RNA genes, histone genes, bithorax complex sequences,
and the entire male Y chromosome of *Drosophila* belong in this category.

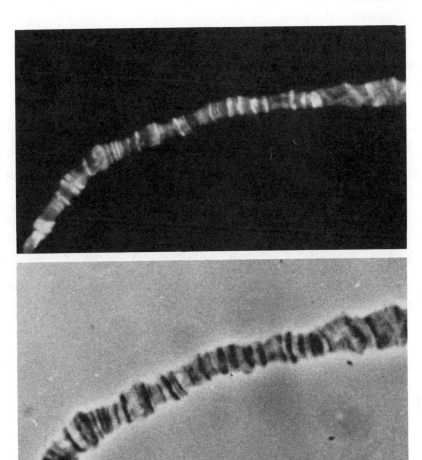

FIGURE 3. Localization of Hl protein on *Drosophila* polytene chromosomes. Upper panel shows localization of histone Hl using indirect immunofluorescence; bottom panel shows the corresponding phase contrast picture of the same chromosome. (From Jamrich, M., Greenleaf, A. L., and Bautz, E. K. F., *Proc. Natl. Acad. Sci. U.S.A.*, 74, 2079, 1977. With permission.)

Many highly repetitive DNA sequences are located close to the centromeres of chromosomes. The centromeres of polytene chromosomes aggregate to form the chromocenter (see Figure 2). The chromocenter has a less prominent banding pattern and can be divided into two parts.[36] The central, most condensed, and optically dark region is called the α-heterochromatin. This region contains a high proportion of underreplicated satellite DNA sequences and is transcriptionally silent.[38,39] The less condensed region, with a subtle banding pattern, is called the β-heterochromatin and contains repetitive sequences as well as active genes. The poor banding pattern is thought to be due to the influence of α-heterochromatin. In males the α-heterochromatic region also contains the most striking example of underreplication — the entire Y chromosome.[40-42] Highly repetitive DNA sequences are also present at the ends of the chromosome, the telomeres.[43-45] Underreplication of some of these sequences has also been demonstrated.[44]

The genes for ribosomal RNA are underreplicated in *Drosophila*[46-48] but in *Chironomus tentans* they are proportionally replicated. Other examples of underreplication are the histone genes[32] and bithorax complex genes.[49] Analysis of the cases of underreplication has shown that they are either heterochromatic or they are in proximity to such sequences. The presence of heterochromatic sequences can be frequently recognized by the absence of a banding pattern, such as that observed in the centromeric heterochromatin, or by constrictions in the case of intercalary heterochromatin. The underreplicated histone and bithorax genes are located close to such constrictions.

Overreplication can be also observed on polytene chromosomes but seems to be limited to Sciaridae. Examples of overreplication were reported in so called "DNA puffs" of *Sciara* and *Rhyncosciara*.[50-52] These examples show that the degree of replication in a given region is regulated. All euchromatic regions overreplicate to the same degree, heterochromatic sequences will not replicate, while others preferentially replicate. All these sequences seem to have an influence on the degree of replication of neighboring sequences. Translocation of euchromatic sequences into chromocentric heterochromatin can cause their severe underreplication.[53-56] Sequences controlling the rate of replication have been identified in the chorion genes of *Drosophila*.[57-59] These sequences appear to act in a stage- and tissue-specific manner. Thus the chorion genes are overreplicated only in the follicle cells of *Drosophila*, and there is evidence that the sequences which are underreplicated in polytene chromosomes of *Calliphora* bristle-forming cells are proportionally replicated in the nurse cells of the same species (Ribbert and Dover, as cited in Korge[60]). The detailed analysis of replication controlling sequences, and for that matter any sequences regulating the structural or functional characteristics of different regions of polytene chromosomes, should be forthcoming in the near future. The development of the *Drosophila* transformation system by Spradling and Rubin[61] makes it relatively easy to introduce novel sequences or reintroduce modified sequences into various positions of the polytene chromosomes. The effects of these sequences upon the structural or functional features of neighboring regions can be so studied. Identification of sequences which delineate the individual bands and interbands would be of specific interest.

III. BANDS AND GENES

The basic banding pattern of polytene chromosomes is constant within a species but differs among species. This feature stimulated many attempts to correlate bands/interbands with genetic units — the genes. The most popular hypothesis suggested that each band corresponded to a single gene.[5,9,62] This was supported by the fact that the number of estimated genes in *Drosophila*, 5000, roughly corresponded to the number of bands visible in the light microscope. The one band-one gene hypothesis was also suggested by genetic analysis of defined chromosomal regions.[63-72] In these studies the number of bands in the analyzed region was identical to the number of complementation groups. More recent evidence, however, has shown that the one band-one gene hypothesis is, in all likelihood, incorrect.

It has been shown that certain bands contain more than one complementation group.[73-75] Biochemical evidence has also suggested that some bands contain more than one transcription unit. Hall et al.[76] found that one band in the 87E region contains four transcription units. Furthermore, the 165 5S genes[77-79] and 100-300 histone genes,[80-82] which occur in tandemly repeated arrays, certainly do not occupy the corresponding number of bands. The *Dunce* gene of *Drosophila* occupies five bands and adjacent interbands, and includes two additional genes encoded in one of its introns.[83] Another such example is the GART locus, in which a cuticle protein gene is present in the intron of another gene.[84] All these results make it unlikely that bands or interbands have a one-to-one correlation with genes. It is more likely that the bands are structural units which, strictly speaking, do not represent genes, transcription, or replication units. Each band, a chromosomal domain, might contain one, a few, or a part of a gene (transcription unit), much in the way the loops of amphibian lampbrush chromosomes do. This packaging might have some influence on transcription or replication (or vice versa) but does not represent either of them. It therefore appears that the pattern of chromosomal domains is determined by a mechanism unknown to us which is unrelated to transcription or replication. Transcription and replication act upon this pattern of chromosomal domains resulting in a variety of local morphological changes which were either discussed above (constriction, chromocenter) or will be discussed below (puffs).

IV. PUFFS, INTERBANDS, AND TRANSCRIPTION

As already mentioned, the basic banding pattern of polytene chromosomes is constant. Nevertheless, specific local modulations of the basic pattern can be observed. Occasionally, certain bands will decondense and form structures called puffs.[85] Puffs are the loci of RNA synthesis[86,87] and contain high levels of RNA polymerase.[88-92] The size of the puffs seems to be proportional to the degree of transcriptional activity.[93,94] Puffs can be formed from a single band or multiple bands. A puff can contain one or more genes.[95-97] Their appearance is tissue specific[98-107] and is developmentally regulated.[108-113] Proteins have been identified which are products of RNA transcription at certain puffs.[114-119] Puffs can also be induced by environmental changes such as elevated temperature.[120] Heat shock of larvae will induce formation of new puffs, redistribution of RNA polymerase, and the appearance of heat shock RNA.[89,92,120,123] Foreign DNA containing a strong promoter will puff when introduced into *Drosophila* by the germline transfer technique.[124] Reduction in size of the newly introduced gene results in a reduction of puff size.[125] Transformation of a transcriptionally inactive gene into *Drosophila* will not result in a puff formation.[125-128]

However, not all transcriptionally active genes will form a puff. The 5S genes and histone genes are known to be transcribed, but the chromosomal regions corresponding to these genes do not puff. Autoradiographic analysis of RNA synthesis revealed that in addition to obvious puffs, other loci show ³H-uridine incorporation.[129,130] These loci were identified as the interbands. Localization studies using antibodies against RNA polymerase[89-92] revealed that many, if not all, interbands contain RNA polymerase molecules. These results were later confirmed using a monospecific antibody to RNA polymerase[131,132] in combination with ³H-uridine incorporation.[133,134] Ribonucleoproteins were localized in the interband regions as well.[135-137]

So why do some transcribed loci puff while others don't? The final answer is not known but it appears that the morphological demonstration of transcription depends on a number of factors. One of them is the location of the transcription unit within the band or interband. Other factors are likely to be the level of transcription, the length of the transcription unit, and the proximity of the next active transcription unit. A combination of these factors could give a different morphological demonstration of transcription at various sites on the chromosomes. So, for example, transposition of the *Drosophila* Sgs-3 protein gene to a new chromosomal site showed that, even though this gene is appropriately regulated at the insertion, the 1.2 kb long transcription unit does not result in formation of a puff.[138-140] However, if the transposed DNA contains the two additional transcription units of genes Sgs-7 and Sgs-8, the resulting 8.1 kb long DNA can support formation of a new puff. These results agree well with data of Simon et al.[125] who showed that an internal deletion in the transposed gene results in a corresponding reduction of puff size. The situation might be similar to that of amphibian lampbrush chromosomes. On these chromosomes every loop (which is possibly equivalent to a band on the polytene chromosomes) consists of one or few transcription units transcribed in the same or opposite direction. Very small genes, like the 5S genes, do not form loops — most likely because the loop is too small to be visible in the light microscope. The larger the size of the transcription unit, the larger the loop.

As already mentioned, the 5S genes on polytene chromosomes do not puff, and the largest known puffs in *Chironomus* generate a huge 75S RNA. In addition, it is possible that only a subset of the thousand or so homologous genes at the same locus are activated at any given time. The influence of activation of only a certain percentage of the genes at a particular locus on chromosome morphology remains to be investigated. Another interesting question is whether the puffing can be separated from transcriptional activity. Under natural conditions, puffing accompanies transcription.[93,121,129] Inhibition of transcription does inhibit the formation of a puff,[141,142] but termination of transcription and removal of RNA polymerase during puffing does not result in the immediate collapse of a puff.[89,113]

Surprisingly, studies of *Drosophila* mutants 1(1) su(f)[ts67g] and 1(1) npr-1 show that the mutated loci puff but do not make any RNA,[139,143-145] suggesting that there are at least two separable elements involved in the process of transcription — one responsible for puffing and the other responsible for transcription.

V. DISTRIBUTION OF PROTEINS ON POLYTENE CHROMOSOMES

The basic composition of polytene chromosomes does not differ in any significant way from other chromatin. DNA is complexed with histones to form a repeated nucleosomal structure, which appears in electron microscopy as a "beaded" structure.[146] The analysis of histone composition on polytene chromosomes shows that histone distribution closely follows the distribution of DNA.[89,147-151] This is visualized in Figure 3 using the indirect immunofluorescence technique. This technique utilizes the fact that the antibodies recognize in cytological preparations the antigen they were raised against. The antigenic proteins are usually visualized indirectly by reacting fluorescent secondary antibodies to the primary antigen-antibody complex. This technique was first used with polytene chromosomes by Desai et al.[148] and has been utilized for a large number of distribution studies of histone and non-histone proteins as well as other chromosomal components. Figure 3 shows a segment of *D. melanogaster* chromosomes reacted with antibodies against histone Hl. The highest concentration of histone Hl is in the bands. This distribution pattern is typical for all five histones and little if any redistribution of histone proteins can be detected during the process of transcription.[89,142,152] A large number of non-histone proteins have been localized on polytene chromosomes using indirect immunofluorescence techniques. The antibodies used were obtained against nuclear extracts; therefore, most of these proteins are of unknown function. Probably the best studied non-histone protein of known function is RNA polymerase. It was shown that the enzyme RNA polymerase I, which is responsible for transcription of ribosomal genes, is almost exclusively present in the nucleolus — the location of the majority of the ribosomal genes.[90] Antibodies against RNA polymerase II, which transcribes the majority of other genes, stain a large number of sites on polytene chromosomes which were identified as puffs and interbands.[89,131,133] Other non-histone proteins were localized[153-156] and deduced to be involved in transcription processes on the basis of their staining pattern. The antibodies against these proteins usually stain a variable number of puffs during the process of transcription. A group of non-histone proteins which bind to RNA have also been identified on polytene chromosomes. Antibodies against RNA binding proteins either react with virtually all transcription sites[137,157] or they stain only few loci, indicating that they are associated with specific transcripts.[158,159] In addition, there are proteins of unknown function which do not show clear association with transcriptional events. Such proteins are, for example, HMG (high mobility group) proteins which are preferentially associated with bands[160] and proteins which bind to centromeric heterochromatin.[161,162]

Indirect immunofluorescence is an extremely powerful technique and the study of protein distribution on polytene chromosomes will bring, in the future, more insight into the processes of transcription and replication. This technique can also be used to identify other components of chromosomes such as Z-DNA regions. Z-DNA can assume a left-handed helical conformation[163,164] and might play a role in gene regulation. Numerous attempts were made to localize Z-DNA regions on polytene chromosomes.[165-171] Unfortunately, the distribution of Z-DNA in polytene chromosomes is dependent of the fixation technique used, so very little has been learned about the Z-DNA distribution *in vivo*.

ACKNOWLEDGMENT

I would like to thank Dr. Kathleen A. Mahon for critical reading of the manuscript.

REFERENCES

1. **Balbiani, E. G.,** Sur la structure du noyau des cellules salivaries chez les larves de *Chironomous, Zool. Anz.,* 4, 637, 1881.
2. **Rambousek, F.,** Cytologische Verhaltnisse der Speicheldrusen der *Chironomus*-Larvae, *Sitzungsber., Konig. bohm. Ges. Wiss., math-maturw. Klasse.,* (cited after Beerman 1962), 1912.
3. **Heitz, E. and Bauer, H.,** Beweise fur die Chromosomennatur der Kernschleifen in den Knauelkernen von *Bibio hortulanus, Z. Zellforsch.,* 17, 67, 1933.
4. **Painter, T. S.,** A new method for the study of chromosome rearrangements and the plotting of chromosome maps, *Science,* 78, 585, 1933.
5. **Painter, T. S.,** Salivary chromosomes and the attack on the gene, *J. Hered.,* 25, 465, 1934.
6. **King, R. L. and Beams, H. W.,** Somatic synapsis in *Chironomus* with special reference to the individuality of the chromosomes, *J. Morphol.,* 56, 577, 1934.
7. **Koltzoff, N. K.,** The structure of the chromosomes in the salivary glands of *Drosophila, Science,* 80, 312, 1934.
8. **Bauer, H.,** Der Aufbau der Chromosomen aus den Speicheldrusen von *Chironomus Thummi* Kiefer (Untersuchungen an den Riesenchromosomen der Dipteren I), *Z. Zellforsch.,* 23, 280, 1935.
9. **Bridges, C. B.,** Salivary chromosome map with a key to the banding of the chromosomes of *D. melanogaster, J. Hered.,* 26, 60, 1935.
10. **Caspersson, T.,** Die Eiweissverteilung in den Strukturen des Zellkerns, *Chromosoma,* 1, 562, 1940.
11. **Alverdes, F.,** Die Kerne in den Speicheldrusen der *Chironomus*-Larve, *Arch. Exp. Zellforsch.,* 9, 168, 1912.
12. **Tanzer, E.,** Die Zellkerne einiger Dipterenlarven und ihre Entwicklung, *Z. Zool.,* 119, 114, 1922.
13. **Metz, C. W.,** Structure of salivary gland chromosomes, *Cold Spring Harbor Symp. Quant. Biol.,* 9, 23, 1941.
14. **Alfert, M.,** Composition and structure of giant chromosome, *Int. Rev. Cytol.,* 3, 131, 1954.
15. **Beerman, W.,** *Riesenchromosomen. Protoplasmatologia, Handbuch der Protoplasmaforschung,* Band VI, D. Springer-Verlag, Vienna, 1962.
16. **Nagl, W.,** *Endopolyploidy and Polyteny in Differentiation and Evolution,* North-Holland, Amsterdam, 1978.
17. **Agard, D. A. and Sedat, J. W.,** Three-dimensional architecture of a polytene nucleus, *Nature (London),* 302, 676, 1983.
18. **Mathog, D., Hochstrasser, M., Gruenbaum, Y., Saumweber, H., and Sedat, J.,** Characteristic folding pattern of polytene chromosomes in *Drosophila* salivary gland nuclei, *Nature (London),* 308, 414, 1984.
19. **Hochstrasser, M. and Sedat, J. W.,** Three-dimensional organization of *Drosophila melanogaster* interphase nuclei. I. Tissue-specific aspects of polytene nuclear architecture, *J. Cell Biol.,* 104, 1455, 1987.
20. **Hochstrasser, M. and Sedat, J. W.,** Three-dimensional organization of *Drosophila melanogaster* interphase nuclei. II. Chromosome spatial organization and gene regulation, *J. Cell Biol.,* 104, 1471, 1987.
21. **Beerman, W. and Pelling, C.,** ³H-Thymidin Markierung einzelner Chromatiden in Riesenchromosomen, *Chromosoma,* 16, 1, 1965.
22. **Kavenoff, R. and Zimm, B. H.,** Chromosome sized DNA molecules from *Drosophila, Chromosoma,* 41, 1, 1973.
23. **Rudkin, G. T., Corlette, S. L., and Schultz, J.,** The relations of the nucleic acid content in salivary gland chromosome bands, *Genetics,* 41, 657, 1956.

24. **Beerman, W.,** in *Results and Problems in Cell Differentiation,* Vol. 4, Beerman, W., Reinert, J., and Ursprung, H., Eds., Springer-Verlag, Berlin, 1972.

25. **Laird, C. D.,** Structural paradox of polytene chromosomes, *Cell,* 22, 869, 1980.

26. **DuPraw, E. J. and Rae, P. M. M.,** Polytene chromosome structure in relation to the "folded fibre" concept, *Nature (London),* 212, 598, 1966.

27. **Sorsa, V.,** Organization of replicative units in salivary gland chromosome bands, *Hereditas,* 78, 298, 1974.

28. **Laird, C. D., Wilkinson, L., Johnson, D., and Sandstrom, C.,** in *Chromosomes Today,* Vol. 7, Bennett, M., Bobrow, M., and Hewitt, G., Eds., George, Allen and Unwin, London, 1980, 74.

29. **Southern, E. M.,** Detection of specific sequences among DNA fragments separated by gel electrophoresis, *J. Mol. Biol.,* 98, 503, 1975.

30. **Spierer, A. and Spierer, P.,** Similar level of polyteny in bands and interbands of *Drosophila* giant chromosomes, *Nature (London),* 307, 176, 1984.

31. **Spierer, P., Spierer, A., Bender, W., and Hogness., D. S.,** Molecular mapping of genetic and chromomeric units in *Drosophila melanogaster, J. Mol. Biol.,* 168, 35, 1983.

32. **Lifschytz, E.,** Sequence replication and banding organization in the polytene chromosomes of *Drosophila melanogaster, J. Mol. Biol.,* 164, 17, 1983.

33. **Sass, H.,** Hierarchy of fibrillar organization levels in the polytene interphase chromosomes of *Chironomus, J. Cell Sci.,* 45, 269, 1980.

34. **Mortin, L. I. and Sedat, J. W.,** Structure of *Drosophila* polytene chromosomes. Evidence for a toroidal organization of the bands, *J. Cell Sci.,* 57, 73, 1982.

35. **Heitz, E.,** Uber α-und β-Heterochromatin sowie Konstanz und Bau der Chromomeren bei *Drosophila, Biol. Zb.,* 54, 588, 1933.

36. **Heitz, E.,** Die somatische Heteropyknose bei *Drosophila melanogaster* und ihre genetische Bedeutung, *Z. Zellforsch. Mikrosk. Anat.,* 20, 237, 1934.

37. **Hinton, T.,** A comparative study of certain heterochromatic regions in the mitotic and salivary gland chromosomes of *Drosophila melanogaster, Genetics,* 27, 119, 1942.

38. **Berendes, H. D. and Keyl, H. G.,** Distribution of DNA in heterochromatin and euchromatin of polytene nuclei *D. hydei, Genetics,* 57, 1, 1967.

39. **Gall, J. G., Cohen, E. H., and Polan, M. L.,** Repetitive DNA sequences in *Drosophila, Chromosoma,* 33, 319, 1971.

40. **Lakhotia, S. C.,** EM autoradiographic studies on polytene nuclei of *D. melanogaster.* III. Localization of nonreplicating chromatin in the chromocentre heterochromatin, *Chromosoma,* 46, 145, 1974.

41. **Lakhotia, S. C. and Jacob, J.,** EM autoradiographic studies on polytene nuclei of *D. melanogaster.* II. Organization and transcriptive activity of the chromocentre, *Exp. Cell Res.,* 86, 253, 1974.

42. **Hilliker, A. J., Appels, R., and Schalet, A.,** The genetic analysis of *D. melanogaster* heterochromatin, *Cell,* 21, 607, 1980.

43. **Rubin, G. M.,** Isolation of a telomeric DNA sequence from *Drosophila melanogaster, Cold Spring Harbor Symp. Quant. Biol.,* 62, 1041, 1978.

44. **Young, B. S., Pession, A., Traverse, K. L., French, C., and Pardue, M. L.,** Telomere regions in *Drosophila* share complex DNA sequences with pericentric heterochromatin, *Cell,* 34, 85, 1983.

45. **Renkawitz-Pohl, R. and Bialojan, S.,** A DNA sequence of *Drosophila melanogaster* with a differential telomeric distribution, *Chromosoma,* 89, 206, 1984.

46. **Henning, W. and Meer, B.,** Reduced polyteny of rRNA cistrons in giant chromosomes of *D. hydei, Nature New Biol.,* 23, 70, 1971.

47. **Spear, B. B. and Gall, J. G.,** Independent control of ribosomal gene replication in polytene chromosomes of Drosophila melanogaster, *Proc. Natl. Acad. Sci. U.S.A.,* 70, 1359, 1973.

48. **Brutlag, D.,** Molecular arrangement and evolution of heterochromatic DNA, *Annu. Rev. Genet.,* 14, 121, 1980.

49. **Spierer, P.,** A molecular approach to chromosome organization, *Dev. Genet.,* 4, 333, 1984.

50. **Breuer, M. E. and Pavan, C.,** Behavior of polytene chromosomes of *Rhynchosciara angelae* at different stages of larval development, *Chromosoma,* 7, 371, 1955.

51. **Crouse, H. V. and Keyl, H. G.,** Extra replication in the "DNA puffs" of *Sciara coprophila, Chromosoma,* 25, 357, 1968.

52. **Glover, D. M., Zaha, A., Stocker, A. J., Santelli, R. V., Pueyo, M. T., Toledo, S. M., and de Lara, F. J. S.,** Gene amplification in *Rhynchosciara* salivary gland chromosomes, *Proc. Natl. Acad. Sci. U.S.A.,* 79, 2947, 1982.

53. **Hartmann-Goldstein, I. J.,** On the relationship between heterochromatinization and variegation in *Drosophila* with special reference to temperature sensitive periods, *Genet. Res.,* 10, 143, 1967.

54. **Ananiev, E. V. and Gvozdev, V. A.,** Changed pattern of transcription and replication in polytene chromosomes of *Drosophila melanogaster* resulting from eu-heterochromatin rearrangement, *Chromosoma,* 45, 173, 1974.

55. **Hartmann-Goldstein, I. J. and Cowell, J.,** in *Current Chromosome Research,* Jones, K. and Brandhaus, P. E., Eds., Elsevier, Amsterdam, 1976, 43.

56. **Cowell, J. and Hartmann-Goldstein, I. J.,** Contrasting response of euchromatin and heterochromatin to translocation in polytene chromosomes of *Drosophila melanogaster, Chromosoma,* 79, 329, 1980.

57. **Spradling, A. C. and Mahowald, A. P.,** Amplification of genes for chorion proteins during oogenesis in *Drosophila melanogaster, Proc. Natl. Acad. Sci. U.S.A.,* 77, 1096, 1980.

58. **Spradling, A. C.,** The organization and amplification of two chromosomal domains containing *Drosophila* chorion genes, *Cell,* 27, 193, 1981.

59. **DeCicco, D. V. and Spradling, A. C.,** Localization of a cis-acting element responsible for the developmentally regulated amplification of *Drosophila* chorion genes, *Cell,* 38, 45, 1984.

60. **Korge, G.,** in *Results and Problems in Cell Differentiation,* Vol. 14, Henning, W., Ed., Springer Verlag, Berlin, 1987, 27.

61. **Spradling, A. C. and Rubin, G. M.,** Transposition of cloned P elements into *Drosophila* germ line chromosomes, *Science,* 218, 341, 1982.

62. **Muller, H. J. and Prokofyeva, A. A.,** The individual gene in relation to the chromomere and the chromosome, *Proc. Natl. Acad. Sci. U.S.A.,* 21, 16, 1935.

63. **Judd, B. H., Shen, M. W., and Kaufmann, T. C.,** The anatomy and function of a segment of the X chromosome of *Drosophila melanogaster, Genetics,* 71, 139, 1972.

64. **Lefevre, G., Jr.,** The relationship between genes and polytene chromosome bands, *Annu. Rev. Genet.,* 8, 51, 1974.

65. **Woodruff, R. C. and Ashburner, M.,** The genetics of a small autosomal region of *Drosophila melanogaster* containing the structural gene for alcohol dehydrogenase. II. Lethal mutations in the region, *Genetics,* 92, 133, 1979.

66. **Gausz, J., Benze, G., Gyurkovics, H., Ashburner, M., Ish-Horowitz, D., and Holden, J. J.,** Genetic characterization of the 87C region of the third chromosome of *Drosophila melanogaster, Genetics,* 93, 917, 1979.

67. **Gausz, J., Gyurkovics, H., Benze, G., Awad, A. A. M., Holden, J. J., and Ish-Horowitz, D.,** Genetic characterization of the region between 86F1,2 and 87B15 on chromosome 3 of *Drosophila melanogaster, Genetics,* 98, 775, 1981.

68. **Lifschytz, E. and Falk, R.,** Fine structure analysis of a chromosome segment in *Drosophila melanogaster.* Analysis of X-ray induced lethals, *Mutat. Res.,* 6, 235, 1968.

69. **Lifschytz, E. and Falk, R.,** Fine structure analysis of a chromosome segment in *Drosophila melanogaster.* Analysis of ethyl methane sulfonate-induced lethals, *Mutat. Res.,* 8, 147, 1969.

70. **Hochman, B.,** Analysis of chromosome 4 in *Drosophila melanogaster.* II. Ethyl methane sulfonate induced lethals, *Genetics,* 67, 235, 1971.

71. **Hochmann, B.,** Analysis of a whole chromosome in *Drosophila, Cold Spring Harbor Symp. Quant. Biol.,* 38, 581, 1973.

72. **Hilliker, A. J., Clark, S. H., Chovnick, A., and Gelbart, W. M.,** Cytogenetic analysis of the chromosomal region immediately adjacent to the rosy locus in *Drosophila melanogaster, Genetics,* 95, 95, 1980.

73. **Young, M. W. and Judd, M. H.,** Nonessential sequences, genes, and the polytene chromosomes bands of *Drosophila melanogaster, Genetics,* 88, 723, 1978.

74. **Wright, T. R. F., Beerman, W., Marsh, J. L., Bishop, C. P., Steward, R., Black, B. C., Tomsett, A. D., and Wright, E. Y.,** The genetics of dopa decarboxylase in *Drosophila melanogaster.* IV. The genetics and cytology of the 37B10-37D1 region, *Chromosoma,* 83, 45, 1981.

75. **Zhimulev, I. F., Pokholkova, G. V., Bgatov, A. V., Semeshin, V. F., and Belyaeva, E. S.,** Fine cytogenetical analysis of the band 10A1-2 and the adjoining regions in the *Drosophila melanogaster* chromosome. II. General analysis, *Chromosoma,* 82, 25, 1981.

76. **Hall, L. M., Mason, P. J., and Spierer, P.,** Transcripts, genes and bands in 315,000 base-pairs of *Drosophila* DNA, *J. Mol. Biol.,* 169, 83, 1983.

77. **Procunier, J. D. and Tartof, K. D.,** Genetic analysis of the 5S genes in *Drosophila melanogaster, Genetics,* 81, 515, 1975.

78. **Procunier, J. D. and Dunn, R. J.,** Genetic and molecular organization of the 5S locus and mutants in *Drosophila melanogaster, Cell,* 15, 1087, 1978.

79. **Wimber, D. E. and Steffensen, D. M.,** Localization of 5S RNA genes on *Drosophila* chromosomes by RNA-DNA hybridization, *Science,* 70, 639, 1970.

80. **Pardue, M. L., Kedes, L. H., Weinberg, E. S., and Birnstiel, M. L.,** Localization of sequences coding for histone messenger RNA in the chromosomes of *Drosophila melanogaster, Chromosoma,* 63, 135, 1977.

81. **Lifton, R. P., Goldberg, M. L., Karp, R. W., and Hogness, D. S.,** The organization of the histone genes in *Drosophila melanogaster:* functional and evolutionary implications, *Cold Spring Harbor Symp. Quant. Biol.,* 42, 1047, 1978.

82. **Chernyshev, A. I., Barshkirov, V. N., Leibovitch, B. A., and Khesin, R. B.,** Increase in the number of histone genes in the case of their deficiency in *Drosophila melanogaster, Mol. Gen. Genet.,* 178, 663, 1980.

83. **Chen, C. N., Malone, T., Beckendorf, S. K., and Davis, R. L.,** At least two genes reside within a large intron of the dunce gene of *Drosophila, Nature (London),* 329, 721, 1987.

84. **Henikoff, S., Keene, M. A., Fechtel, K., and Fristrom, J. W.,** Gene within a gene: nested *Drosophila* genes encode unrelated proteins on opposite DNA strands, *Cell,* 44, 32, 1986.

85. **Poulson, D. F. and Metz, C. W.,** Studies on the structure of nucleolus forming regions and related structures in the giant salivary gland chromosomes of Diptera, *J. Morphol.,* 63, 363, 1938.

86. **Pelling, C.,** Chromosomal synthesis of ribonucleic acid as shown by the incorporation of uridine labelled with tritium, *Nature (London),* 184, 655, 1959.

87. **Sirlin, J. L.,** Cell sites of RNA and protein synthesis in the salivary gland of *Smittia* (Chironomidae), *Exp. Cell Res.,* 19, 177, 1960.

88. **Plagens, U., Greenleaf, A. L., and Bautz, E. K. F.,** Distribution of RNA polymerase on *Drosophila* polytene chromosomes as studied by indirect immunofluorescence, *Chromosoma,* 59, 1517, 1976.

89. **Jamrich, M., Greenleaf, A. L., and Bautz, E. K. F.,** Localization of RNA polymerase in polytene chromosomes of *Drosophila, Proc. Natl. Acad. Sci. U.S.A.,* 74, 2079, 1977.

90. **Jamrich, M., Greenleaf, A. L., and Bautz, E. K. F.,** Functional organization of polytene chromosomes, *Cold Spring Harbor Symp. Quant. Biol.,* 42, 389, 1977.

91. **Jamrich, M., Haars, R., Wulf, E., and Bautz, F. A.,** Correlation of RNA polymerase B and transcriptional activity in the chromosomes of *Drosophila melanogaster, Chromosoma,* 64, 319, 1977.

92. **Greenleaf, A. L., Plagens, U., Jamrich, M., and Bautz, E. K. F.,** RNA polymerase B (or II) in heat induced puffs of *Drosophila* polytene chromosomes, *Chromosoma,* 65, 127, 1978.

93. **Pelling, C.,** Ribonukleinsauresynthese der Riesenchromosomen. Autoradiographische Unterosuchungen an *Chironomus tentans, Chromosoma,* 15, 71, 1964.

94. **Edstrom, J. and Daneholt, B.,** Sedimentation properties of the newly synthesized RNA from isolated nuclear components of *Chironomus tentans* salivary gland cells, *Exp. Cell Res.,* 57, 205, 1967.

95. **Crowley, T. E., Bond, M. W., and Meyerowitz, E. M.,** The structural genes for three *Drosophila* glue proteins reside at a single polytene chromosome puff locus, *Mol. Cell Biol.,* 3, 623, 1983.

96. **Semechin, V. F., Baricheva, E. M., Belyaeva, E. S., and Zhimulev, I. F.,** Electron microscopical analysis of *Drosophila* polytene chromosomes. II. Development of complex puffs, *Chromosoma,* 91, 210, 1985.

97. **Semechin, V. F., Baricheva, E. M., Belyaeva, E. S., and Zhimulev, I. F.,** Electron microscopical analysis of *Drosophila* polytene chromosomes. III. Mapping of puffs developing from one band, *Chromosoma,* 91, 234, 1985.

98. **Beerman, W.,** Chromosomenkonstanz und spezifische Modifikationen der Chromosomenstruktur in der Entwicklung von *Chironomus tentans, Chromosoma,* 5, 139, 1952.

99. **Beerman, W.,** Ein Balbiani-Ring als Locus einer Speicheldrusemutation, *Chromosoma,* 12, 1, 1961.

100. **Mechelke, F.,** Reversible Strukturmodifikationen der Speicheldrusenchromosomen von *Acricotopus lucidus, Chromosoma,* 5, 511, 1953.

101. **Mechelke, F.,** The timetable of physiological activity of several loci in the salivary gland chromosomes of *Acricotopus lucidus, Proceedings of the 10th International Congress on Genetics,* Vol. II., University of Toronto Press, Toronto, 1958.

102. **Mechelke, F.,** Spezielle Funktionszustande des genetischen Materials, *Wissenschaftliche Konferenz der Gesselschaft der Deutschen Naturforscher und Artzte,* Springer-Verlag, Berlin, Gottingen-Heidelberg, 1963, 15.

103. **Becker, H. J.,** Die Puffs der Speicheldrusenchromosomen von *Drosophila melanogaster.* 1. Mitteilung, Beobachtungen zum Verhalten des Puffmusters in Normalstamm und bei zwei Mutanten, Giant und Lethal-Giant-Larvae, *Chromosoma,* 10, 654, 1959.

104. **Becker, H. J.,** Die Puffs der Speicheldrusenchromosomen von *Drosophila melanogaster.* II. Die Auslosung der Puffbildung, ihre Spezifitat und ihre Beziehung zur Funktion der Ringdruse, *Chromosoma,* 13, 341, 1962.

105. **Berendes, H. D.,** Salivary gland functions and chromosomal puffing patterns in *D. hydei, Chromosoma,* 17, 35, 1965.

106. **Berendes, H. D.,** The induction of changes in chromosomal activity in different polytene types of cell in *D. hydei, Dev. Biol.,* 11, 371, 1965.

107. **Berendes, H. D.,** The hormone ecdysone as effector of specific changes in the pattern of gene activities of *Drosophila hydei, Chromosoma,* 29, 118, 1967.

108. **Clever, U. and Karlson, P.,** Induktion von Puff-Veranderungen in den Speicheldrusenchromosomen von *C. tentants* durch Ecdysone, *Exp. Cell Res.,* 20, 623, 1960.

109. **Panitz, R.,** Innersekretorische Wirkung auf Strukturmodifikationen der Speicheldrusenchromosomen von *Acricotopus lucidus, Naturwissenschaften,* 47, 383, 1960.

110. **Panitz, R.,** Hormonkontrollierte Genaktivitaten in den Riesenchromosomen von *Acricotopus lucidus, Biol. Zb.,* 83, 187, 1964.

111. **Clever, U.,** Genaktivitaten in den Riesenchromosomen von *Chironomus tentans* und ihre Beziehung zur Entwicklung. I. Genaktivierungen durch Ecdysone, *Chromosoma,* 12, 607, 1961.

112. **Ashburner, M.,** Autosomal puffing patterns in a laboratory stock of *Drosophila melanogaster, Chromosoma,* 21, 398, 1967.

113. **Ashburner, M.,** Ecdysone induction of puffing in polytene chromosomes of *D. melanogaster.* Effects of inhibitors of RNA synthesis, *Exp. Cell. Res.,* 71, 433, 1972.

114. **Baudisch, W. and Panitz, R.,** Kontrolle eines biochemischen Merkmals in den Speicheldrusen von *Acricotopus lucidus* durch einen Balbiani ring, *Exp. Cell. Res.,* 49, 470, 1968.

115. **Grossbach, U.,** Chromosomen-Aktivitat und biochemische Zelldifferenzierung in den Speicheldrusen von *Camptechironomus, Chromosoma,* 28, 136, 1969.

116. **Grossbach, U.,** Chromosome puffs and gene expression in polytene cells, *Cold Spring Harbor Symp. Quant. Biol.,* 38, 619, 1973.

117. **Korge, G.,** Chromosome puff activity and protein synthesis in larval salivary glands of *D. melanogaster, Proc. Natl. Acad. Sci. U.S.A.,* 72, 4550, 1975.

118. **Korge, G.,** Direct correlation between a chromosome puff and the synthesis of a larval salivary protein in *Drosophila melanogaster, Chromosoma,* 62, 155, 1977.

119. **Akam, M. E., Roberts, D. B., Richards, G. P., and Ashburner, M.,** *Drosophila:* the genetics of two major larval proteins, *Cell,* 13, 215, 1978.

120. **Ritossa, F.,** A new puffing pattern induced by temperature shock and DNP in *Drosophila, Experientia,* 18, 571, 1962.

121. **Berendes, H. D.,** Factors involved in the expression of gene activity in polytene chromosomes, *Chromosoma,* 24, 418, 1968.

122. **Tissieres, A., Mitchell, H. K., and Tracy, U. M.,** Protein synthesis in salivary glands of *Drosophila melanogaster.* Relation to chromosome puffs, *J. Mol. Biol.,* 84, 389, 1974.

123. **Ashburner, M. and Bonner, J. J.,** The induction of gene activity in *Drosophila* by heat shock, *Cell,* 17, 241, 1979.

124. **Lis, J. T., Simon, J. A., and Sutton, C. A.,** New heat shock puffs and beta-galactosidase activity resulting from transformation of *Drosophila* with an hsp70-lacZ hybrid gene, *Cell,* 35, 403, 1983.

125. **Simon, J. A., Sutton, C. A., Lobell, R. B., Glaser, R. L., and Lis, J. T.,** Determinants of heat shock-induced chromosome puffing, *Cell,* 40, 805, 1985.

126. **Bonner, J. J., Parks, C., Parks-Thornburg, J., Martin, M. A., and Pelham, H. R. B.,** The use of promoter fusion in *Drosophila* genetics: isolation of mutants affecting the heat shock response, *Cell,* 37, 979, 1984.

127. **Cohen, R. S. and Meselson, M.,** Inducible transcription and puffing in *Drosophila melanogaster* transformed with *hsp*70-phage λ hybrid heat shock genes, *Proc. Natl. Acad. Sci. U.S.A.,* 81, 5509, 1984.

128. **Dudler, R. and Travers, A. A.,** Upstream elements necessary for optimal function of the hsp 70 promoter in transformed flies, *Cell,* 38, 391, 1984.

129. **Zhimulev, I. F. and Belyaeva, F. A.,** [3]H uridine labeling patterns in the *Drosophila melanogaster* salivary gland chromosomes X, 2R, and 3L, *Chromosoma,* 49, 219, 1975.

130. **Semeshin, V. F., Zhimulev, I. F., and Belyaeva, E. S.,** Electron microscope autoradiographic study on transcriptional activity of *Drosophila melanogaster* polytene chromosomes, *Chromosoma,* 73, 163, 1979.

131. **Kramer, A., Haars, R., Kabish, R., Will, H., Bautz, F. A., and Bautz, E. K. F.,** Monoclonal antibody directed against RNA polymerase II of *Drosophila melanogaster, Mol. Gen. Genet.,* 180, 193, 1980.

132. **Sass, H. and Bautz, E. K. F.,** Immunoelectron microscopic localization of RNA polymerase B on isolated polytene chromosomes of *Chironomus tentans, Chromosoma,* 85, 633, 1982.

133. **Sass, H.,** RNA polymerase B in polytene chromosomes: immunofluorescent and autoradiographic analysis during stimulated and repressed RNA synthesis, *Cell,* 28, 269, 1982.

134. **Sass, H. and Bautz, E. K. F.,** Interbands of polytene chromosomes: binding sites and start points for RNA polymerase B(II), *Chromosoma,* 86, 77, 1982.

135. **Skaer, R. J.,** Interband transcription in *Drosophila, J. Cell Sci.,* 26, 251, 1977.

136. **Saumweber, H., Symmons, P., Kabisch, R., Will, H., and Bonhoefer, F.,** Monoclonal antibodies against chromosomal proteins of *Drosophila melanogaster, Chromosoma,* 80, 253, 1980.

137. **Christensen, M. E., LeStourgeon, M., Jamrich, M., Howard, G. C., Seruinan, L. A., Silver, L. M., and Elgin, S. C.,** Distribution studies on polytene chromosomes using antibodies directed aginst hnRNP, *J. Cell Biol.,* 90, 18, 1981.

138. **Richards, G., Caasab, A., Bourouis, M., Jarry, B., and Dissous, C.,** The normal developmental regulation of a cloned *Sgs-3* glue gene chromosomally integrated in *Drosophila melanogaster* by P element transformation, *EMBO J.,* 2, 2137, 1983.

139. **Meyerowitz, E. M., Crosby, M. A., Garfinkel, M. D., Martin, C. H., Mathers, P. H., and Raghavan, V.,** The 68C gene puff of *Drosophila, Cold Spring Harbor Symp. Quant. Biol.,* 50, 347, 1985.

140. **Crosby, M. A. and Meyerowitz, E. M.,** *Drosophila* glue gene *Sgs-3*: sequences required for puffing and transcriptional regulation, *Dev. Biol.,* 118, 593, 1986.
141. **Beerman, W.,** Control of differentiation at the chromosomal level, *J. Exp. Zool.,* 157, 49, 1964.
142. **Beerman, W.,** Effect of α-amanitin on puffing and intranuclear RNA synthesis in *Chironomus* salivary glands, *Chromosoma,* 34, 152, 1971.
143. **Hansson, L., Lineruth, K., and Lambertsson, A.,** Effect of the *l(1)su(f)*[ts67g] mutation of *Drosophila melanogaster* on the glue protein synthesis, *Wilheim Roux Arch. Dev. Biol.,* 192, 308, 1982.
144. **Hansson, L. and Lambertsson, A.,** The role of su(f) gene function and ecdysterone in transcription of glue polypeptide mRNAs in *Drosophila melanogaster, Mol. Gen. Genet.,* 192, 395, 1983.
145. **Crowley, T. E., Mathers, P. H., and Meyerowitz, E. M.,** A trans-acting regulatory product necessary for expression of the *Drosophila melanogaster* 68C glue gene cluster, *Cell,* 39, 149, 1984.
146. **Olins, A. L. and Olins, D. E.,** Spheroid chromatin units (ν bodies), *Science,* 183, 330, 1974.
147. **Swift, H.,** The histones of polytene chromosomes, in *The Nucleohistones,* Tso, P. and Bonner, J., Eds., Holden Day, San Francisco, 1964.
148. **Desai, L. S., Pothier, L., Foley, G. E., and Adams, R. A.,** Immunofluorescent labeling of chromosomes with antisera to histones and histone fraction, *Exp. Cell Res.,* 70, 468, 1972.
149. **Alfageme, C. R., Rudkin, G. T., and Cohen, L. H.,** Locations of chromosomal proteins in polytene chromosomes, *Proc. Natl. Acad. Sci. U.S.A.,* 73, 2038, 1976.
150. **Silver, L. M. and Elgin, S. C. R.,** Distribution patterns of three subfractions of *Drosophila* nonhistone chromosomal proteins: possible correlation with gene activity, *Cell,* 11, 971, 1977.
151. **Kurth, P. D., Moudrianakis, E. N., and Bustin, M.,** Histone localization in polytene chromosomes by immunofluorescence, *J. Cell Biol.,* 78, 910, 1978.
152. **Silver, L. M. and Elgin, S. C. R.,** A method for determination of the *in situ* distribution of chromosomal proteins, *Proc. Natl. Acad. Sci. U.S.A.,* 73, 423, 1976.
153. **Mayfield, J. E., Serunian, L. A., Silver, L. M., and Elgin, S. C. R.,** A protein released by DNase I digestion of *Drosophila* nuclei is preferentially associated with puffs, *Cell,* 14, 539, 1978.
154. **Steiner, E. K., Eissenberg, J. C., and Elgin, S. C. R.,** A cytological approach to the ordering of events in gene activation using the Sgs-4 locus of *Drosophila melanogaster, J. Cell Biol.,* 99, 233, 1984.
155. **Silver, L. M. and Elgin, S. C. R.,** Production and characterization of antisera against three individual NH proteins: a case of a generally distributed NH protein, *Chromosoma,* 68, 101, 1978.
156. **Howard, G. C., Abmayr, S. M., Shinefeld, L. A., Sato, V. L., and Elgin, S. C. R.,** Monoclonal antibodies against a specific nonhistone chromosomal protein associated with active genes, *J. Cell Biol.,* 88, 219, 1981.
157. **Kabisch, R. and Bautz, E. K. F.,** Differential distribution of RNA polymerase B and nonhistone proteins in polytene chromosomes of *Drosophila melanogaster,* EMBO J., 2, 395, 1983.
158. **Dangli, A., Grond, C., Kloetzel, P., and Bautz, E. K. F.,** Heat-shock puff 93D from *Drosophila melanogaster:* accumulation of a RNP-specific antigen associated with giant particles of possible storage function, *EMBO J.,* 2, 1747, 1983.
159. **Risau, W., Symmons, P., Saumweber, H., and Frash, M.,** Nonpackaging and packaging protein of hnRNA in *Drosophila melanogaster, Cell,* 33, 529, 1983.
160. **Kurth, P. D. and Bustin, M.,** Localization of chromosomal protein HMG-1 in polytene chromosomes of *Chironomus thummi, J. Cell. Biol.,* 89, 70, 1981.
161. **Will, H. and Bautz, E. K. F.,** Immunological identification of a chromocenter associated protein in polytene chromosomes of *Drosophila, Exp. Cell Res.,* 125, 401, 1980.
162. **James, T. C. and Elgin, S. C.,** Identification of a nonhistone chromosomal protein associated with heterochromatin in *Drosophila melanogaster* and its gene, *Mol. Cell. Biol.,* 6, 3862, 1986.
163. **Wang, A. H. J., Quigley, G. J., Kolpak, F. J., van der Marel, G., van Boom, J. H., and Rich, A.,** Left-handed double helical DNA: variation in the backbone conformation, *Science,* 211, 171, 1981.
164. **Drew, H. R. and Dickerson, R. E.,** Conformation and dynamics in a Z-DNA tetramer, *J. Mol. Biol.,* 152, 723, 1981.
165. **Nordheim, A., Pardue, M. L., Lafer, E. M., Moller, A., Stollar, B. D., and Rich, A.,** Antibodies to left-handed Z-DNA bind to interband regions of *Drosophila* polytene chromosomes, *Nature (London),* 294, 417, 1981.
166. **Lemeunier, F., Derbin, C., Malfoy, B., Leng, M., and Taillandier, E.,** Identification of left-handed Z-DNA by indirect immunofluorescence in polytene chromosomes of *Chironomus thummi thummi, Exp. Cell Res.,* 141, 508, 1982.
167. **Arndt-Jovin, D. J., Robert-Nicoud, M., Zarling, D. A., Greider, C., Weimer, E., and Jovin, T. M.,** Left-handed Z-DNA in bands of acid-fixed polytene chromosomes, *Proc. Natl. Acad. Sci. U.S.A.,* 80, 4344, 1983.
168. **Arndt-Jovin, D. J., Robert-Nicoud, M., Baurschmidt, P., and Jovin, T. M.,** Immunofluorescence localization of Z-DNA in chromosomes: quantitation by scanning microphotometry and computer-assisted image analysis, *J. Cell. Biol.,* 101, 1422, 1985.

169. **Hill, R. J. and Stollar, B. D.,** Dependence of Z-DNA antibody binding to polytene chromosomes on acid fixation and DNA torsional strain, *Nature (London),* 305, 338, 1983.
170. **Lancillotti, F., Lopez, M. C., Alonso, C., and Stollar, B. D.,** Locations of Z-DNA in polytene chromosomes, *J. Cell. Biol.,* 100, 1759, 1985.
171. **Lancillotti, F., Lopez, M. C., Arias, P., and Alonso, C.,** Z-DNA in transcriptionally active chromosomes, *Proc. Natl. Acad. Sci. U.S.A.,* 84, 1560, 1987.

Chapter 5

SACCHAROMYCES CEREVISIAE: STRUCTURE AND BEHAVIOR OF NATURAL AND ARTIFICIAL CHROMOSOMES

Virginia A. Zakian

TABLE OF CONTENTS

I. INTRODUCTION

The single-celled eukaryote *Saccharomyces cerevisiae* (baker's yeast) can be readily manipulated by some of the most sophisticated genetic and molecular biological techniques. Therefore, yeast has become a favorite model system for analysis of a wide variety of eukaryotic-specific processes. Yeast also provides unique benefits for the analysis of chromosome structure and function. Like other eukaryotes, yeast contains multiple chromosomes consisting of linear DNA molecules. However, the sixteen yeast chromosomes are relatively small, ranging in size from 260 kb (chromosome I) to 3100 kb (chromosome XII)[1,164] with a mean size of about 900 kb. In contrast, human chromosomes are from ten (chromosome 21) to 100 (chromosome 1) times the size of the largest yeast chromosome.[2] Although their small size precludes conventional cytological studies, intact yeast chromosomal DNA molecules can be readily separated and manipulated by pulsed field gel electrophoresis[3,4,5] (Figure 1).

A major advantage of yeast for studies on chromosome behavior is that, unlike many other eukaryotes, it has a high tolerance for aneuploidy. Its normal life cycle includes haploid (n) and diploid (2n) phases. Moreover, both disomic (n + 1) and monosomic (2n − 1) strains can be constructed and stably propagated.[6,7] These properties make it relatively easy to monitor the rate at which individual chromosomes are lost. Studies on different chromosomes indicate that yeast chromosomes are replicated and segregated with high fidelity: mitotic loss rates are typically 10^{-5} to 10^{-6} per division.[8-13] In meiosis, the rate of chromosome loss is about 2×10^{-4} per viable spore.[14]

Probably the most important property of yeast for analysis of chromosome structure is that it is the only system in which there are functional assays for all *cis*-acting elements known to be required for replication and segregation of eukaryotic chromosomes. Three such structures have been identified: (1) replication origins distributed throughout the length of chromosomal DNA molecules that are necessary for initiation of DNA synthesis, (2) centromeres, the sites of attachment for spindle microtubules which serve to promote accurate segregation of replicated chromosomes, and (3) telomeres, the physical ends of linear chromosomes, which are necessary for complete duplication and stability of DNA termini. These same three structures are probably required for the stable maintenance of all eukaryotic chromosomes from yeast to humans. Functional assays have made it possible to construct artificial yeast chromosomes that can be maintained in mitosis and meiosis and which can be manipulated easily both *in vitro* and *in vivo*.

In this article, data will be reviewed that describe the structure and behavior of yeast replication origins, centromeres, and telomeres. The chapter will end with a description of the properties of artificial yeast chromosomes and a discussion of factors that influence their stability.

II. REPLICATION ORIGINS

A. BACKGROUND

As is true for most eukaryotes, chromosomal DNA replication in yeast occurs during a discrete portion of the cell cycle (S phase), and, based on electron microscopic and density transfer studies, is limited to a single round of semi-conservative replication per S phase.[15-18] Moreover, even the smallest yeast chromosome is replicated via the activation of multiple initiation sites.[19] In contrast, the *Escherichia coli* chromosome, which is more than ten times larger than yeast chromosome I, is replicated from a single unique origin that can be activated more than once per cell cycle (reviewed in Reference 20). Electron microscopic analysis of small yeast chromosomal DNA molecules indicates that these multiple replication origins are spaced, on the average, 36 kb apart.[19] Density transfer experiments

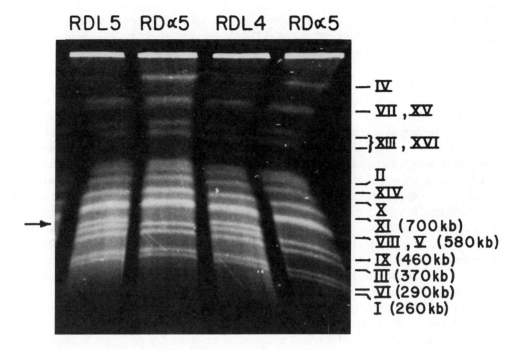

FIGURE 1.   Yeast DNAs were prepared by modifications[1] of the gel insert method of Schwartz and Cantor.[3] These DNAs were subjected to orthogonal field gel electrophoresis (OFAGE[4]) in a 1.5% agarose gel for 21 hours at 14° using a switching interval of 80 s. After electrophoresis, the gel was stained with ethidium bromide. Assignment of ethidium bromide staining bands to individual chromosomes and sizes of the smallest chromosomes are based on data presented in Carle and Olson.[1] Under OFAGE conditions, chromosome XII, the largest yeast chromosome, often does not enter the gel.[1] Although chromosome XII is underrepresented in this gel relative to other chromosomes, hybridization analysis indicated that the molecules of chromosome XII that entered the gel comigrated with chromosome IV (data not shown). The strains used in this gel are RDα5 (lanes 2 and 4; *MATα ade2*-101 *lys2*-901 *ura-52Δtrp1*-901; obtained from P. Heiter) and RDL5 (lane 1) and RDL4 (lane 3). RDL4 and RDL5 were derived by transformation of RDα5 with a 9 kb artificial chromosome carrying the centromere region of chromosome IV.[158] In both strains, recombination between the 9 kb artificial chromosome and authentic chromosome IV resulted in the splitting of chromosome IV into two telocentric derivatives: IVLt, which carries the left arm of intact chromosome IV, migrates with an apparent molecular weight of 520 kb (indicated by arrow) and IVRt, which carries the right arm of intact chromosome IV, comigrates with chromosomes VII and XV.[158]

carried out with cells undergoing a synchronous S phase demonstrate that specific regions on chromosomes are replicated during discrete subintervals of S phase.[18]

B. ISOLATION AND CHARACTERIZATION OF AUTONOMOUSLY REPLICATING SEQUENCES (*ARSs*) OF *SACCHAROMYCES CEREVISIAE*

Molecular evidence for specificity of chromosomal origins comes from transformation studies. Yeast can be transformed with recombinant DNA plasmids containing a selectable gene.[21] Two different modes of transformation have been described.[22] Circular plasmids containing most fragments of yeast chromosomal DNA transform at low frequency (1 to 10 transformations per microgram plasmid DNA) via homologous integration of the plasmid into the chromosome. However, a subset of DNA fragments causes high frequency transformation of plasmids (10^3 to 10^5 transformants per microgram.)[22,23] In these cases, the plasmid is maintained in multiple extrachromosomal copies in the transformed cells. In the absence of selection, these extrachromosomal plasmids are unstable as a result of their inefficient segregation into the nucleus of the daughter cell.[24]

DNA fragments that support high frequency transformation and extrachromosomal replication of plasmids are called *ARS*s for autonomously replicating sequences.[25] The behavior of plasmids containing yeast *ARS*s is very similar to that of plasmids containing known origins of replication from bacterial and viral DNAs. By analogy, it was hypothesized that in their normal chromosomal location *ARS*s are origins of DNA replication.[22,23] A large body of evidence consistent with this hypothesis has been collected. For example, the number[26,27] and spacing[28] of *ARS*s on yeast chromosomes are roughly similar to the number and spacing of replication origins as estimated from electron microscopic studies on small yeast chromosomes.[19] Moreover, electron microscopic analysis indicates that replication of the tandemly repeated ribosomal RNA genes often begins in the nontranscribed spacer,[29] a region that also acts as a weak *ARS*.[30,31] Data from *in vitro* replication studies have also been used to support the hypothesis that *ARS*s act as replication origins. *ARS*-specific initiation of DNA replication on plasmid DNA *in vitro* has been reported by a number of groups.[32-35] However, these systems have been criticized for their low efficiency and poor reproducibility. In one case, the apparent *ARS*-specificity of initiation of DNA replication has been shown to be due to prepriming of plasmid templates in *E. coli*.[36] Finally, the replication behavior of plasmids containing *ARS1*, an *ARS* found adjacent to *TRP1* on chromosome IV, is similar to that of chromosomal DNA, a fact consistent with the possibility that *ARS1* is a chromosomal origin of DNA replication. These similarities include S-phase limited replication and dependence on the same gene products required for chromosomal replication.[18,37] Moreover, replication of an *ARS1* plasmid occurs only once in each cell cycle and early in the S phase, as does replication of the chromosomal copy of *ARS1*.[18]

Recent data have provided the first unequivocal evidence that *ARS*s act as origins of DNA replication *in vivo*, at least for plasmid DNA.[38,39] The model system used was 2 μm DNA, an endogenous multiple copy plasmid whose replication is normally controlled in a manner similar to that of chromosomal DNA.[17,40] This 6.3 kb plasmid contains a single *ARS* that has been delimited to a region of about 75 bp.[41,42] The *in vivo* origin of DNA replication for 2 μm DNA was mapped to a position indistinguishable from that of its *ARS* by two completely different two-dimensional gel electrophoretic methods.[38,39] These data confirm earlier electron microscopic studies that suggested that the majority of 2 μm DNA molecules being replication near the *ARS*.[43] One of these groups also demonstrated that a recombinant DNA plasmid containing *ARS1* also begins replication near (or at) the *ARS*.[38] It should now be possible to use these origin-mapping techniques to determine if *ARS*s serve as initiation sites in their normal chromosomal context.

C. SEQUENCE REQUIREMENTS FOR *ARS* FUNCTION

For a variety of technical and intellectual reasons, it has been difficult to determine which DNA sequences are necessary for *ARS* function. First, there is no single criterion for *ARS* activity. High frequency transformation is a convenient, but clearly inadequate, assay for *ARS* function. Rates of loss (stability) of acentric plasmids are now known to reflect primarily defects in segregation,[24] not replication.[44] Currently, the most useful and accurate method to assay *ARS* function is to insert a putative *ARS* into a plasmid containing both a centromere and a marker that permits visual determination of rates of plasmid loss[45,46] (see Centromere section for more details). By using these plasmid systems, it has been shown that different *ARS*s can produce rates of plasmid loss that differ by more than 50-fold.[45] It is also clear that plasmid sequences adjacent to the *ARS* can influence *ARS* activity.[47,48] Thus, the *ARS* phenotype is not an all or none phenomenon.

Despite these problems, a number of generalizations can be made about the sequence and structure requirements for *ARS*s. Sequence analysis reveals that all *ARS*s carry one or more copies (or near copies) of an eleven bp sequence called the core consensus (also called domain A) that can be represented as $^{A}_{T}TTTAT^{A}_{G}TTT^{A}_{T}$.[49-51] The essentiality of the core consensus is demonstrated by its ubiquity in Ars$^+$ DNAs and by the fact that single base pair changes in the consensus can abolish *ARS* function.[52] However, a single copy of the core consensus is not sufficient for full *ARS* activity. For example, although a 19 bp region from *ARS1* that contains the 11 bp core consensus can support the autonomous replication of a centromere plasmid, cells carrying this plasmid grow extremely slowly.[53] Moreover, there are regions of yeast DNA that contain exact matches to the consensus but which nonetheless do not display *ARS* activity.[54,55]

Sequencing studies reveal no sequences other than the core consensus that are common to all *ARS*s. Indeed, by the criterion of Southern hybridization, most yeast *ARS*s are single copy DNAs[25] (see discussion of X and Y' for examples of repetitive sequences that act as *ARS*s). However, it is clear that DNA in addition to the core consensus is necessary for full *ARS* function although the amount and sequence of this additional DNA varies for different *ARS*s. For example, for the *ARS* adjacent to the *HO* gene, as little as 46 bps are necessary for full *ARS* function,[52] whereas 300 to 400 bps are necessary for full activity of *ARS1*.[53,56] In the case of the *HO ARS*, as well as many other *ARS*s, essentially no sequences 5' to the T-rich strand of the core consensus are necessary.[47,52,55] However, for *ARS1*, deletions or insertions into the 5' flanking region can adversely affect plasmid stability.[46,53,57] All *ARS*s require at least some sequences flanking the 3' side of the T-rich strand of the core consensus for full activity. In some, but clearly not all *ARS*s, the 3' flanking region contains one or more copies (or near copies) of a second 11 bp consensus sequence that can be represented as $CTTTTAGC_{TTT}^{AAA}$.[55] Not only do some *ARS*s lack this 3' consensus element, but its removal from one *ARS* did not affect its activity.[58] For *ARS1* the 3' flanking region has been shown to contain bent DNA[59] and to be devoid of nucleosomes,[60] characteristics that are also features of, for example, the SV40 origin of DNA replication.[61,62]

Recently, a synthetic *ARS* was constructed that consists solely of multiple copies of the core consensus sequence with some copies of the consensus inverted relative to other copies.[58] This synthetic *ARS* works as efficiently as a strong *ARS* from chromosome III. This result raises the possibility that even for authentic *ARS*s, the requirement for sequences other than the core consensus might be fulfilled by additional copies or near copies of this element.

The size and sequence requirements for *ARS* function probably vary for different *ARS*s. Assuming that chromosomal origins of DNA replication act as *ARS*s when carried on a plasmid, these differences may simply indicate that plasmid systems are inadequate to assay the behavior of sequences that normally function in a chromosomal environment. Alternatively, these differences may reflect *in vivo* differences among different *ARS*s. *ARS*s could differ in their time or frequency of activation. For example, different classes of *ARS*s might be used in early versus late S phase or in mitotic versus premeiotic S phase. Some sequences defined as *ARS*s by the transformation assay might never function as origins of replication when situated within chromosomal DNA. Finally, all *ARS*s on a given chromosome or in a common subchromosomal environment might share common structural features. For example, telomere proximal regions often carry middle repetitive sequences called X and Y', both of which contain *ARS*s[63,64] (see Telomere section for more details). Density transfer experiments done with synchronized cells indicate that many Y', and at least one X, replicate late in S phase,[165] a fact consistent with the possibility that telomere-associated *ARS*s are a specialized set of replication origins.

D. PROTEINS THAT INTERACT WITH *ARS*s

If *ARS*s are chromosomal origins of DNA replication, analogy with bacterial and viral systems predicts that there are also likely to be *ARS* binding proteins whose presence is required for initiation of DNA synthesis. If there are multiple classes of *ARS*s (as discussed above), there might also be multiple initiation proteins. Alternatively, all origins might be activated by the binding of a common initiator protein. If so, the binding site for such a protein is likely to be the core consensus since it is the only sequence common to all *ARS*s. Consistent with this hypothesis is the observation that single base pair changes in this element can abolish *ARS* activity.[52] Both biochemical and genetic schemes are being used to identify yeast proteins with the properties expected for origin-specific initiation proteins.

The absence of an efficient *ARS*-specific *in vitro* DNA replication system has meant that biochemical approaches have had to rely solely on *ARS*-specific DNA binding as an assay for the presence of initiator proteins. Both gel retardation[65-69] and nitrocellulose filter binding[70] have been used to monitor these activities. The *ARS* fragments that have been used for these studies include *ARS1*,[65-67] a telomere adjacent *ARS* called *ARS120*,[68,69] and an *ARS* associated with HMRE, a site also involved in repression of transcription at one of the silent mating type loci.[70,71] In each case, the investigators have identified proteins that protect 20-30 bps of DNA in the regions 3' to the T-rich strand of the core consensus (domain B). None of these activities appears to interact directly with the core consensus. The protein purified by its binding to *ARS1*[66,67] is probably the same as that identified in crude extracts by binding to HMRE.[70,71] This protein, called ABFI, has an estimated molecular weight of 135 kDa.[66,67] Both ABFI[66,67,70] and the activity described by Eisenberg and Tye[68] bind only to a subset of *ARS*s. To date, there is no evidence that ABFI or any other of the identified *ARS* binding proteins plays a role in DNA replication. Indeed, deletion from *ARS120* of the binding site for the protein has no effect on *ARS* function,[68] and deletion of the ABFI binding site on *ARS1* has very modest effects on plasmid stability.[67]

E. GENES THAT AFFECT *ARS* FUNCTION

Attempts to identify genes encoding *trans*-acting factors that interact with *ARS*s have relied on changes in plasmid stability as a phenotype for putative mutants.[72-74,166] In each case, the investigators monitored the stability of a circular plasmid containing both an *ARS* and a centromere. In the most extensive study, 40 minichromosome maintenance defective mutants (mcm⁻) representing 16 complementation groups were identified.[72] Most of these mutants affect the stability of each plasmid tested regardless of the identity of its *ARS* (nonspecific mutants). However, mutants in five complementation groups had different effects on plasmids containing different *ARS*s.[72,75] Surprisingly, each specific mutant affected the stability of the same subset of *ARS*s. At least one of the *mcm* mutations identifies an essential gene.[75]

Mutations in many different types of genes are likely to influence the stability of circular centromere-bearing plasmids. Therefore, it is essential that the mutant phenotype be correlated directly with *ARS* behavior. For the *mcm* mutants, *ARS* specificity is supported by the observation that all of them decrease the stability of plasmids containing a variety of centromeres. More importantly, *mcm2* increases loss rates of acentric circular and linear plasmids as well as circular and linear derivatives of chromosome III carrying three *ARS*s.[76] Consistent with its having a role in DNA replication, *mcm2* increases rates of mitotic recombination. (Faulty replication often leaves DNA lesions which are substrates for recombination.[77]) However, *mcm2* has only small effects on loss rates of authentic chromosomes.[76]

Other investigators have used visual assays to identify mutants that stabilize circular plasmids containing a centromere and a partially functional (or weak) *ARS*.[73,166] In one study, nine complementation groups were identified that enhance the stability of a plasmid containing a mutant HO *ARS*.[73] At least one complementation group defines an essential gene: DNA sequencing demonstrates that this gene encodes a 48 kDa protein.[74] In many cases, these mutants also improve the stability of plasmids containing other weak *ARS*s. In another study, mutations in *TUP1*, a gene involved in dTMP uptake, were found to influence differentially the stability of plasmids containing different *ARS*s.[166] The relationship of any of these gene products to either *ARS* binding proteins or to authentic initiation proteins remains to be determined.

III. CENTROMERES

A. BACKGROUND

Yeast centromeres, like those of all eukaryotes, mediate attachment of the chromosome to the mitotic and meiotic spindles, thereby ensuring its proper segregation. In addition, during meiosis, the centromere presumably prevents separation of sister chromatids until the time of the meiosis II division. The position of the centromere as inferred from genetic analysis on each of the 16 yeast chromosomes indicates that all yeast chromosomes are metacentric.[78] Electron microscopic analysis suggests that in mitosis a single microtubule attaches to each centromere.[79]

B. ISOLATION AND CHARACTERIZATION OF CENTROMERIC (CEN) DNA

Yeast centromeres were first isolated by cloning genes known to be linked tightly to their centromeres and testing this centromere proximal DNA for the ability to influence the stability of ARS plasmids. The first centromere to be cloned was from chromosome III.[80] A contiguous region of about 25 kb containing both LEU2 and CDC10, two genes tightly linked to the centromere of chromosome III, was isolated on overlapping plasmids.[81,82] Because these plasmids also contained ARS1, they transformed yeast at high frequency and were maintained in extrachromosomal form in transformed cells. However, plasmids containing the CDC10 gene were much more stable than other ARS1 plasmids;[80] whereas ARS1 plasmids are typically lost at rates of 0.2 to 0.3 per cell division, these plasmids were lost at rates of 0.01 to 0.03 per cell division.[83] Moreover, plasmids containing CDC10 were transmitted much more efficiently in meiosis than plasmids containing only ARSs, and during meiosis these plasmids often demonstrated first division segregation, as expected for a plasmid containing a centromere.[80] The stabilizing sequence was delimited to a fragment of 1.6 kb that was separable from both LEU2 and CDC10.[80] These and later studies demonstrated that centromeres (CENs) are not simply more efficient ARSs since fragments containing a CEN either have no ARS[84] or only very weak ARS activity.[80] The centromeres from chromosome IV,[84] chromosome VI,[85] and chromosome XI[86] were cloned by methods similar to those used to isolate CEN3.

Definitive proof that the 1.6 kb fragment from chromosome III contains the bona fide chromosome III centromere came from experiments in which deletion of this region caused enormous increases in the loss rate of chromosome III.[87] Replacement of the deleted region with a 624 bp subclone from the 1.6 kb fragment sufficed to restore normal stability to chromosome III in both mitosis and meiosis.[87] Experiments were also done in which CEN3 was replaced by CEN11. Since this CEN11 bearing chromosome III displayed normal chromosome III behavior in both mitosis and meiosis, centromeres are unlikely to mediate homologue pairing during meiosis I.[87]

The ability of centromeric DNA to stabilize ARS plasmids in mitosis was used as the basis of a general selection scheme to isolate yeast centromeres.[88] Fragments of yeast DNA were inserted into an ARS1 vector and used to transform yeast. Individual transformants were transferred to nonselective plates and subsequently replica plated to both selective and nonselective media. Plasmids were isolated from colonies with a large number of cells still able to grow on selective media. Two types of stabilizing sequences were identified, centromeres and portions of 2 μm DNA. During meiosis, plasmids containing centromeres most often segregated $2^+:2^-$, whereas those containing 2 μm DNA segregated predominantly $4^+:0^-$. The centromere from chromosome V was also isolated by its ability to stabilize an ARS plasmid.[89]

Plasmids containing a CEN (in addition to an ARS) also differ from ARS plasmids in their copy number. Whereas ARS plasmids are typically found in 10 to 20 copies per cell, CEN plasmids are normally maintained at 1 to 2 copies per cell (reviewed[83]). This difference can be attributed, at least in part, to the fact that although *ARS* plasmids are replicated efficiently,[18,37,44] they often fail to segregate to the daughter cell.[24] Therefore, the copy number of *ARS* plasmids in plasmid-bearing cells increases. In contrast, after replication *CEN* plasmids usually segregate to both progeny cells.[24] In addition, cells containing multiple copies of *CEN* plasmids grow poorly compared to normal cells,[90] suggesting that proteins or structures that interact with centromeric DNA are present in limited amounts in the yeast cell and that centromere plasmids can compete with authentic chromosomes for these factors.

The ability of *CEN* DNA to reduce copy number of *ARS* plasmids has been exploited to develop another general scheme to select yeast centromeres.[91] This scheme takes advantage of the fact that multiple copies of the ochre-suppressing tRNA gene *SUP11* are lethal to yeast cells. Fragments of yeast DNA were inserted into a plasmid containing *SUP11*, *ARS1*, and a selectable marker and used to transform yeast. Eleven fragments that allowed extra-chromosomal maintenance of *SUP11* were isolated of which ten represented centromeric sequences (the eleventh was derived from 2 μm DNA). Taken together, the various centromere isolation schemes have resulted in the identification of most of the 16 yeast centromeres.

C. VECTORS FOR VISUAL MONITORING OF *CEN* PLASMIDS

Two different vector systems have been developed that permit visual analysis of the rate and mechanism of loss of circular plasmids containing a *CEN* and an *ARS*.[45,46] Both systems rely on the fact that defects in the purine biosynthetic pathway can affect the color of yeast colonies. Cells blocked relatively late in the pathway (e.g., *ade1* or *ade2* cells) produce red colonies on plates containing low amounts of adenine because the substrate on which the *ADE2* protein acts, amino imidazole ribotide (AIR), is converted to a red pigment.[92] Cells blocked prior to production of AIR (e.g., *ade3* cells) do not produce the precursor and, like wild-type cells, produce white colonies. In one system, a plasmid-borne copy of the *ade3-2p* allele is used to affect colony color of an *ade3 ade2* strain.[46] The *ade3-2p* allele is a partially functioning allele of *ADE3*. A single copy of *ade3-2p* is sufficient to allow an *ade3* cell to grow in the absence of adenine, but two copies of *ade3-2p* are needed in *ade2 ade3* cells to produce the same amount of red pigment produced in an *ade2* strain. Therefore, cells carrying one copy of *ade3-2p* produce pink colonies in the same amount of time that a red colony forms in an *ade2* strain. The rate at which red pigment accumulates is dependent on the copy number of *ade3-2p* such that *ade3 ade2* cells transformed by an *ade3-2p* bearing plasmid can produce red (2 copies of plasmid), pink (one copy), or white (no copies) colonies. A similar scheme utilizes a plasmid containing the ochre suppressing tRNA gene *SUP11*.[45] This plasmid is studied in a diploid strain homozygous for an ochre-suppressible mutation in *ADE2*, a strain that normally produces red colonies. Cells containing one and two copies of the plasmid produce, respectively, pink and white colonies.

Both plasmid systems can be used to determine rates of plasmid loss from the frequency of half-sectored colonies. For example, for cells carrying the *ade3-2p* allele, a pink:white half-sectored colony is indicative of 1:0 segregation, a pattern most consistent with loss being due to a failure of plasmid replication. This interpretation is supported by the fact that plasmids with mutant *ARS*s display increased numbers of 1:0 events.[46] For cells carrying the *ade3-2p* allele, a red:white half-sectored colony is indicative of 2:0 segregation, a pattern most consistent with nondisjunction. Again, this interpretation is supported by the fact that plasmids carrying mutant *CEN*s display elevated rates of 2:0 events.[46] An analysis of the frequency and types of half-sectored colonies produced in cells carrying small circular *CEN* plasmids demonstrated that most (70 to 80%) loss events result from 1:0 events and are therefore probably due to replication failure.[45,46] These analyses also confirmed earlier studies demonstrating that small circular *CEN* plasmids are four to five orders of magnitude less stable than natural yeast chromosomes. Systems sensitive to the copy number of centromere plasmids and allowing a visual determination of rates of plasmid loss are being exploited by many labs to identify genes that influence chromosome behavior.

A. Consensus Sequence for Yeast Centromeres

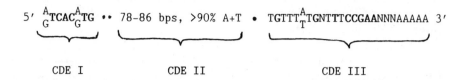

CDE I   CDE II   CDE III

B. Effects of Single Base Pair Changes in CDE III from <u>CEN3</u>.

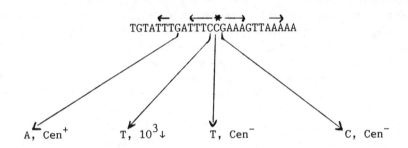

FIGURE 2.   (A) Sequence represents bases that occur at a minimum of 6 of 11 sequenced centromeres. Invariant bases are in bold print. N, any base. Based on data summarized in Fitzgerald-Hayes.[98] (B) Single bp changes were made as indicated in CDE III of *CEN3*. Loss rates of altered chromosome III with various *CEN*s are indicated[100,104]: Cen[+], essentially wild-type loss rate; Cen[−], no centromere function; 10^3 ↓ , loss rate ~10^3 greater than wild-type. Arrows indicate position of partial dyad. Asterik indicates center of dyad.

D. SEQUENCE REQUIREMENTS FOR *CEN* FUNCTION

Although centromeres appear to be unique sequence DNAs by the criterion of Southern hybridization, comparison of the sequences of eleven different *CEN*s demonstrates common structural features.[85,91,93-96] Three elements called centromere DNA elements (CDE) I, II, and III have been found on all sequenced *CEN*s (Figure 2). CDE I is an 8 bp element with the sequence 5′ᴬGTCACᴬGTG 3′. CDE II is a region characterized both by its length (78 to 86 bps) and its high A + T content (>90%), although its actual sequence varies from centromere to centromere. In CDE II, As or Ts are often found in blocks of up to seven in a row. Fragments of DNA containing CDE II display anomalously slow mobility in polyacrylamide gels, a characteristic of bent DNA.[97] The distance between CDE I and CDE II varies from 1 to 8 bps.[98] CDE III is a 26 bp region with partial dyad symmetry (Figure 2B) whose left most boundary is CDE II.

The roles of CDE I, CDE II, and CDE III have been studied using both site directed mutagenesis and deletion analysis. The phenotypes of deleted or mutated *CEN* fragments have been determined in two ways. The easier method is to determine if the mutation affects the stability of a small circular *ARS* plasmid (e.g., Reference 95). Alternatively mutated *CEN*s have been substituted for wild type *CEN3* on chromosome III[87] by fragment mediated transformation[99] and their phenotypes analyzed by determining the loss rate of chromosome III.[100-102] Although clearly more laborious, the latter method is more likely to reflect normal chromosome behavior.

Deletion analysis has demonstrated that full *CEN3* function can be carried on restriction fragments as small as 211 bp.[100] Moreover, a 136 bp synthetic *CEN3* containing only 10 bps to the left of CDE I and 9 bps to the right of CDE III appears to provide full *CEN* function, at least on circular plasmids (unpublished data cited in Reference 98). Single base pair changes in CDE I have little or even no effect on plasmid and chromosome stability in mitosis.[95,98,102] Indeed, complete deletion of CDE I reduces but does not abolish mitotic *CEN* function: the loss rates of chromosome III lacking CDE I is only ~20-fold greater than that of chromosome III with wild-type *CEN3*.[102] However, the presence of an intact CDE I is essential for proper segregation of chromosomes in meiosis.[102] Surprisingly, the meiotic effects of CDE I mutations are only manifest in strains in which both homologs are defective. The meiotic behavior of plasmids with CDE I deletions suggests that the absence of CDE I causes precocious first division segregation of replicated plasmids.

The presence of CDE II is essential for wild-type *CEN* function since deletion of most of CDE II causes a 10,000-fold increase in the mitotic loss rate of chromosome III.[101] Deletions in CDE II can be partially compensated for by insertions of equal amounts of noncentromeric sequences.[101,102] Insertion of random sequences of high A + T content restores chromosome stability to rates comparable to that of chromosomes carrying wild-type *CENs* (loss rates ~4× wild-type) whereas substitution of DNA of lower A + T content (e.g., 44% A + T) restores chromosome stability to a lesser extent (loss rate 40× wild-type).[102] Small increases (10 and 45 bps) in the size of CDE II have only small effects on plasmid stability[95] while increases of 100 bps cause about a 100-fold increase in the mitotic loss rate of chromosome III.[101] In meiosis, the presence but not the exact sequence of CDE II is important.[102] Taken together, these data indicate that although the presence of CDE II is required for centromere function, both its sequence and size are somewhat flexible. Nonetheless, substantial changes in the size or A + T content of CDE II have negative effects on centromere activity.

Like CDE II, the presence of CDE III is essential for mitotic centromere function: even small deletions from its right hand boundary can abolish *CEN* activity.[103] Moreover, single base-pair changes at some of the conserved bases within CDE III eliminate *CEN* function[100,104] (Figure 2B). For example, changing the C to a T at the center of symmetry of the partial dyad destroys mitotic *CEN* activity on both plasmids and chromosomes. The fact that single base pair changes in CDE III can eliminate *CEN* activity is consistent with the possibility that CDE III is a binding site for a protein important for *CEN* function. In contrast, other invariant bases within CDE III can be altered without measurable effects on *CEN* function. These studies illustrate the importance of studying *CEN* mutations within a chromosomal context since one mutation that had small effects on plasmid stability caused a 1000-fold increase in the loss rate of chromosome III.[100]

E. PROTEINS THAT INTERACT WITH *CEN* DNA

Micrococcal nuclease digestion was used to demonstrate that yeast centromeres on both chromosomes and plasmids assume a unique chromatin organization.[105] On chromosomes III and XI, a region of 220 to 250 bps centered about CDE III is in a protected, nonnucleosomal configuration. This region is flanked on both sides by DNase hypersensitive sites. Nucleosomes are positioned at precise intervals to either side of this core centromere region over a total distance of about 3.5 kb. A similar chromatin structure is seen for plasmid-borne *CENs* except that in these cases the precisely positioned nucleosomes occur only in *CEN* DNA, not in adjacent plasmid sequences. Like histones, the proteins that associate with *CEN* DNA are removed by 1 *M* NaCl. That this chromatin structure might be related to *CEN* function is supported by the observation that mutant *CENs* assume altered chromatin configurations *in vivo* in which the extent of nuclease protection within the core centromere region correlates with the amount of centromere function.[106]

Proteins that interact with centromeres are being identified. CP1 (centromere-binding protein 1), assayed by nitrocellulose filter binding, is a relatively abundant, heat stable protein of ~60 kDa that binds to CDE I as well as to many sites upstream of polymerase II transcribed genes.[107] The abundance of CP1 ($\geqslant$500 molecules per cell), the relatively small contribution of CDE I to mitotic *CEN* function, and the existence of numerous copies of CDE I at noncentromeric locations all argue that CP1 is more likely to play a role in transcription than in chromosome segregation.

Two *in vitro* assays, exonuclease III protection and gel-fragment retardation, have been used to support the hypothesis that CDE III serves as a protein recognition site.[104] For exonuclease III experiments, a 633 bp *CEN3* fragment was incubated with whole cell extracts. Two specific exonuclease III stop points were mapped to 10 and 24 bases to the right side of CDE III (Figure 2A). (Stop points to the left of CDE III were not determined.) This result was corroborated by the fact that gel retardation assays demonstrate that *CEN3* fragments can be specifically retarded after incubation with yeast extracts. To detect specificity for *CEN* DNA, it was necessary to fractionate whole cell extracts on heparin-agarose and use the flow through as the source of proteins, thereby eliminating most DNA binding proteins. Competition experiments with a variety of mutant and wild-type *CEN*s provide strong support for the argument that the activity being detected binds to CDE III. This activity might be the same as that detected by binding to a 39 bp oligonucleotide containing CDE III from *CEN6*.[103]

A number of laboratories are using genetic screens to identify chromosomally borne mutations that affect the maintenance of *CEN* plasmids. These mutations identify genes that are candidates for genes encoding centromere-binding proteins. Types of mutations indicative of defects in *CEN*-binding proteins include those that increase 2:0 (rather than 1:0) segregation events and those that increase the stability of chromosomes or plasmids containing mutant centromeres. These considerations are important given that mutations in genes encoding *ARS* binding proteins are also expected to affect plasmid and chromosome stability.

IV. TELOMERES

A. BACKGROUND

The genetic map derived for yeast indicates that, as in other eukaryotes, individual chromosomes are linear.[78] The physical ends of linear chromosomes are called telomeres; they appear to be required for complete duplication and stability of chromosomes. Telomeres are also believed to mediate interactions among both homologous and nonhomologous chromosomes and to contribute to the three-dimensional organization of the nucleus (reviewed in Reference 108).

The sequences at natural DNA termini were first studied in ciliated protozoans where the small size and large number of linear macronuclear DNA molecules result in a relatively large abundance of DNA ends. The ribosomal RNA genes (rDNA) from *Tetrahymena* macronuclei can be readily isolated due to their high copy number and small size (~10^4 copies of a 21 kb palindromic DNA molecule[109]). Terminal restriction fragments from isolated rDNA were sequenced directly and found to contain 20 to 70 copies of the simple repeat C_4A_2/T_2G_4, where the C_4A_2 strand runs 5' to 3' from the end of the rDNA molecule towards its interior.[110] In addition, single-strand discontinuities were detected within the C_4A_2 strand. The structure and sequence of DNA termini from several species of hypotrichous ciliates, including *Oxytricha,* were determined by direct sequencing of total macronuclear DNA.[111-113] Again the analysis was facilitated by the small size of macronuclear DNA molecules which in *Oxytricha* average only 2.5 kb.[114] Each *Oxytricha* DNA terminus bears 20 bps of C_4A_4/T_4G_4 repeats. In addition, at each end the G_4T_4 strand is extended as a 16 base single-stranded tail with the sequence $(G_4T_4)_2$. As in *Tetrahymena,* the CA-rich strand runs 5' to 3' from the end towards the center of the DNA molecule.

B. FUNCTIONAL ASSAYS FOR TELOMERES

Because ciliate termini can be readily obtained, it was possible to test terminal restriction fragments from both *Tetrahymena* rDNA and total *Oxytricha* macronuclear DNA for the ability to support maintenance of linear plasmids in yeast. Linearized *ARS* plasmids containing a selectable gene were ligated to terminal restriction fragments from either *Tetrahymena* rDNA[115,116] or total *Oxytricha* macronuclear DNA[117] and introduced by transformation into yeast. Many transformants contained linear plasmids. Linearity was established by (1) the presence of a single hybridizing species, rather than the multiple forms characteristic of circular plasmids,[115-117] (2) the number of fragments produced by restriction digestion of plasmid DNA,[115,116] or (3) behavior of plasmids during two-dimensional agarose gel electrophoresis under conditions that permit separation of circular and linear DNA molecules.[117] Although the structure of linear plasmids was essentially that predicted from the size of the input fragments, the terminal restriction fragments were about 200 to 1000 bps longer after propagation in yeast than when taken directly from *Tetrahymena* or *Oxytricha*.[115,117] Southern hybridization[117,118] and DNA sequence analysis[119] indicated that the added DNA has the same sequence as that found at the ends of yeast chromosomes (discussed further below). The addition of yeast sequences on ciliate termini occurs in the absence of the *RAD52* gene product which is required for the majority of mitotic recombination events in yeast.[120,121]

The ability of certain sequences to support maintenance of linear plasmids provides a functional assay for telomeres in yeast. However, unlike *ARS*s and *CEN*s, there is no direct selection available for telomeres: individual transformants must be screened by Southern hybridization for the presence of linear DNA molecules. To date, no fragments other than those bearing sequences found on natural eukaryotic DNA termini have been shown to be capable of acting as telomeres in yeast. Linear *ARS* plasmids without telomeric sequences can only transform by self-ligation (thereby forming a circular plasmid) or by recombination with chromosomal DNA.

C. ISOLATION AND CHARACTERIZATION OF TELOMERIC DNA

Authentic yeast telomeres were also identified by the functional cloning assay.[115] Total DNA from a strain carrying a linear plasmid with terminal fragments derived from *Tetrahymena* rDNA was digested with *Pvu*I, a restriction enzyme with a single site within the linear plasmid. The DNA was then ligated and introduced by transformation into yeast, selecting for a gene carried on the original plasmid and screening transformants for the presence of linear plasmids. Three transformants contained linear plasmids carrying a ~2.6 kb *Pvu*I fragment from yeast in place of one of the *Tetrahymena* rDNA fragments. An internal portion of this fragment was cloned in *E. coli* and then used to probe restriction digests of total yeast DNA. These studies demonstrated that the *Pvu*I fragment acting as a telomere on the linear plasmid was a middle repetitive yeast sequence. Mapping with a number of restriction enzymes demonstrated that one end of the chromosomal copies of this fragment behaves as if it carries a recognition site for many restriction enzymes (*Xho*I, *Hind*III, *Cla*I, and *Bam*HI). Thus, on a number of chromosomes, one end of the *Pvu*I fragment is either a hypersensitive site for nucleases or the physical end of the chromosome. The fact that this *Pvu*I fragment supports maintenance of a linear plasmid provides compelling evidence for the second alternative.

In addition to natural termini, cloned telomeric repeats from yeast and ciliates can support formation of linear plasmids in yeast.[167,168] To serve as a telomere, the cloned telomeric repeats must be oriented in the same manner as they are on natural chromosomes (i.e., CA-rich strand running 5′ to 3′ from the terminus towards the center of the plasmid). These studies demonstrate that the CA-rich sequences alone are sufficient for formation of new telomeres in yeast.

The telomeric end of the *Pvu*I fragment from yeast chromosomal DNA that acted as a telomere on a recombinant DNA plasmid[115] was subcloned in *E. coli* after S1 digestion to remove the extreme molecular end of the linear plasmid, a procedure that deleted ~400 bps.[119] Sequence analysis of this cloned DNA revealed 81 bps of an irregular CA-rich satellite-like DNA whose sequence can be represented as $5'[C_{2-3}A(CA)_{1-3}]3'$ (often abbreviated $C_{1-3}A$). Like the CA-rich terminal repeats in ciliates, the yeast CA-rich strand runs $5'$ to $3'$ from the end of the chromosome towards the centromere. An *ARS* consensus sequence was detected about 200 bps centromere proximal to the $C_{1-3}A$ repeats or ~700 bps from the physical end of the chromosome. A *Tetrahymena* rDNA fragment used to support telomere formation in yeast, cloned and sequenced by similar methods, was found to contain 228 bps of $C_{1-3}A$ DNA added directly to the C_4A_2 *Tetrahymena* repeats.[119]

The lengths of terminal tracts of $C_{1-3}A$ repeats vary among different laboratory strains of yeast.[122,123] For example, the average lengths of terminal $C_{1-3}A$ tracts in strains A364a and 2262 are, respectively, 360 bps and 620 bps.[122] The meiotic segregation patterns of terminal restriction fragments produced when two strains of different telomere lengths are crossed suggest that telomere length is controlled by a small number of genes with naturally occurring allelic variants.[122,123] In addition to differences among different strains, individual telomeres can vary in size by as much as 200 bps among different clonal isolates of the same strain, a heterogeneity that increases with increased numbers of cell divisions.[124] Intrastrain variation in telomere lengths is not coordinate: individual telomeres vary independently of other telomeres in the cell. It is likely that this intrastrain heterogeneity in telomere lengths is a reflection of telomere-specific duplication processes, such as nontemplated DNA replication, recombination, or both (see below for further discussion).

Mutations in genes that interact with telomeric DNAs might be expected to change the average lengths of terminal tracts of $C_{1-3}A$ repeats. For example, a leaky mutation in a telomere specific replication enzyme might decrease the lengths of these tracts. After screening 200 highly mutagenized strains by Southern hybridization, three strains were identified with telomeres 200 to 300 bps shorter than the starting strain.[125] In each case, the short telomere phenotype was recessive and segregated $2^+:2^-$ in meiosis. The three mutants identified two genes called *TEL1* and *TEL2*. No other phenotype, such as temperature sensitive growth, cosegregated with short telomeres.

Mutations in *CDC17*, an essential gene, can also influence telomere length.[126] Temperature-sensitive alleles of *CDC17* cause a cell cycle specific arrest phenotype at 36°C. Arrest occurs prior to nuclear division but after replication of ≥50% of chromosomal DNA.[127] When *cdc17* cells are grown at semipermissive temperatures, increased recombination occurs in telomere proximal regions of the chromosome and $C_{1-3}A$ tracts on chromosome ends increase in size by 11 to 18 bps per generation.[126] Moreover, *cdc17* cells grown at permissive temperatures can have telomeres up to 900 bps longer than congenic strains. It is now clear that *CDC17* encodes the gene for the catalytic core subunit of DNA polymerase I.[128] Why mutations in a gene presumably required for replication of the entire chromosome should have telomere-specific effects is not known.

The physical structure of yeast chromosomal termini is unknown. Earlier experiments demonstrated that large molecular weight chromosomal DNA molecules rapidly renature after denaturation. These and other results suggested that the ends of yeast chromosomes are cross-linked.[129,130] However, reinvestigation of terminal cross-linking using cloned telomere probes did not support this possibility.[131] There is some evidence that yeast telomeric fragments are substrates for nick-translation with DNA polymerase I,[115] which might mean that the $C_{1-3}A$ repeats, like the *Tetrahymena* C_4A_2 repeats, have single-strand discontinuities.

D. TELOMERE-ASSOCIATED SEQUENCES

In addition to $C_{1-3}A$ repeats, many yeast telomeres carry copies of two middle repetitive sequences called X and Y'[63,64] (see Figure 3). Y' is a highly conserved element of 6.7 kb.

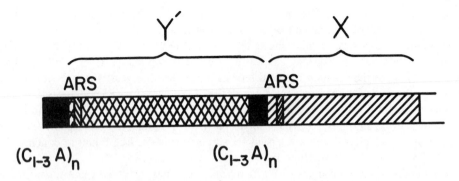

FIGURE 3.   Structure of yeast telomeres. As drawn, the end of the chromosome is towards the left and the centromere is towards the right. Two middle repetitive sequences Y' and X are found on many yeast telomeres. Y' exists in zero to four copies[64,122,123,132,134]. Most, but not all, telomeres carry X[134]. Blocks of C_{1-3} A repeats, ■; *ARSs*, ▨ .

It has been inferred from hybridization studies that some telomeres lack Y' whereas others contain up to four tandem copies of the element.[63,64,122,123,132] The *Pvu*I fragment identified as a telomere by the functional cloning experiment was derived from a Y' element.[115] X, a less conserved element than Y', varies in size from 0.3 to 3.75 kb. On telomeres carrying both X and Y', Y' is found telomere proximal to X.[64]

Pulsed field gel electrophoresis has been used to determine the distribution of X and Y' on individual chromosomes in different yeast strains.[133,134] The patterns of hybridization varied from strain to strain, a conclusion also deduced from the number and size of Y'-hybridizing bands in restriction enzyme digested DNA from different strains.[123] One or two of the smallest chromosomes lacked Y' and in some strains, chromosome I, the smallest yeast chromosome, lacked detectable amounts of both X and Y'.[134] These data argue that X and Y' are not essential for maintenance of at least small chromosomes, although more subtle effects on chromosome behavior are still possible. This conclusion is in agreement with the observation that a deletion derivative of chromosome III that lacks X and Y' is as stable as the natural chromosome III from which it was derived.[135] However, it is not known whether the original chromosome III carried X, Y', or both.

Internal tracts of 25 to 140 bp of C_{1-3}A repeats have been found between adjacent Y' elements and between X and Y' repeats.[136] Since telomere cloning experiments have demonstrated that telomeric sequences in the proper orientation are sufficient to support formation of new telomeres, a yeast chromosome broken within or near an X or Y' element should be a suitable substrate for telomere formation. Consistent with this possibility, a linear plasmid broken within a Y' element can be efficiently "healed".[120] However, in this case healing occurred by *RAD52*-dependent recombination with chromosomal telomeres, rather than by *de novo* addition of C_{1-3}A repeats.

E. TELOMERE REPLICATION

Telomeres are believed to facilitate the complete duplication of DNA termini. Typical DNA polymerases, which require a 3' primer and are able to synthesize DNA only in the 5' to 3' direction, would leave 5' gaps at the ends of newly replicated strands.[137] Thus, it is likely that special replication enzymes or mechanisms (or both) are involved in telomere duplication thereby avoiding loss of genetic information with each round of DNA replication. In addition to tracts of C_{1-3}A repeats, X and Y' both carry *ARSs*.[26,64] Given the special replication requirements of telomeric DNA and the fact that many Y' and at least some X elements replicate late in S phase,[165] it is tempting to speculate that these X and Y' *ARSs* are specialized replication origins necessary for telomere duplication. Consistent with this hypothesis, the terminal restriction fragments derived from *Tetrahymena* rDNA that support telomere formation in yeast also act as an *ARS* in yeast.[138,139] However, if this hypothesis were true, telomeres that lack both X and Y'[134] would have to carry an as yet unidentified class of telomeric *ARS*s. In any case, at least some of the *Oxytricha* fragments that support telomere formation in yeast do not contain *ARS*s.[168] Moreover, linear artificial chromosomes with terminal *ARS*s are no more stable in mitosis than similar constructs without terminal *ARS*s.[169] Taken together, these results argue against an important telomere-specific function for the X and Y' *ARS*s during chromosome replication.

A number of telomere replication models postulate an inverted repeat (palindrome) as an intermediate in telomere replication.[148-151] After replication, these inverted repeats must be resolved in order to yield individual chromosomes. When yeast is transformed with circular plasmids containing an inverted repeat consisting of C_4A_2 sequences, the circular plasmid is resolved relatively efficiently to form a linear molecule.[151] Although this result is consistent with the possibility that inverted repeats are intermediates in telomere replication, there is no evidence for fused telomeres during the yeast cell cycle. Moreover, an inverted repeat model of telomere replication does not predict the addition of $C_{1-3}A$ repeats during telomere formation. The activity detected in these experiments might play a role in resolving rare end-to-end fusions of chromosomes (although there is as yet no evidence that such events occur in yeast).

F. PROTEINS THAT INTERACT WITH TELOMERES

A number of different approaches are being used to identify proteins that interact with yeast telomeres. Using a gel retardation assay, an activity has been detected in yeast extracts that binds to tracts of $C_{1-3}A$ repeats.[140] The minimum length for binding is between 7 and 25 bps. Binding *in vitro* does not require any specialized telomere structure since $C_{1-3}A$ tracts from internal regions of yeast chromosomes or on recombinant DNA plasmids are bound as efficiently as terminal $C_{1-3}A$ tracts. In contrast, tracts of C_4A_2 or C_4A_4 repeats are not suitable substrates for binding even though both sequences can support telomere formation *in vivo*.

The activity that binds to telomeric sequences may be the same protein as that isolated by its ability to bind both to sequences involved in inactivation of transcription at the silent mating type loci[70,71] and to sequences necessary for activation of transcription of ribosomal protein genes.[141,142] This protein called TUF,[141,142] RAPl,[143] or GRFl[70] will be called RAPl (repressor/activator site binding protein) herein. RAPl has been purified to homogeneity by DNA affinity chromatography, and the gene encoding it cloned and sequenced.[143] Sequence analysis predicts a highly acidic protein of 92.5 kDa. Gene disruption experiments demonstrate that RAPl is essential for growth. From nitrocellulose filter binding assays using synthetic oligonucleotides as competitors, a consensus sequence for high affinity RAPl binding sites was deduced that can be represented as 5' $^{AA}_{GC}$ACCCANNCA$^{TT}_{CC}$ 3'.[70] Some versions of this consensus are identical to the consensus sequence 5' $C_{1-3}A$ 3' deduced for yeast telomeres. Indeed yeast telomeric repeats act as high affinity binding sites for RAPl *in vitro*.[70] It is not yet clear if RAPl is the only protein that binds to telomeric $C_{1-3}A$ repeats as assayed by gel retardation techniques or if the essential role of RAPl is related to a function carried out at yeast telomeres.

Other proteins that interact with telomeres have been identified in ciliated protozoans. Two proteins have been purified from *Oxytricha* that bind tenaciously but noncovalently to DNA termini and in so doing protect them against exonucleolytic degradation *in vitro*.[144] Unlike the $C_{1-3}A$ binding activity detected in yeast, the *Oxytricha* proteins require both the sequence and structure of DNA termini for their most efficient binding. Although no comparable terminus-specific protein has been detected biochemically in yeast, the phenotype of yeast cells carrying large numbers of DNA termini is consistent with the existence of such proteins.[170] Cells were transformed by circular or linear plasmids containing *leu2-d*, a promotor defective allele of *LEU2* that can complement a *leu2* cell only when present in 100 to 200 copies per cells. The circular plasmid, which contained an insert of 81 bp of $C_{1-3}A$ repeats, had no detectable effect on the telomeres of cells containing it. However, cells carrying the linear plasmid (and therefore 200 to 400 extra DNA termini) had telomeres about $2\times$ longer than the parent strain. This result is consistent with the possibility that there are proteins that specifically interact with the ends of yeast chromosomes and that these proteins are present in limiting amounts. Since the $C_{1-3}A$ binding activity monitored by *in vitro* gel retardation assays exhibits no preference for terminal $C_{1-3}A$ tracts,[140] it is unlikely to be the activity titrated out *in vivo* by excess telomeres. Unlike the *in vitro* behavior of the *Oxytricha* termini binding proteins, the putative yeast protein appears to protect against processes that increase telomere length, such as replication or recombination.

An excellent candidate for a telomere-specific replication enzyme has been identified in *Tetrahymena*. This activity adds tandem T_2G_4 repeats onto the single stranded DNA oligonucleotide $(T_2G_4)_4$.[145,146] *In vivo*, such an activity would be capable of extending the 3' strand of DNA termini in a template independent manner. Moreover, the *Tetrahymena* enzyme can extend oligonucleotides bearing the GT-strand of telomeric sequences from a variety of organisms, including yeast. If a comparable enzyme exists in yeast, it would explain both the ability of ciliate termini to support telomere formation in yeast and their modification by the addition of $G_{1-3}T$ repeats. Alternatively, the yeast results could be explained by recombination models of telomere duplication (e.g., Reference 147) if ciliate termini have sufficient homology to $C_{1-3}A$ repeats to permit base pairing with yeast telomeres *in vivo*.[136]

Mutations in the three genes known to affect telomere length, *TEL1*, *TEL2*, and *CDC17*, all demonstrate long phenotypic lags before telomere length differences are manifest.[125,126] Likewise, intrastrain heterogeneity in telomere lengths can only be seen after many generations of growth.[124] These results probably indicate that only the very ends of chromosomes are replicated by telomere-specific mechanisms: most of the DNA in the terminal restriction fragments whose size is monitored in telomere length experiments is presumably replicated by standard semiconservative DNA replication.

V. ARTIFICIAL CHROMOSOMES

Although centromeres improve the segregation of circular *ARS* plasmids in both mitosis and meiosis, even *CEN* plasmids are lost at very high rates compared to natural chromosomes. One possible explanation for this result is that instability is a direct consequence of circularity. For example, sister chromatid exchange between circular *CEN* plasmids can produce dicentric plasmids[152,153] which, like dicentric chromosomes,[154-156] are unstable. To test this possibility, small (10 to 15 kb) artificial linear chromosomes, containing an *ARS*, a *CEN*, *Tetrahymena* rDNA termini, and a selectable marker, were constructed *in vitro* and introduced by transformation into yeast.[116,157] These minichromosomes transform yeast efficiently, but, surprisingly, are three to four times less stable than circular *CEN* plasmids of similar size and structure and four or five orders of magnitude less stable than real chromosomes. Since small linear chromosomes, like *ARS* (but not *CEN*) plasmids, are present in multiple copies per cell,[157] their low stability presumably reflects faulty segregation rather than inefficient replication.

The first artificial chromosomes differed from authentic yeast chromosomes in four major ways: (1) they are telocentric (centromere very close to telomere) whereas in natural yeast chromosomes, the centromere is always located >50 kb from both telomeres; (2) they bear surrogate telomeres derived from ciliate DNA; (3) they are about twenty times smaller than the smallest yeast chromosome, and (4) they lack most of the sequences normally found on a yeast chromosome. Since the close (3.5 to 12 kb) spacing of centromeres and telomeres on chromosome-sized DNA molecules has only small effects on chromosome stability, this factor is unlikely to account for the low stability of small artificial chromosomes.[8,135,158] Likewise, derivativies of natural chromosomes in which telomere function is supplied by yeast modified ciliate termini are about as stable as natural chromosomes,[13,135,158] and small artificial chromosomes bearing $C_{1-3}A$ repeats with or without a Y' ARS are no more stable than artificial chromosomes carrying *Tetrahymena* termini.[169]

Unlike close centromere-telomere spacing and ciliate termini, neither of which markedly affects chromosome loss rates, small size has large negative effects on chromosome stability. The lengths of artificial chromosomes can be increased by integration of one or more copies of bacteriophage lambda DNA which is presumed to add no positive (or negative) sequences that influence stability. When this is done, the stability of both linear and circular chromosomes increases for sizes up to ~100 kb.[45,157] In this size range, circles are still more stable than linears.[45] Above ~100 kb, circles decrease in stability, possibly because the frequency of sister chromatid exchange increases with size.[45] In contrast, linear artificial chromosomes continue to get more stable as they get longer: the loss rate per division of a 137 kb linear artificial chromosome, the largest studied to date, is $\sim 1.5 \times 10^{-3}$.[45] This 137 kb artificial chromosome is about half the size of chromosome I, the smallest yeast chromosome, but is still two to three orders of magnitude less stable than chromosome III (loss rate, 10^{-5} to 10^{-6} per division;[9,13] size, 370 kb[1]).

Size also influences stability of natural chromosomes as demonstrated by the effects of deletions on the stability of chromosome III.[9,13] However, deletion derivatives of chromosome III that are in the size range of artificial chromosomes are much more sensitive to size changes than chromosomal-sized DNA molecules. For example, down to 70 kb, even large deletions have relatively small effects on the stability of chromosome III derivatives whereas for chromosomes from 70 to 29 kb, small deletions have large deleterious effects.[9] Likewise, for DNA molecules in the size range of chromosomes, differences in size probably have only modest effects on stability. For example, chromosome V[10] (estimated size, ~580 kb[1]) is lost at the same rate as chromosome VII[11] (estimated size ~1100 kb[171]). Thus, not surprisingly, even the smallest yeast chromosomes are very stable during mitosis. However, size-related stabilities are apparent under certain conditions. For example, when nuclear fusion is prevented after zygote formation, chromosomes are often passed between the two nuclei in the cell; and, in this situation, the frequency of transfer for different chromosomes is inversely proportional to their size.[7]

In addition to the correlation between size of small chromosomes and their stability, there appears to be a sharp size cutoff below which stable propagation of linear chromosomes is no longer possible.[158] A 9 kb linear minichromosome constructed *in vitro* transforms yeast efficiently but is propagated so poorly that the only stable transformants are those produced by events that fundamentally alter its size or structure. In contrast, artificial chromosomes only somewhat larger (13 and 15 kb) but in other ways similar to the 9 kb minichromosome are stably maintained.[116] A similar size limitation does not apply to other ARS plasmids since small (1.4 to 3 kb) circles with or without centromeres and small (3 kb) acentric linear plasmids are maintained efficiently.[37,120,159]

It is not yet clear if there are *cis*-acting sequences other than ARSs, CENs, and telomeres important for chromosome stability. Presumably, any such sequences are nonessential since artificial chromosomes can be maintained in their absence. An example of such a sequence is a transcription terminator since it has been shown that transcription from a nearby, strong promotor can eliminate ARS[48] and CEN activity.[160,161] (However, most artificial chromosomes are constructed in such a way that the CEN and ARS are flanked by terminators; thus, their instability is unlikely to be due to transcription induced loss.)

One argument for the existence of additional stability elements is the observation that deletion derivatives of chromosome III appear to be more stable than artificial chromosomes of similar size.[9] For example, 150 kb deletion derivatives of chromosome III (loss rates $\sim 10^{-4}$ to 10^{-5} per division[9,13]) are 10 to 100 times more stable than a 137 kb artificial chromosome.[45] However, it is also possible that differences in stability between natural and artificial chromosomes of similar sizes are more apparent than real: chromosome loss values obtained in different laboratories are difficult to compare directly (for example, literature values for the loss rate of chromosome III vary as much as tenfold[9,13]); and, in most cases, sizes of artificial chromosomes and chromosome III derivatives were not determined by pulsed field gel electrophoresis and are therefore not precise.

In addition to providing important insights into the behavior of eukaryotic chromosomes, work with artificial yeast chromosomes is of practical significance. Recently, >400 kb fragments from human DNA have been cloned as parts of artificial chromosomes in yeast.[162] Although it is not yet practical to use yeast to generate large amounts of cloned DNA, cloning in yeast does circumvent certain problems inherent with *E. coli*, such as maintenance of repeated sequences. It has also been shown recently that chromosomes from the fission yeast *Schizosaccharomyces pombe* can be maintained in mouse tissue culture cells.[163] Since yeasts are currently the only eukaryotes in which chromosomal DNA can be easily and precisely manipulated, these results raise the possibility that yeast could serve as intermediaries in the manipulation of the mammalian genome.

ACKNOWLEDGMENTS

I thank my colleagues who communicated preprints and results prior to publication and M. Conrad, D. Gottschling, R. Levis, A. Pluta, K. Runge, and R. Wellinger for comments on this manuscript. Special thanks to A. Taylor for help with some of the fine points of word processing. Work in my lab is supported by grant NP-574 from the American Cancer Society and grant GM26938 from the National Institutes of Health.

REFERENCES

1. **Carle, G. F. and Olson, M. V.**, An electrophoretic karyotype for yeast, *Proc. Natl. Acad. Sci. U.S.A.*, 82, 3756, 1985.
2. **Therman, E.**, *Human Chromosomes*, Springer-Verlag, New York, 1986, 1.
3. **Schwartz, D. C. and Cantor, C. R.**, Separation of yeast chromosome-sized DNAs by pulsed field gradient gel electrophoresis, *Cell*, 37, 67, 1984.
4. **Carle, G. F. and Olson, M. V.**, Separation of chromosomal DNA from yeast by orthogonal-field-alternation gel electrophoresis, *Nucleic Acids Res.*, 12, 5647, 1984.
5. **Chu, G., Vollrath, D., and Davis, R. W.**, Separation of large DNA molecules by contour-clamped homogeneous electric fields, *Science*, 234, 1582, 1986.
6. **Bruenn, J. and Mortimer, R. K.**, Isolation of monosomics in yeast, *J. Bact.*, 102, 548, 1970.
7. **Dutcher, S. K.**, Internuclear transfer of genetic information in kar1-1/KAR1 heterokaryons in *Saccharomyces cerevisiae*, *Mol. Cell. Biol.*, 1, 245, 1981.
8. **Surosky, R. T. and Tye, B.-K.**, Construction of telocentric chromosomes in *Saccharomyces cerevisiae*, *Proc. Natl. Acad. Sci. U.S.A.*, 82, 2106, 1985.
9. **Surosky, R. T., Newlon, C. S., and Tye, B.-K.**, The mitotic stability of deletion derivatives of chromosome III in yeast, *Proc. Natl. Acad. Sci. U.S.A.*, 83, 414, 1986.
10. **Hartwell, L. and Smith, D.**, Altered fidelity of mitotic chromosome transmission in cell cycle mutants of *S. cerevisiae*, *Genetics*, 110, 131, 1985.
11. **Hartwell, L. H., Dutcher, S. K., Wood, J. S., and Garvik, B.**, The fidelity of mitotic chromosome reproduction in *Saccharomyces cerevisiae*, *Rec. Adv. Yeast Mol. Biol.*, 1, 28, 1982.
12. **Esposito, M. S., Maleas, D. T., Bjornstad, K. A., and Bruschi, C. V.**, Simultaneous detection of changes in chromosome number, gene conversion and intergenic recombination during mitosis of *Saccharomyces cerevisiae*: spontaneous and ultraviolet light induced events, *Curr. Genet.*, 6, 5, 1982.

13. **Murray, A. W., Schultes, N. P., and Szostak, J. W.,** Chromosome length controls mitotic chromosome segregation in yeast, *Cell,* 45, 529, 1986.
14. **Sora, S., Lucchini, G., and Magni, G. E.,** Meiotic diploid progeny and meiotic non-disjunction in *Saccharomyces cerevisiae, Genetics,* 101, 17, 1982.
15. **Newlon, C. S., Petes, T. D., Hereford, L. M., and Fangman, W. L.,** Replication of yeast chromosomal DNA, *Nature (London),* 247, 32, 1974.
16. **Brewer, B. J., Zakian, V. A., and Fangman, W. L.,** Replication and meiotic transmission of yeast ribosomal RNA genes, *Proc. Natl. Acad. Sci. U.S.A.,* 77, 6739, 1980.
17. **Zakian, V. A., Brewer, B. J., and Fangman, W. L.,** Replication of each copy of the yeast 2 micron DNA plasmid occurs during the S-phase, *Cell,* 17, 923, 1979.
18. **Fangman, W. L., Hice, R. H., and Chlebowicz-Sledziewska, E.,** ARS replication during the S phase, *Cell,* 32, 831, 1983.
19. **Newlon, C. S. and Burke, W.,** Replication of small chromosomal DNAs in yeast, in *Mechanistic Studies on DNA and Genetic Recombination, ICN-UCLA Symp. Mol. Cell. Biol.,* Vol. 19, Alberts, B., Ed., Alan R. Liss, New York, 1980, 399.
20. **Zyskind, J. W. and Smith, D. W.,** The bacterial origin of replication, oriC, *Cell,* 46, 489, 1986.
21. **Hinnen, A., Hicks, J. B., and Fink, G. R.,** Transformation of yeast, *Proc. Natl. Acad. Sci. U.S.A.,* 75, 1929, 1978.
22. **Struhl, K., Stinchcomb, D. T., Scherer, S., and Davis, R. W.,** High-frequency transformation of yeast: autonomous replication of hybrid DNA molecules, *Proc. Natl. Acad. Sci. U.S.A.,* 76, 1035, 1979.
23. **Hsiao, C.-L. and Carbon, J.,** High frequency transformation of yeast by plasmid containing the cloned ARG4 gene, *Proc. Natl. Acad. Sci. U.S.A.,* 76, 3829, 1979.
24. **Murray, A. W. and Szostak, J. W.,** Pedigree analysis of plasmid segregation in yeast, *Cell,* 34, 961, 1983.
25. **Stinchcomb, D. T., Struhl, K., and Davis, R. W.,** Isolation and characterization of a yeast chromosomal replicator, *Nature (London),* 282, 39, 1979.
26. **Chan, C. S. M. and Tye, B.-K.,** Autonomously replicating sequences in *Saccharomyces cerevisiae, Proc. Natl. Acad. Sci. U.S.A.,* 77, 6329, 1980.
27. **Beach, D., Piper, M., and Shall, S.,** Isolation of chromosomal origins of replication in yeast, *Nature (London),* 284, 185, 1980.
28. **Newlon, C. S., Green, R. P., Hardeman, K. J., Kim, K. E., Lipchitz, L. R., Palzkill, T. G., Synn, S., and Woody, S. T.,** Structure and organization of yeast chromosome III, in *Yeast Cell Biology, UCLA Symp. Mol. Cell. Biol.,* New Series, Vol. 33, Hicks, J., Ed., Alan R. Liss, New York, 1986, 211.
29. **Saffer, L. D. and Miller, O. L., Jr.,** Electron microscopic study of *Saccharomyces cerevisiae* rDNA chromatin replication, *Mol. Cell. Biol.,* 6, 1148, 1986.
30. **Szostak, J. W. and Wu, R.,** Insertion of a genetic marker into the ribosomal DNA of yeast, *Plasmid,* 2, 536, 1979.
31. **Kouprina, N. Y. and Larinonov, V. L.,** The study of a rDNA replicator in *Saccharomyces, Curr. Genet.,* 7, 433, 1983.
32. **Scott, J. F.,** Preferential utilization of a yeast chromosomal replication origin as template for enzymatic DNA synthesis, in *Mechanistic Studies of DNA Replication and Genetic Recombination, ICN-UCLA Symp. on Mol. Cell Biol.,* Vol. 19, Alberts, B., Ed., Alan R. Liss, New York, 1980, 379.
33. **Kojo, H., Greenberg, B. D., and Sugino, A.,** Yeast 2-μm plasmid DNA replication in vitro: origin and direction, *Proc. Natl. Acad. Sci. U.S.A.,* 78, 7261, 1981.
34. **Celniker, S. E. and Campbell, J. L.,** Yeast DNA replication *in vitro:* initiation and elongation events mimic *in vivo* processes, *Cell,* 31, 201, 1982.
35. **Jazwinski, S. M. and Edelman, G. M.,** Protein complexes from active replicative fractions associate *in vitro* with the replication origins of yeast 2-μm DNA plasmid, *Proc. Natl. Acad. Sci. U.S.A.,* 79, 3428, 1982.
36. **Jong, A. Y. S. and Scott, J. F.,** DNA synthesis in yeast cell-free extracts dependent on recombinant DNA plasmids purified from *Escherichia coli, Nucleic Acids Res.,* 13, 2943, 1985.
37. **Zakian, V. A. and Scott, J. F.,** Construction, replication and chromatin structure of TRP1 RI circle, a multiple-copy synthetic plasmid derived from *Saccharomyces cerevisiae* chromosomal DNA, *Mol. Cell. Biol.,* 2, 221, 1982.
38. **Brewer, B. J. and Fangman, W. L.,** The localization of replication origins on ARS plasmids in *Saccharomyces cerevisiae, Cell,* 51, 463, 1987.
39. **Huberman, J. A., Spotila, L. D., Nawotka, K. A., El-Assouli, S. M., and Davis, L. R.,** The *in vivo* replication origin of the yeast 2 μm plasmid, *Cell,* 51, 473, 1987.
40. **Livingston, D. M. and Kupfer, D. M.,** Control of *Saccharomyces cerevisiae* 2 μm DNA replication by cell division cycle genes that control nuclear DNA replication, *J. Mol. Biol.,* 116, 249, 1977.
41. **Broach, J. R. and Hicks, J. B.,** Replication and recombination functions associated with the yeast plasmid, 2 μ circle, *Cell,* 21, 501, 1980.

42. **Jayaram, M., Ki, Y.-Y., and Broach, J. R.**, The yeast plasmid 2 μ circle encodes components required for its high copy propagation, *Cell*, 34, 95, 1983.

43. **Newlon, C. S., Devenish, R. J., Suci, P. A., and Roffis, C. J.**, Replication origins used *in vivo* in yeast, in *Initiation of DNA Replication, ICN-UCLA Symp. Mol. Cell. Biol.*, Vol. 22, Ray, D. S., Ed., Alan R. Liss, New York, 1981, 501.

44. **Zakian, V. A. and Kupfer, D. M.**, Replication and segregation of an unstable plasmid in yeast, *Plasmid*, 8, 15, 1982.

45. **Hieter, P., Mann, C., Snyder, M., and Davis, R. W.**, Mitotic stability of yeast chromosomes: a colony color assay that measures nondisjunction and chromosome loss, *Cell*, 40, 381, 1985.

46. **Koshland, D., Kent, J. C., and Hartwell, L. H.**, Genetic analysis of the mitotic transmission of mini-chromosomes, *Cell*, 40, 393, 1985.

47. **Bouton, A. H. and Smith, M.**, Fine-structure analysis of the DNA sequence requirements for autonomous replication of *Saccharomyces cerevisiae* plasmids, *Mol. Cell. Biol.*, 6, 2354, 1986.

48. **Snyder, M., Sapolsky, R. J., and Davis, R. W.**, Transcription interferes with elements important for chromosome maintenance in *Saccharomyces cerevisiae*, *Mol. Cell. Biol.*, 8, 2184, 1988.

49. **Stinchcomb, D. T., Mann, C., Selker, E., and Davis, R. W.**, DNA sequences that allow the replication and segregation of yeast chromosomes, in *Initiation of DNA Replication, ICN-UCLA Symp. Mol. Cell. Biol.*, Vol. 22, Ray, D. S., Ed., Alan R. Liss, New York, 1981, 473.

50. **Broach, J. R., Li, Y.-Y., Feldman, J., Jayaram, M., Abraham, J., Nasmyth, K. A., and Hicks, J. B.**, Localization and sequence analysis of yeast origins of DNA replication, *Cold Spring Harbor Symp. Quant. Biol.*, 47, 1165, 1982.

51. **Williamson, D. H.**, The yeast ARS element, six years on, a progress report, *Yeast*, 1, 1, 1985.

52. **Kearsey, S.**, Structural requirements for the function of a yeast chromosomal replicator, *Cell*, 37, 299, 1984.

53. **Srienc, F., Bailey, J. E., and Campbell, J.**, Effect of ARS1 mutations on chromosome stability in *Saccharomyces cerevisiae*, *Mol. Cell. Biol.*, 5, 1676, 1985.

54. **Blanc, H.**, Two modules from the hypersuppressive *rho⁻* mitochondrial DNA are required for plasmid replication in yeast, *Gene*, 30, 47, 1984.

55. **Palzkill, T. G., Oliver, S. G., and Newlon, C. S.**, DNA sequence analysis of ARS elements from chromosome III of *Saccharomyces cerevisiae*, identification of a new conserved sequence, *Nucleic Acids Res.*, 14, 6245, 1986.

56. **Celniker, S. E., Sweder, K., Srienc, F., Bailey, J. E., and Campbell, J. L.**, Deletion mutations affecting autonomously replicating sequence ARS1 of *Saccharomyces cerevisiae*, *Mol. Cell. Biol.*, 4, 2455, 1984.

57. **Strich, R., Woontner, M., and Scott, J. F.**, Mutations in ARS1 increase the rate of simple loss of plasmids in *Saccharomyces cerevisiae*, *Yeast*, 2, 169, 1986.

58. **Palzkill, T. G. and Newlon, C. S.**, A yeast replication origin consists of multiple copies of a small conserved sequence, *Cell*, 53, 441, 1988.

59. **Snyder, M., Buchman, A. R., and Davis, R. W.**, Bent DNA at a yeast autonomously replicating sequence, *Nature (London)*, 324, 87, 1986.

60. **Thoma, F. and Simpson, R. T.**, Local protein-DNA interactions may determine nucleosome positions on yeast plasmids, *Nature (London)*, 315, 250, 1985.

61. **Deb, S., De Lucia, A., Koff, A., Tsui, S., and Tegtmeyer, P.**, The adenine-thymine domain of the simian virus 40 core origin directs DNA bending and coordinately regulates DNA replication, *Mol. Cell. Biol.*, 6, 4578, 1986.

62. **Varshavsky, A. J., Sundin, O., and Bohn, M.**, A stretch of "late" SV40 viral DNA about 400 bp long which includes the origin of replication is specifically exposed in SV40 minichromosomes, *Cell*, 16, 453, 1979.

63. **Chan, C. S. M. and Tye, B.-K.**, A family of *Saccharomyces cerevisiae* repetitive autonomously replication sequences that have very similar genomic environments, *J. Mol. Biol.*, 168, 505, 1983.

64. **Chan, C. S. M. and Tye, B.-K.**, Organization of DNA sequences and replication origins at yeast telomeres, *Cell*, 33, 563, 1983.

65. **Budd, M., Gordon, C., Sitney, K., Sweder, K., and Campbell, J. L.**, Yeast DNA polymerases and ARS binding proteins, in *Cancer Cells, Eukaryotic DNA Replication*, Vol. 6, Cold Spring Harbor Symp., Cold Spring Harbor, NY, in press.

66. **Diffley, J. F. X., and Stillman, B.**, Interactions between purified cellular proteins and yeast origins of DNA replication, in *Cancer Cells, Eukaryotic DNA Replication*, Vol. 6, Cold Spring Harbor Symp., Cold Spring Harbor, NY, in press.

67. **Diffley, J. F. X. and Stillman, B.**, Purification of a yeast protein that binds to origins of DNA replication and a transcriptional silencer, *Proc. Natl. Acad. Sci. U.S.A.*, 85, 2120, 1988.

68. **Eisenberg, S. and Tye, B.-K.**, Identification of an ARS DNA-binding protein in *Saccharomyces cerevisiae*, in *DNA Replication and Recombination, UCLA Symp. Mol. Cell. Biol.*, New Series, Vol 47, McMacken, R. and Kelly, T. J., Eds., Alan R. Liss, New York, 1987, 391.

69. **Eisenberg, S., Civalier, C., and Tye, B.-K.,** Specific interaction between a *Saccharomyces cerevisiae* protein and a DNA element associated with certain autonomously replicating sequences, *Proc. Natl. Acad. Sci. U.S.A.,* 85, 743, 1988.

70. **Buchman, A. R., Kimmerly, W. J., Rine, J., and Kornberg, R. D.,** Two DNA-binding factors recognize specific sequences at silencers, upstream activating sequences, autonomously replicating sequences, and telomeres in *Saccharomyces cerevisiae, Mol. Cell. Biol.,* 8, 210, 1988.

71. **Shore, D., Stillman, D. J., Brand, A. H., and Nasmyth, K. A.,** Identification of silencer binding proteins from yeast, possible roles in SIR control and DNA replication, *EMBO J.,* 6, 461, 1987.

72. **Maine, G. T., Sinha, P., and Tye, B.-K.,** Mutants of *Saccharomyces cerevisiae* defective in the maintenance of mini-chromosomes, *Genetics,* 106, 365, 1985.

73. **Kearsey, S.,** Mutations which enhance minichromosome stability in *S. cerevisiae,* in *DNA Replication and Recombination, UCLA Symp. Mol. Cell. Biol.,* New Series, Vol. 47, McMacken, R. and Kelly, T. J., Eds., Alan R. Liss, New York, 1987, 355.

74. **Kearsey, S. E. and Edwards, J.,** Mutations that increase the mitotic stability of minichromosomes in yeast, characterization of RAR1, *Mol. Gen. Genet.,* 210, 509, 1987.

75. **Gibson, S., Surosky, R., Sinha, P., Maine, G., and Tye, B.-K.,** Complexity of the enzyme system for the initiation of DNA replication in yeast, in *DNA Replication and Recombination, UCLA Symp. Mol. Cell. Biol.,* New Series, Vol. 47, McMacken, R. and Kelly, T. J., Eds., Alan R. Liss, New York, 1987, 341.

76. **Sinha, P., Chang, V., and Tye, B.-K.,** A mutant that affects the function of autonomously replicating sequences in yeast, *J. Mol. Biol.,* 192, 805, 1986.

77. **Broker, T. and Doermann, A. H.,** Molecular and genetic recombination of bacteriophage T4, *Annu. Rev. Genet.,* 9, 213, 1975.

78. **Mortimer, R. K. and Schild, D.,** Genetic map of *Saccharomyces cerevisiae,* edition 9, *Micro. Rev.,* 49, 181, 1985.

79. **Peterson, J. B. and Ris, H.,** Electron microscopic study of the spindle and chromosome movement in the yeast *Saccharomyces cerevisiae, J. Cell Sci.,* 22, 219, 1976.

80. **Clarke, L. and Carbon, J.,** Isolation of a yeast centromere and construction of functional small circular chromosomes, *Nature (London),* 287, 504, 1980.

81. **Ratzkin, B. and Carbon, J.,** Functional expression of cloned yeast DNA in *Escherichia coli, Proc. Natl. Acad. Sci. U.S.A.,* 74, 487, 1977.

82. **Clarke, L. and Carbon, J.,** Isolation of the centromere-linked CDC10 gene by complementation in yeast, *Proc. Natl. Acad. Sci. U.S.A.,* 77, 2173, 1980.

83. **Clarke, L. and Carbon, J.,** The structure and function of yeast centromeres, *Annu. Rev. Genet.,* 19, 29, 1985.

84. **Stinchcomb, D. T., Mann, C., and Davis, R. W.,** Centromeric DNA from *Saccharomyces cerevisiae, J. Mol. Biol.,* 158, 157, 1982.

85. **Panzeri, L. and Philippsen, P.,** Centromeric DNA from chromosome VI in *Saccharomyces cerevisiae* strains, *EMBO J.,* 1, 1605, 1982.

86. **Fitzgerald-Hayes, M., Buhler, J-M., Cooper, T. G., and Carbon, J.,** Isolation and subcloning analysis of functional centromere DNA (*CEN 11*) from *Saccharomyces cerevisiae* chromosome XI, *Mol. Cell. Biol.,* 2, 1982, 82.

87. **Clarke, L. and Carbon, J.,** Genomic substitutions of centromeres in *Saccharomyces cerevisiae, Nature (London),* 305, 23, 1983.

88. **Hsiao, C-L. and Carbon, J.,** Direct selection procedure for the isolation of functional centromeric DNA, *Proc. Natl. Acad. Sci. U.S.A.,* 78, 3760, 1981.

89. **Maine, G. T., Surosky, R. T., and Tye, B.-K.,** Isolation and characterization of the centromere from chromosome V(*CEN5*) of *Saccharomyces cerevisiae, Mol. Cell. Biol.,* 4, 86, 1984.

90. **Futcher, B. and Carbon, J.,** Toxic effects of excess cloned centromeres, *Mol. Cell. Biol.,* 6, 2213, 1986.

91. **Hieter, P., Pridmore, D., Hegemann, J. H., Thomas, M., David, R. W., and Philippsen, P.,** Functional selection and analysis of yeast centromeric DNA, *Cell,* 42, 913, 1985.

92. **Silver, J. M. and Eaton, N. R.,** Functional blocks of the AD_1 and AD_2 mutants of *Saccharomyces cerevisiae, Biochem. Biophys. Res. Comm.,* 3, 301, 1969.

93. **Fitzgerald-Hayes, M., Clarke, L., and Carbon, J.,** Nucleotide sequence comparisons and functional analysis of yeast centromere DNAs, *Cell,* 29, 235, 1982.

94. **Neitz, M. and Carbon, J.,** Identification and characterization of the centromere from chromosome XIV in *Saccharomyces cerevisiae, Mol. Cell Biol.,* 5, 2887, 1985.

96. **Panzeri, L., Landonio, L., Stotz, A., and Philippsen, P.,** Role of conserved sequence elements in yeast centromere DNA, *EMBO J.,* 4, 1867, 1985.

96. **Mann, C. and Davis, R. W.,** Structure and sequence of the centromeric DNA of chromosome 4 in *Saccharomyces cerevisiae, Mol. Cell. Biol.,* 6, 241, 1986.

97. **Ng, R., Ness, J., and Carbon, J.,** Structural studies on centromeres in the yeast, *Saccharomyces cerevisiae,* in *Extrachromosomal Elements in Lower Eukaryotes,* Wickner, R. B., Hinnebusch, A., Gunsalus, I. C., Lambowitz, A. M., and Hollaender, A., Eds., Plenum Press, New York, 1986, 479.

98. **Fitzgerald-Hayes, M.,** Yeast centromeres, *Yeast,* 3, 187, 1987.
99. **Rothstein, R. J.,** One step gene disruption in yeast, *Methods Enzymol.,* 101C, 202, 1983.
100. **McGrew, J., Diehl, B., and Fitzgerald-Hayes, M.,** Single base-pair mutations in centromere element III cause aberrant chromosome segregation in *Saccharomyces cerevisiae, Mol. Cell. Biol.,* 6, 530, 1986.
101. **Gaudet, A. and Fitzgerald-Hayes, M.,** Alterations in the adenine-plus-thymine-rich region of *CEN3* affect centromere function in *Saccharomyces cerevisiae, Mol. Cell. Biol.,* 7, 68, 1987.
102. **Cumberledge, S. and Carbon, J.,** Mutational analysis of meiotic and mitotic centromere function in *Saccharomyces cerevisiae, Genetics,* 117, 203, 1987.
103. **Hegemann, J. H., Pridmore, R. D., Schneider, R., and Philippsen, P.,** Mutations in the right boundary of *Saccharomyces cerevisiae* centromere 6 lead to nonfunctional or partially functional centromeres, *Mol. Gen. Genet.,* 205, 305, 1986.
104. **Ng, R. and Carbon, J.,** Mutational and *in vitro* protein-binding studies on centromere DNA from *Saccharomyces cerevisiae, Mol. Cell. Biol.,* 7, 4522, 1987.
105. **Bloom, K. S. and Carbon, J.,** Yeast centromere DNA is in a unique and highly ordered structure in chromosomes and small circular minichromosomes, *Cell,* 29, 305, 1982.
106. **Saunders, M., Fitzgerald-Hayes, M., and Bloom, K.,** Chromatin structure of altered yeast centromeres, *Proc. Natl. Acad. Sci. U.S.A.,* 84, 175, 1988.
107. **Bram, R. J. and Kornberg, R. D.,** Isolation of a *Saccharomyces cerevisiae* centromere DNA-binding protein, its human homolog, and its possible role as a transcription factor, *Mol. Cell. Biol.,* 7, 403, 1987.
108. **Blackburn, E. H. and Szostak, J. W.,** The molecular structure of centromeres and telomeres, *Annu. Rev. Biochem.,* 53, 163, 1984.
109. **Yao, M-C.,** Amplification of ribosomal RNA genes, in *Molecular Biology of Ciliated Protozoa,* Gall, J. G., Ed., Academic Press, New York, 1986, 179.
110. **Blackburn, E. H. and Gall, J. G.,** A tandemly repeated sequence at the termini of the extrachromosomal ribosomal RNA genes in *Tetrahymena, J. Mol. Biol.,* 120, 33, 1978.
111. **Oka, Y., Shiota, S., Nakai, S., Nishida, Y., and Okubo, S.,** Inverted terminal repeat sequence in the macronuclear DNA of *Stylonychia pustulata, Gene,* 10, 301, 1980.
112. **Klobutcher, L. A., Swanton, M. T., Donini, P., and Prescott, D. M.,** All gene-sized DNA molecules in four species of hypotrichs have the same terminal sequence and an unusual 3' terminus, *Proc. Natl. Acad. Sci. U.S.A.,* 78, 3015, 1981.
113. **Pluta, A. F., Kaine, B. P., and Spear, B. B.,** The terminal organization of macronuclear DNA in *Oxytrich fallax, Nucleic Acids Res.,* 10, 8145, 1982.
114. **Klobutcher, L. A. and Prescott, D. M.,** The special case of the hypotrichs, in *Molecular Biology of Ciliated Protozoa,* Gall, J. G., Ed., Academic Press, New York, 1986, 111.
115. **Szostak, J. W. and Blackburn, E. H.,** Cloning yeast telomeres on linear plasmid vectors, *Cell,* 29, 245, 1982.
116. **Dani, G. M. and Zakian, V. A.,** Mitotic and meiotic stability of linear plasmids in yeast, *Proc. Natl. Acad. Sci. U.S.A.,* 80, 3406, 1983.
117. **Pluta, A. F., Dani, G. M., Spear, B. B., and Zakian, V. A.,** Elaboration of telomeres in yeast: recognition and modification of termini from *Oxytricha* macronuclear DNA, *Proc. Natl. Acad. Sci. U.S.A.,* 81, 1475, 1984.
118. **Walmsley, R. M., Szostak, J. W., and Petes, T. D.,** Is there left-handed DNA at the ends of yeast chromosomes?, *Nature (London),* 302, 84, 1983.
119. **Shampay, J., Szostak, J. W., and Blackburn, E. H.,** DNA sequences of telomeres maintained in yeast, *Nature (London),* 310, 154, 1984.
120. **Dunn, B., Szauter, P., Pardue, M. L., and Szostak, J. W.,** Transfer of yeast telomeres to linear plasmids by recombination, *Cell,* 39, 191, 1984.
121. **Zakian, V. A., Blanton, H. B., and Dani, G.,** Formation and stability of linear plasmids in a recombination deficient strain of yeast, *Curr. Gen.,* 9, 441, 1985.
122. **Walmsley, R. M. and Petes, T. D.,** Genetic control of chromosome length in yeast, *Proc. Natl. Acad. Sci. U.S.A.,* 82, 506, 1985.
123. **Horowitz, H., Thorburn, P., and Haber, J. E.,** Rearrangements of highly polymorphic regions near telomeres of *Saccharomyces cerevisiae, Mol. Cell. Biol.,* 4, 2509, 1984.
124. **Shampay, J. and Blackburn, E. H.,** Generation of telomere-length heterogeneity in *Saccharomyces cerevisiae, Proc. Natl. Acad. Sci. U.S.A.,* 85, 534, 1988.
125. **Lustig, A. J. and Petes, T. D.,** Identification of yeast mutants with altered telomere structure, *Proc. Natl. Acad. Sci. U.S.A.,* 83, 1398, 1986.
126. **Carson, M. J. and Hartwell, L.,** *CDC17:* an essential gene that prevents telomere elongation in yeast, *Cell,* 42, 249, 1985.
127. **Wood, J. S. and Hartwell, L. H.,** A dependent pathway of gene functions leading to chromosome segregation in *Saccharomyces cerevisiae, J. Cell Biol.,* 94, 718, 1982.

128. **Carson, M.,** *CDC17* the structural gene for DNA polymerase I of yeast: mitotic hyperrecombination and effects on telomere metabolism, Ph.D. thesis, University of Washington, Seattle, 1987.

129. **Forte, M. A. and Fangman, W. L.,** Naturally occurring cross-links in yeast chromosomal DNA, *Cell,* 8, 425, 1976.

130. **Forte, M. A. and Fangman, W. L.,** Yeast chromosomal DNA molecules have strands which are cross-linked at their termini, *Chromosoma,* 72, 131, 1979.

131. **Szostak, J. W.,** Evolutionary conservation of the structure of eucaryotic telomeres, *Rec. Adv. Yeast Mol. Biol.,* 1, 76, 1982.

132. **Button, L. L. and Astell, C. R.,** The *Saccharomyces cerevisiae* chromosome III left telomere has a type X, but not a type Y, ARS region, *Mol. Cell. Biol.,* 6, 1352, 1986.

133. **Zakian, V. A., Blanton, H. M., and Wetzel, L.,** Distribution of telomere-associated sequences in yeast, in *Extrachromosomal Elements in Lower Eukaryotes,* Wickner, R. B., Hinnebusch, A., Gunsalus, I. C., Lambowitz, A. M., and Hollaender, A., Eds., Plenum Press, New York, 1986, 493.

134. **Zakian, V. A. and Blanton, H. M.,** Distribution of telomere-associated sequences on natural chromosomes in *Saccharomyces cerevisiae, Mol. Cell. Biol.,* 8, 2257, 1988.

135. **Murray, A. W. and Szostak, J. W.,** Construction and behavior of circularly permuted and telocentric chromosomes in *Saccharomyces cerevisiae, Mol. Cell. Biol.,* 6, 3166, 1986.

136. **Walmsley, R. W., Chan, C. S. M., Tye, B.-K., and Petes, T. D.,** Unusual DNA sequences associated with the ends of yeast chromosomes, *Nature (London),* 310, 157, 1984.

137. **Watson, J. D.,** Origin of concatemeric DNA, *Nature New Biol.,* 239, 197, 1972.

138. **Kiss, G. B., Amin, A. A., and Pearlman, R. E.,** Two separate regions of the extrachromosomal ribosomal deoxyribonucleic acid of *Tetrahymena thermophila* enable autonomous replication of plasmids in *Saccharomyces cerevisiae, Mol. Cell. Biol.,* 1, 535, 1981.

139. **Amin, A. A. and Pearlman, R. E.,** Autonomously replicating sequences from the non transcribed spacers of *Tetrahymena thermophila* ribosomal DNA, *Nucleic Acids Res.,* 13, 2647, 1985.

140. **Berman, J., Tachibana, C. Y., and Tye, B.-K.,** Identification of a teleomere-binding activity from yeast, *Proc. Natl. Acad. Sci. U.S.A.,* 83, 3713, 1986.

141. **Vignais, M. L., Woudt, L. P., Wassenaar, G. M., Mager, W. H., Sentenac, A., and Planta, R. J.,** Specific binding of TUF factor to upstream activation sites of yeast ribosomal protein genes, *EMBO J.,* 6, 1451, 1987.

142. **Huet, J. and Sentenac, A.,** TUF, the yeast DNA-binding factor specific for UAS$_{rpg}$ upstream activating sequences: identification of the protein and its DNA-binding domain, *Proc. Natl. Acad. Sci. U.S.A.,* 84, 3648, 1987.

143. **Shore, D. and Nasmyth, K.,** Purification and cloning of a DNA binding protein from yeast that binds to both silencer and activator elements, *Cell,* 51, 721, 1987.

144. **Gottschling, D. E. and Zakian, V. A.,** Telomere proteins: specific recognition and protection of the natural termini of *Oxytricha* macronuclear DNA, *Cell,* 47, 195, 1986.

145. **Greider, C. W. and Blackburn, E. H.,** Identification of a specific telomere terminal transferase activity in *Tetrahymena* extracts, *Cell,* 43, 405, 1985.

146. **Greider, C. W. and Blackburn, E. H.,** The telomere terminal transferase of *Tetrahymena* is a ribonucleoprotein enzyme with two kinds of primer specificity, *Cell,* 51, 887, 1987.

147. **Heumann, J. M.,** A model for replication of the ends of linear chromosomes, *Nucleic Acid Res.,* 3, 3167, 1976.

148. **Bateman, A. J.,** Simplification of palindromic telomere theory, *Nature (London),* 253, 379, 1975.

149. **Cavalier-Smith, T.,** Palindromic base sequences and replication of eukaryotic chromosomes, *Nature (London),* 250, 467, 1974.

150. **Dancis, B. M. and Holmquist, G. P.,** Telomere replication and fusion in eukaryotes, *J. Theor. Biol.,* 78, 211, 1979.

151. **Szostak, J. W.,** Replication and resolution of telomeres in yeast, *Cold Spring Harbor Symp. Quant. Biol.,* 47, 1187, 1983.

152. **Mann, C. and Davis, R. W.,** Instability of dicentric plasmids in yeast, *Proc. Natl. Acad. Sci. U.S.A.,* 80, 228, 1983.

153. **Koshland, D., Rutledge, L., Fitzgerald-Hayes, M., and Hartwell, L.,** A genetic analysis of dicentric minichromosomes in *Saccharomyces cerevisiae, Cell,* 48, 801, 1987.

154. **Haber, J. E., Thorburn, P. C., and Rogers, D.,** Meiotic and mitotic behavior of dicentric chromosomes in *Saccharomyces cerevisiae, Genetics,* 106, 185, 1984.

155. **Haber, J. E. and Thorburn, D. C.,** Healing of broken linear dicentric chromosomes in yeast, *Genetics,* 106, 207, 1984.

156. **Surosky, R. T. and Tye, B.-K.,** Resolution of dicentric chromosomes by Ty-mediated recombination in yeast, *Genetics,* 110, 397, 1985.

157. **Murray, A. W. and Szostak, J. W.,** Construction of artificial chromosomes in yeast, *Nature (London),* 305, 189, 1983.

158. **Zakian, V. A., Blanton, H. M., Wetzel, L., and Dani, G. M.,** A size threshold for yeast chromosomes: generation of telocentric chromosomes from an unstable mini-chromosome, *Mol. Cell. Biol.,* 6, 925, 1986.

159. **Bloom, K., Amaya, E., and Yeh, E.,** Centromeric DNA structure in yeast chromatin, in *Molecular Biology of the Cytoskeleton,* Borisy, G. G., Cleveland, D. W., and Murphy, D. B., Eds., 1984, 1975.

160. **Chlebowicz-Sledziewska, E. and Sledziewski, A. Z.,** Construction of multicopy yeast plasmids with regulated centromere function, *Gene,* 39, 25, 1985.

161. **Hill, A. and Bloom, K.,** Genetic manipulation of centromere function, *Mol. Cell. Biol.,* 7, 2397, 1987.

162. **Burke, D. T., Carle, G. F., and Olson, M. V.,** Cloning of large segments of exogenous DNA into yeast by means of artificial chromosome vectors, *Science,* 236, 806, 1987.

163. **Allshire, R. C., Cranston, G., Gosden, J. R., Maule, J. C., Hastie, N. D., and Fantes, P. A.,** A fission yeast chromosome can replicate autonomously in mouse cells, *Cell,* 50, 391, 1987.

164. **Carle, G. F. and Olson, M. V.,** personal communication, 1988.

165. **McCarroll, R. and Fangman, W. L.,** Time of replication of yeast centromeres and telomeres, *Cell,* 54, 505, 1988.

166. **Thrash-Bingham, C. and Fangman, W. L.,** *Mol. Cell. Biol.,* 9, 809, 1989.

167. **Murray, A. F., Claus, T. E., and Szostak, J. W.,** Characterization of two telomeric DNA processing reactions in *Saccharomyces cerevisiae, Mol. Cell. Biol.,* 8, 4642, 1988.

168. **Pluta, A. F. and Zakian, V. A.,** *Nature (London),* 337, 429, 1989.

169. **Wellinger, R. and Zakian, V. A.,** *Proc. Natl. Acad. Sci. U.S.A.,* 86, 973, 1989.

170. **Runge, K. and Zakian, V. A.,** *Mol. Cell Biol.,* 9, 1488, 1989.

171. **Olson, M.,** personal communication, 1988.

Chapter 6

DNA REPLICATION IN HIGHER PLANTS

Sabine Heinhorst, Gordon C. Cannon, and Arthur Weissbach

TABLE OF CONTENTS

I. INTRODUCTION

The plant cell has three compartments, each of which contains its own genetic material. DNA in the plant nucleus is complexed with histones to form nucleosomes, the basic structural repeat units of chromatin. The nucleosome fiber, in turn, is folded to generate yet a higher order of DNA structure in the eukaryotic nucleus, the chromosome. The complexity of plant nuclear DNA ranges from 7×10^7 to 10^{11} base pairs (bp) and is characterized by a large amount of repeated sequences. Another feature of plant nuclear DNA is its extremely high degree of methylation. Animal nuclear DNA, on the other hand, is methylated to a lesser degree.

The second DNA complement in the plant cell is mitochondrial DNA (mtDNA). In contrast to animal mtDNA, the plant mitochondrial genome (chondriome) is quite large. Size ranges of several hundred to a few thousand kilobase pairs (kbp) have been reported for almost all species so far examined. Besides its large size, the plant chondriome is characterized by size heterogeneity which is presumably generated by recombinational events that fragment the master molecule into species of various sizes and shapes.

The organelle characteristic for plant cells, the plastid, contains a DNA species the size and structure of which is remarkably conserved among higher plants. With few exceptions, the circular plastid genome (plastome) is between 120 and 200 kbp in size and contains a large inverted repeat unit. Unlike nuclear DNA, neither chondriome nor plastid DNA (ctDNA) is methylated or complexed with histones to form nucleosomes.

Nuclear, mitochondrial, and plastid DNAs each code for components that are part of the distinct protein synthesis apparatus in the respective cellular compartments. In addition, there is an extensive sharing of genetic information between nuclear DNA and organellar genomes. Generally, multisubunit proteins that are located and function in either plastids or mitochondria are only in part encoded by the organellar genome. The genes for other subunits of these proteins are located in nuclear DNA, expressed in the cytoplasm and subsequently imported into the organelle, where they join the organelle DNA encoded subunits to form functional holoenzymes. Another form of information sharing between the three genomes in the plant cell is an interorganelle DNA exchange. Fragments of the plastome seem to be integrated at various sites in nuclear as well as mtDNA. Neither the function nor the method of DNA transfer between compartments is known.

The different genome sizes and structures in the three plant cell compartments call for different modes, mechanisms, and controls of DNA replication. We will outline in this review what is currently known about replication of nuclear, mitochondrial, and chloroplast DNA in the higher plant cell. Since the information currently available on this subject is somewhat sparse, we will, when necessary, refer to animal or algal systems to complement the data.

II. NUCLEAR DNA

A. COMPLEXITY AND PLOIDY

The amount of nuclear DNA (C-value) varies greatly between species of higher plants. Among angiosperms, more than 600-fold differences in C-values have been reported. By far the smallest nuclear genome has been found in the crucifer *Arabidopsis thaliana* with a nuclear DNA content of 0.2 pg per unreplicated haploid genome representing 7×10^7 bp, whereas most higher plants have genome sizes of 10^9 and some *Liliaceae* of up to 10^{11} bp.[1,2] While the significance of these large differences in C-values of higher plant genomes is not well understood, a correlation appears to exist between nuclear DNA content and the duration of the cell cycle.[3-5] Length of mitosis in turn influences the minimum generation time and seems to be correlated with the type of life cycle (annual or perennial) and the adaptability of a plant species to a climatic zone.[6]

The vast differences in C-values among higher plant genomes cannot be explained by differences in genetic complexity of similar proportions, but are rather due to the presence of large amounts of repetitive sequences.[7,8] Estimates of the amount of single copy DNA range from a few to 50% of the total DNA content, which would leave 50 to over 90% of the DNA as repeated sequences that are interspersed with short unique regions in various ways.[8] Flavell[8] postulates that the interspersion pattern in plant nuclear DNA might be a prerequisite for the maintenance of large amounts of repetitive DNA. Short, single-copy sequences would make repeated DNA segments unique, which would enable the cell to maintain these segments as euchromatin. Consistent with this hypothesis is the preferential association of tandem arrays of highly repeated sequences with constitutive heterochromatin. Of the single-copy DNA, only a fraction seems to be transcriptionally active.[8-10] Most of the reiterated sequences have no known regulatory function and are not transcribed.

It is clear from the facts mentioned above that there has been a tendency for extensive changes in the plant genome during evolution. More rapid genome changes occur during plant development. In addition to large differences in C-values, higher plants exhibit a wide range of ploidy levels.[1,11] Within a species, meristematic tissue is diploid, while cells in differentiating tissue tend to increase the chromosome number per nucleus without a subsequent mitosis. The extent of endopolyploidy varies between tissue and cell types.[12] It is not known which factors regulate the polyploidization process, but there is evidence that phytohormones may play a role.[13,14] The increase in ploidy of somatic cells is often characterized by a differential replication of certain portions of DNA, leading to over- and underrepresentations of certain sequences,[14] but the significance of this selectivity is not known.

Walbot and Cullis[15] postulate that the apparent great tolerance of plants for alterations in their genome is a prerequisite for adaptability of the plant to rapidly changing environmental conditions. An example supporting this hypothesis is the occurrence of new, stable phenotypes in flax after just one generation of growth under unfavorable conditions. The phenotypic changes were accompanied by changes in nuclear DNA content and amplification of most of the repetitive DNA, including rRNA and 5S RNA genes.[15]

B. CHROMATIN STRUCTURE

Nuclear DNA is tightly associated with basic proteins, histones, to form higher order structures, termed nucleosomes. In higher plants, chromatin is organized in much the same way as in animals. The 10-nm nucleosome fibers fold up to generate a higher order of chromatin packaging, the 30-nm fiber, which in turn must condense in a yet ill defined way to account for the packaging of the long DNA fibers into chromosomes. The basic repeat length (170 to 200 bp) of plant nucleosomes seems to be invariable, regardless of tissue type or developmental stage in most cases studied,[16] though there is evidence that the chromatin structure changes in *Phaseolus vulgaris* phaseolin genes are tissue specific and are related to expression of the genes in these tissues.[17] Plant and animal histones seem to be functionally identical as judged by their apparent exchangeability *in vitro* and the similarity of interactions between the histone classes in the nucleosome.[18-20] Furthermore, antibodies raised against chicken erythrocyte nucleosomes cross-react with those isolated from plants.[21] Other integral chromatin components are the nonhistone proteins, a gorup of proteins that is not very conserved evolutionarily and defined mainly by its nonhistone character. Among these proteins a subgroup, termed high-mobility group proteins, seems to be associated with transcriptionally active genes, but is not very well characterized in higher plants.[16] Certain nonhistone proteins seem to play a crucial role in maintaining the overall scaffold structure of the chromosome, as evidenced by electron microscopic examination of salt-extracted metaphase chromosomes devoid of histones.[22]

C. CELL CYCLE, SYNCHRONIZATION

The eukaryotic cell cycle can be divided into a presynthetic phase (G1), the synthetic

TABLE 1
Synchronization of the Plant Cell Cycle

Plant material	Synchronizing treatment	Arrested stage	Parameters investigated	Ref.
Haplopappus gracilis, SC	5-Aminouracil	None	Mitotic index, cytotoxicity, abnormal mitoses	26
	HU	G1/S		
	FUdR	G1/S		
Nicotiana tabacum, SC	Cytokinin		Mitotic index	27
Acer pseudoplatanus, SC	Starvation	G1 (possibly)	Mitotic index, DNA synthesis, respiration rate	28
Glycine max, SC	FUdR	G1/S	Mitotic index, DNA synthesis	29
Glycine max, SC	Ethylene/CO_2	G2 and S	Mitotic index, cell growth	25
Daucus carota, SC	Auxin	G1	Mitotic index, cell growth	30
Zea mays, SC	HU + colchicine	G1/S	Mitotic index, ploidy	31
Nicotiana tabacum, SC	Aphidicolin	G1/S	Mitotic index, transformation	32
Daucus carota, SC	HU	G1/S	Mitotic index	33
	FUdR	M		
	Fresh medium + colchicine			
Daucus carota, SC	Aphidicolin	G1/S	DNA synthesis	34
Nicotiana tabacum protoplasts	Aphidicolin	G1/S	Mitotic index, transformation	35
Nicotiana tabacum, SC	Aphidicolin	G1/S	DNA synthesis	36
Nicotiana tabacum, SC	HU	G1/S	DNA synthesis	37
	Nalidixic acid			
Datura innoxia, CC	FUdR	G1/S	Nuclear DNA content	38
Triticum monococcum,	HU	G1/S	Mitotic index	39
Petroselinum hortense, SC	HU + colchicine			
Pisum sativum, root tips	FUdR + starvation	G1/S	Mitotic index, DNA synthesis	40
Haplopappus gracilis, root tips	Aphidicolin	G1/S	Mitotic index, DNA synthesis	41

Note: SC = suspension cells, CC = callus culture, HU = hydroxyurea, FUdR = 5-fluorodeoxyuridine.

phase (S) during which the genetic complement is being replicated, a postsynthetic phase (G2), and finally mitosis (M) resulting in the formation of two daughter cells.[23] To study molecular and cellular mechanisms governing this sequence of events, such as the replication of DNA during the S phase, a synchronously growing cell population is desirable. Since there are only a limited number of naturally synchronous systems,[24] methods to induce synchronous divisions in normally asynchronous plant tissue explants or cell cultures have been developed. Generally, this can be achieved by addition of DNA synthesis inhibitors, changes in the plant growth substance complement of the medium, nutrient starvation, and temporary anaerobic conditions. Apart from some previous work on plant tissue explants, plant cell suspension cultures have usually been used as the experimental system. However, despite the obvious advantages of using a population of more or less uniform cells, Constabel et al.[25] have pointed out that besides often long generation times, sensitivity of plant cells to changes in media composition and their habit of growing in aggregates are factors limiting the complete success of this approach. Table 1 summarizes the more recent reports on synchrony in plant cell cultures. Aphidicolin emerges as the synchronizing agent of choice,[42] due to its high specificity as an inhibitor of nuclear DNA synthesis, its low toxicity, and the high degree of synchrony obtained in a variety of plant species. The tendency of some plant cell cultures to degrade aphidicolin can be counteracted by repeated additions of the drug to the medium in order to maintain the minimal inhibitory concentration.[43] Galli and Sala[41] followed the mitotic index in root tip meristems of seedlings and germinating embryos of *Happlopappus gracilis* after treatment with aphidicolin. The authors observed two successive synchronous cell cycles, reaching mitotic indices in the embryos that are close to the total number of cycling cells. In our hands, aphidicolin produced excellent synchrony in a cultured tobacco cell line, resulting in immediate onset of DNA synthesis after removal of the inhibitor from the medium.[36] In a variety of other plant cell cultures, 80 to 95% of the cycling cells were synchronized by a single aphidicolin treatment.[32,34]

D. REPLICONS

The replication of eukaryotic chromosome takes place during the S phase of the cell cycle and is characterized by the sequential activation of a large number of replication units called replicons. Bidirectional extension of replication forks from replicon origins results in merging of neighboring replicons and eventually in a completely duplicated chromosome. Autoradiography of isolated radio-labeled DNA fibers allows determination of replicon size, which is defined as the distance between the initiation points of two neighboring replicons and rate of replication.[44,45] Applied to the study of chromosome replication in higher plants, this technique has led to the compilation of replicon sizes and elongation rates for a variety of monocot and dicot species. Both parameters can vary greatly among and within a species and are dependent on several factors, such as cellular metabolism[46] and temporal position in S phase.[46,47] Among angiosperms, dicots tend to have an average replicon length of 22.2 μm, while monocots seem to have a shorter mean replicon size of 14.4 μm.[5] To determine if differences in replicon size and/or replication rates could account for the differences in length of S phase among plants with different genome sizes, Van't Hof and Bjerknes[48] and Francis, Kidd, and Bennett[5] investigated these parameters in various diploid angiosperms with large differences in C-values. The authors found no clear correlation between length of S phase and either rate of replication or replicon size and concluded that the differences in S phase length with differing C-values are due to the number of replicon families that are activated sequentially. Support for this hypothesis comes from the finding that *A. thaliana* seedlings (C-value = 0.2 pg, S phase duration 2.8 h) have two replicon families that are activated sequentially during S phase with a 36-min interval.[49] DNA fiber autoradiography data on other plant species with much larger C-values are consistent with the less synchronous replication of DNA during S phase that is indicative of a larger number of replicon families.[5,48] Though there are no clear data available on the number of replicon familes in plant species with larger genomes, Van't Hof[50] and Van't Hof and Bjerknes[48] estimated that *Pisum sativum* root meristem cells (C-value = 4.6 pg,[51] S phase duration 5 h[48]) contain a total of 4.3 × 10^4 replicons that are organized into 5 sequentially activated replicon familes.

Van't Hof[52] suggests that there is a heirarchy of control of replicon activation in the higher plant chromosomes. Neighboring replicons in a cluster tend to be replicated in synchrony. Several spatially distinct clusters form a family, the members of which are activated at more or less the same time and replicate during a distinct period within the S phase. A replicon family might in turn (possibly via gene products encoded by members of that family) regulate the replication of other replicon families, leading to an exact temporal order of replicon family activation during S phase. It is not known what sequence or conformation requirements must be met to mark a segment of DNA as an origin of replication in a replicon. An approach to characterize nuclear origins of replication has been to insert DNA fragments into a yeast plasmid unable to replicate and determine which recombinant plasmids are able to transform yeast cells at high frequencies. Cloned sequences promoting replication of the plasmid in yeast are called autonomously replicating seuqences or *ARS*. With this approach, chromosomal ARS from *Nicotiana tabacum*[53] and from *Triticum aestivum*[54] have been isolated and characterized. An *ARS* from tobacco that was sequenced contains an 11 bp consensus sequence that is essential for several yeast *ARS*, as well as several A + T-rich stretches of direct and inverted repeats that are characteristic of *ARS* and replication origins.[55] It is unresolved, however, if *ARS* function as origins of replication *in vivo*. Recently,

Van't Hof and coworkers[56,57] succeeded in identifying the replicon origin of a cluster of repeated rRNA genes from synchronized root meristem cells of *Pisum sativum*. The authors made use of the fact that single-stranded DNA molecules, such as those encountered in replicating intermediates, can be selectively retained on benzoylated naphthoylated DEAE (BND)-cellulose from total restriction-digested, pulse-labeled nuclear DNA.[58] The origin of replication was located in the nontranscribed spacer region between the 25S and 18S rRNA genes, as in other eukaryotic rRNA gene families.

Investigations with inhibitors of protein synthesis in *Allium cepa* root meristems revealed that the initiation of DNA synthesis at the beginning of the S phase, but not the subsequent continuation throughout S, is dependent on concomitant protein synthesis.[59] However, Schvartzman and Van't Hof[60] demonstrated that the joining of nascent replicons is to some extent dependent on concomitant protein synthesis in *P. sativum*. In animal and plant studies it has been shown that nuclear DNA maturation, which is the joining of nascent replicons, occurs in a stepwise manner.[52,61] Replication forks between neighboring replicons in a cluster advance to the ends of each replicon, leaving single-stranded gaps between adjacent replicons. In pea root meristem cells, these gaps are connected to form chromosome-size molecules during late S or early G2 phase of the cell cycle.[61,62]

E. ENZYMOLOGY OF DNA REPLICATION

The enzymology of eukaryotic DNA replication has been studied in lower eukaryotes (yeast) where the availability of conditional mutants defective in distinctive steps of the process are available. In mammals, the lack of such mutants has lead to the development of a few well-defined virus models (SV40, polyoma, adenovirus) that replicate the viral genome in an *in vitro* system. The advantage of these systems, which presumably replicate DNA in a fashion mimicking that in the cell nucleus, is the relative ease of manipulation, allowing the dissection of the DNA replication process with respect to protein, template, and cofactor requirements by selective removal and addition of components. In higher plants, comparable information is not available. A few, presumed replication enzymes that have been isolated and characterized have been the subject of a recent review.[63] We will add some new information, but most of the information in this part of the review will come from data on animal systems.

1. Initiation

As already pointed out above, the initiation of replication must involve the recognition of specific sequences on a replicon that constitute the origin. Localized melting at the origin to expose both DNA template strands is a prerequisite for the subsequent synthesis of daughter strands. To advance the nascent replication fork, an enzyme would have to concomitantly unwind the duplex at the front of the fork. In SV40, the only virus-encoded protein necessary for replication of the SV40 genome is the large T antigen.[64] This protein functions as an origin-specific DNA-binding protein[65,66] and a DNA dependent ATPase that catalyzes localized melting of duplex DNA.[67,68] Recently, the enzyme has been shown to selectively unwind the DNA duplex starting at the SV40 origin of replication *in vitro*.[69] The only report of DNA-dependent ATPase activity in plants comes from Hotta and Stern,[70] who found DNA unwinding activities in meiotic cells of *Lilium*. The most prominent activity was associated with a 130,000-mol wt polypeptide and paralleled closely the events of meiosis. The enzyme is believed to function in recombination events during meiosis, and a possible involvement in DNA replication was not investigated.

Single-stranded DNA-binding or helix-destabilizing proteins bind to single-stranded DNA sections, reducing the melting temperature of the duplex and maintaining the single strands in an extended conformation. It has been shown that these proteins stimulate the replicative DNA polymerase in various *in vitro* systems,[71] though their exact function in eukaryotic DNA replication is not yet clear. In higher plants, single-stranded DNA-binding proteins have been found in meiotic cells of *Lilium*.[72,73] An involvement of these proteins in DNA synthesis was not shown.

DNA polymerases of either prokaryotic or eukaryotic origins are unable to initiate a new DNA strand, but need a free 3'-OH for further addition of deoxynucleotides to the nascent chain.[74] This is particularly critical on the lagging strand of the replication fork, where new Okazaki fragments are initated when more and more DNA template becomes available as the fork advances. An enzyme that catalyzes the formation of oligoribonucleotide primers on a single-stranded DNA template, called DNA primase, has been isolated from a variety of eukaryotic sources. The enzyme is tightly associated with DNA polymerase alpha and copurifies with the latter activity through most purification steps.[75-80] Some evidence suggests that both polymerase and primase activity from calf thymus might reside on the same polypeptide.[81] During SV40 DNA replication *in vivo,* the enzyme presumably generates primers 9 to 11 nucleotides long approximately once every 135 bases, a distance which corresponds closely to the average length of Okazaki fragments in that system.[82,83] The only reports of a primase activity from higher plant sources come from Graveline et al.[84] and Marchesi et al.[85] The enzyme from wheat embryos has a molecular weight of 110,000 and was reported to be associated with a gamma-like DNA polymerase based on the resistance of primase-dependent DNA synthesis to aphidicolin in a poly(dT)-directed assay.[84] Marchesi et al.,[85] using a single-stranded bacteriophage fd DNA template, found that DNA polymerase alpha from cultured rice cells seems to elongate the RNA primers in this system. Further experiments with various synthetic templates demonstrated that the involvement of either DNA polymerase alpha or gamma in primer elongation *in vitro* depends highly on the template used. The primase isolated from rice cells has a molecular weight of 166,000 and is not associated with any DNA polymerase activity during purification. Since total cellular extracts were used to purify the enzyme from both sources, the subcellular location of the primase from plants is not yet known.

2. Chain Elongation

Upon advancement of the growing replication fork by a DNA unwinding protein, the resulting torsional stress in the molecule ahead of the fork must be released. Enzymes which change the topological state of DNA by introducing transient strand breaks are called topoisomerases. They are found in prokaryotes and eukaryotes, and depending on their reaction mechanism, i.e., introduction of single- or double-stranded breaks, are classified as class I and class II, respectively.[86] There is a variety of evidence relating the activity of topoisomerases to the DNA replication process. Recently, yeast mutants deficient in either topoisomerase I or II have been isolated, and studies with these mutants suggest an involvement of topoisomerase I in the maintenance of the chromatin structure during the cell cycle. Topoisomerase II, on the other hand, appears to play a role in the segregation of chromosomes during mitosis.[87,88] A recent report on SV40 DNA replication suggests that topoisomerase I or II can relieve the torsional stress in the replicating molecule, while only the latter enzyme is able to decatenate the daughter molecules.[89] In higher plants, a DNA topoisomerase I has been isolated from wheat germ.[90,91] The enzyme resembles closely topoisomerase I enzymes from other eukaryotic sources and appears to show some specificity for the base sequence at the breakage site. A type II topoisomerase from cauliflower inflorescences, with a molecular weight of 223,000, was characterized by Fukata et al.[92] The authors speculate that the relative abundance of the enzyme in the inflorescence might be related to rapid nuclear divisions, though no results pertaining to this suggestion were reported.

Animals contain three different DNA polymerases that have been classified as alpha, beta and gamma.[93,94] DNA polymerase beta is a low molecular weight (50,000) enzyme, located in the nucleus, where it is tightly associated with chromatin and believed to be involved in DNA repair processes. DNA polymerase gamma is found in nuclei and mitochondria and has been shown to replicate mitochondrial DNA.[95,96] The enzyme located in

TABLE 2
Plant DNA Polymerases Alpha

	Source		
	Oryza sativa[102,103] suspension cells	*Spinacia oleracea*[104] leaves	*Triticum aestivum*[105,106] embryos
Molecular weight	180,000	160,000	110,000
Divalent cation	Mg^{2+}	Mg^{2+}	Mg^{2+}
Preferred template	Activated DNA	Activated DNA	Poly[d(AT)]
Native DNA	No	No	ND
Denatured DNA	No	No	ND
NEM[a]	Sensitive	Sensitive	Sensitive
Ionic strength	Sensitive	Sensitive	Sensitive
Aphidicolin	Sensitive	Sensitive	Sensitive
ddTTP[b]	Resistant	ND	Resistant
Primase	No	ND	ND
3'-5' Exonuclease	Yes	No	Yes

Note: ND = not determined.

[a] *N*-ethylmaleimide.
[b] Dideoxythymidine triphosphate.

the nucleus, DNA polymerase alpha,[97] is the sole enzyme responsible for semidiscontinuous nuclear DNA replciation.[98,99] Additionally, a DNA polymerase delta has been described that resembles DNA polymerase alpha in many respects, but has a 3'-5' exonuclease activity and appears to be immunologically different.[100,101] The function of this enzyme is not known.

With few exceptions (Table 2), DNA polymerase alpha from plants has not been well characterized.[63,107] As judged from the available data, the enzymes from rice[102,103] and spinach[104] are similar in their template preferences, cofactor requirements, and inhibitor sensitivities to DNA polymerase alpha from animals. However, the lack of immunological similarities between the plant and mammalian enzyme[108] points to some structural differences between the two. Support for DNA polymerase alpha being the replicative enzyme in plant cells comes from experiments by Amileni et al.[102] and Sala et al.[103,109] The authors found that DNA polymerase alpha from suspension cells of *Oryza sativa, Acer pseudoplatanus, Parthenocissus tricuspidata* and *Medicago sativa* was the most abundant DNA polymerase activity in dividing cells and responded to changes in growth rates. Moreover, the partially purified DNA polymerase alpha from *O. sativa* cells is sensitive to aphidicolin, which has been shown to be a specific inhibitor of eukaryotic DNA polymerase alpha.[110,111] As visualized by autoradiography, exposure of *O. sativa* cells to the inhibitor affected incorporation of labeled thymidine into nuclei, but had no effect on organellar DNA synthesis.

It should be pointed out that, to date, there is no evidence for a close association of plant DNA polymerase alpha with a primase activity, and the evidence for the presence of a 3'-5' exonuclease activity is not fully convincing, since there is a disparity between the enzymes listed in Table 2.[102-106] The absence of an exonuclease activity in the enzyme preparation from spinach[104] might be explained by the considerably higher degree of purity compared to the enzymes from rice[102,103] and wheat.[105,106] This lack of other enzyme activities associated with DNA polymerase alpha from plant sources distinguishes the enzyme from its animal counterpart. Whether the plant enzyme is associated with a supramolecular replication enzyme complex *in vivo* remains to be seen.

A DNA polymerase activity from plant cells has been characterized that is similar to the gamma class of enzymes. This enzyme seems to be responsible for organellar DNA replication and will be discussed under those headings. To date, no convincing case can be made for the presence of a DNA polymerase beta in plant cells. Some earlier reports[63,112,113] present evidence that a low molecular weight, chromatin-bound DNA polymerase activity with some properties relating it to the beta class of enzymes is present in plant cells. Since only relatively crude preparations were used, this evidence must at best be viewed as circumstantial until a more detailed characterization of these enzymes is available.

3. Maturation of Nascent Chains

In order to allow the DNA polymerase to complete the nascent Okazaki fragments, the RNA primers have to be removed and the resulting gaps filled in the DNA polymerase. Furthermore, the last $5',3'$-phosphodiester bond between adjacent fragments of newly replicated DNA has to be sealed. There is evidence that in mammalian cells the hydrolysis of oligoribunucleotides involves ribonuclease H, an endonuclease that recognizes RNA/DNA hybrid structures, and possibly a $5'$-$3'$ exonuclease.[98] In cultured cells of *Daucus carota*,[114] the activity of RNase H increased parallel to that of DNA polymerase during the exponential growth phase of the cells. A relatively crude preparation of the enzyme was used for these experiments, and the cellular location of the enzyme as well as its involvement in DNA replication were not determined.

A polynucleotide ligase activity has been reported from a variety of plant species.[115-117] The enzyme is similar to its animal counterpart in its requirement for nicked template, ATP and Mg^{2+}, and for a sulfhydryl compound. It is not clear, from the limited data available on ligase activity in plant cells, if and how the enzyme is involved in DNA replication.

F. *IN VITRO* DNA REPLICATION SYSTEM

With few exceptions, attempts to mimic the eukaryotic DNA replication process *in vitro* have not been very successful. Cannon et al.[118] found that in soybean suspension cells that were permeabilized with lysophosphatidyl choline, nuclear DNA synthesis was completely abolished. Roman et al.[119] describe a system based on nuclei isolated from soybean suspension cell protoplasts. The purified nuclei incorporate radio-labeled TTP into deoxyribonuclease-sensitive, trichloroacetic acid precipitable material at an estimated 38% the rate of *in vivo* DNA replication. Furthermore, short replication intermediates could be chased into longer fragments, indicative of the joining of nascent fragments in the system. Experiments with UV-irradiated protoplasts, a treatment that presumably inhibits replicative synthesis and triggers the onset of repair synthesis, showed that the synthesis seen in isolated control nuclei was much higher than in irradiated ones, presumably reflecting true DNA replication in the former. Moreover, fluorodeoxyuridine-synchronized nuclei, upon release from the block, showed an increase in DNA synthesis activity that was in good agreement with that seen *in vivo*. Caboche and Lark[120] reported that the DNA synthesis observed in nuclei isolated from cultured soybean cells is due to the preferential replication of certain repetitive sequences, while others, such as the rRNA genes, are not replicated at all.

III. CHLOROPLAST DNA

A. GENOME STRUCTURE AND CODING CAPACITY

Plastids of eukaryotic algae and higher plants have been known for some time to contain genetically active DNA. The higher plant chloroplast genome exists as a covalently closed, double-stranded circular DNA molecule of approximately 120 to 200 kilobase pairs (kbp).[121] With the exception of a few legumes, nearly all higher plant plastomes are arranged as two inverted repeat units separated by a large and small unique region. The inverted repeat units have been implicated in the evolutionary stability of the plastome.[122] On the ctDNA a number of genes have been identified that code for components of the plastid's own protein synthesizing apparatus. Additionally, the plastome harbors many genes important to photosynthesis.[121,123,124] Previous estimations arrived at a coding capacity of the plastome for approximately 100 proteins and 35 to 43 RNA species. Recent advances in cloning and sequencing techniques have allowed the compilation of sequence data for the entire plastid genomes from *Marchantia polymorpha*[125] and *Nicotiana tabacum*,[126] analysis of which has disclosed 128 and 146 genes, putative genes, and open reading frames in the liverwort and tobacco, respectively.

B. INTERORGANELLE DNA EXCHANGE

As briefly mentioned in the Introduction, organellar DNA sequences are not confined to their respective cellular compartment. Mitochondria and the nucleus have been shown to contain sequences that are homologous to ctDNA,[127-133] and there is evidence for regions on the mitochondrial genome that cross-hybridize with nuclear DNA.[134] In spinach, eleven out of twelve cloned ctDNA fragments showed various degrees of homology to mtDNA.[133] Renaturation kinetics suggest that as many as four copies of the plastome are present per haploid nuclear genome in spinach. The plastome sequences seem to be integrated into nuclear DNA at a number of sites, and there is no indication of transcription products being formed.[130] The same was thought to be true for the widespread extensive homologies observed between the mitochondrial genome and the plastome. Carlson et al.,[135] however, recently reported that *in organello* radiolabeled transcription products from *Brassica napus* chloroplasts hybridize to restriction fragments from mitochondrial DNA and vice versa. The reasons for or functions of this interorganelle distribution of DNA have yet to be explained. Mechanisms that would allow this type of transfer have been proposed and include organelle specific transposons or vectors, as well as organelle fusion,[127] but to date none have been shown to function in this capacity.

C. PLASTOME COPY NUMBERS

In higher plants, each chloroplast contains multiple genome copies. Values as high as 50,000 copies of the plastome per cell have been reported for various species at different stages of tissue development.[136-141] The necessity for multiple copies of the chloroplast genome is still a mystery. In a recent review Bendich[142] has suggested that multiple genome copies might be a means of amplifying gene dosage, thereby allowing for an increased expression of genes required by the chloroplast. The author postulates that in analogy to bacterial systems the protein synthetic capability of the chloroplast is increased by elevated transcription rates of ribosomal RNAs encoded within the inverted repeats of the plastome. Bendich further suggests that the increase in plastid genome copies during leaf development provides enough plastid rRNA genes to balance nuclear and plastid protein synthesis rates. This hypothesis is consistent with our observations in cultured tobacco cells undergoing a transition from photomixotrophic to photoautotrophic conditions. Plastome copy numbers increased in response to a presumably greater demand for plastid-encoded gene products by the developing photosynthesizing cells.[140]

By using microfluorometry to quantitate amounts of 4,6-diamidino-2-phenyl indole (DAPI)-stained DNA in individual chloroplasts, Lawrence and Possingham[141] have determined that plastome levels vary in cells of developing spinach leaves. In general, the authors concluded that there is a 25-fold increase in ctDNA relative to nuclear DNA in spinach leaf cells during development and growth. However, in elongating, nondividing cells of the leaf's distal region, chloroplast division exceeds the rate of plastome replication, resulting in a concomitant decrease in chloroplast ploidy. These conclusions are complementary to other results reported for mesophyll chloroplasts from sugarbeet,[136] spinach leaf disks,[138] and pea leaves.[136] Lawrence and Possingham[141] further report that in spinach leaf epidermal cells, the number of plastomes per organelle remained constant at approximately 50 copies per chloroplast, independent of the position within the leaf. It is not clear if the differences in relative ctDNA content per organelle are important for the differentiation process of leaf tissue, but they are certainly consistent with the aforementioned suggestion[142] that amplification of chloroplast genes is required for an increased photosynthetic activity of the developing leaf cell.

D. PLASTOME REPLICATION DURING THE CELL CYCLE

The very important finding from Lawrence and Possingham's[141] and Rose et al.'s[145] studies, namely, the independence of ctDNA replication from the period of nuclear DNA synthesis during the S phase of the cell cycle, is in agreement with our *in vivo* studies with tobacco suspension cells[36,37] and those reported by Sala et al.[109] for rice cells. Following exposure of cultured tobacco cells to inhibitors of DNA synthesis, plastome replication rates were measured by quantitative molecular hybridization. Chloroplast DNA replication continued in the absence of nuclear DNA synthesis, which had been completely inhibited by aphidicolin, a potent inhibitor of the nuclear replicative DNA polymerase alpha. Removal of the drug from the culture medium resulted in a synchronous start of nuclear DNA replication, while the rate of ctDNA synthesis remained constant.[36] Similar results were obtained with the inhibitors nalidixic acid and hydroxyurea,[37] emphasizing again the independence of nuclear and plastid genome replication.

E. CONTROL OF CHLOROPLAST DNA SYNTHESIS BY THE NUCLEUS

Despite the presence of a large number of genes on the plastome, it is estimated that more than 90% of all chloroplast proteins are nuclear encoded, synthesized in the cytoplasm and transported into the organelle post-translationally.[144] It is notable that there is extensive sharing between nucleus and chloroplast in genetic information content for multisubunit chloroplast proteins. Probably the most widely known example is the enzyme ribulose-1,5-bisphosphate carboxylase, the small subunit of which is nuclear encoded and imported into the organelle, while the gene for the large subunit of the enzyme resides on the plastome. It is evident that the separation of genes for protein subunits between organelle and nucleus provides the latter with a control mechanism for organelle associated processes. One can envision that one such process would be organelle DNA replication, a control of which by the nucleus would guarantee a plastome copy number balanced with the cell's needs. Findings of Leonard and Rose[146] indicate a dependence of both nuclear and chloroplast division on nuclear encoded proteins in spinach leaf disks. Interestingly, inhibition of organellar protein synthesis had no effect on the replication of either organelle or nucleus. This interplay between chloroplast and nucleus on the level of protein synthesis was shown to occur in cultured tobacco cells as well.[36] Nuclear DNA replication declined rapidly to 2 to 3% of its normal rate after the addition of cycloheximide to the culture medium. In contrast, ctDNA synthesis continued with only a minor decrease for 1.5 h after addition of the potent inhibitor of cytoplasmic protein synthesis, at which point it demonstrated a dramatic decrease. Addition of chloramphenicol to the culture had no significant effect on either nuclear or plastid DNA replication, although the drug clearly inhibited organellar protein synthesis.

Further evidence for the exclusive nuclear location of genes required for ctDNA replication comes from experiments with plants whose plastids lack functional ribosomes.[147-149] The presence of plastid DNA has been established in the albino portions of maize[147] and barley[149] leaves, which are known to be deficient in plastid protein synthesis, as well as in heat-treated leaves of rye,[148] where chloroplast ribosome function is impaired. Quantitative studies of ctDNA levels in the barley albino mutant "albostrians" demonstrated that plastome-encoded proteins are not necessary to maintain normal plastome replication.[149] The best explanation of these findings is that all proteins required by the chloroplast to replicate its genome are encoded on nuclear DNA, synthesized in the cytoplasm, and imported into the organelle. From the above mentioned evidence it is apparent that the nucleus has the potential to control the level of organelle genomes, which in turn may have a significant influence on gene expression in the organelle itself.

F. REPLICATION MECHANISM, REPLICATION ORIGINS

As already pointed out, various viral systems have served as models for nuclear DNA replication in animals. Likewise, the rather small (14 kbp) genomes of mitochondria from mammals has facilitated the dissection of the molecular machinery required for DNA replication in that organelle. In contrast, molecular studies on ctDNA replication have been scarce. Electron microscopic observations of replicating ctDNA from pea and maize have indicated that replication of the circular molecule is initiated by the introduction of two displacement (D-) loops on opposite parental strands 7.1 kbp apart.[150] The D-loops are extended past each other to form a Cairns-type replicative intermediate. In addition, structures consistent with the rolling circle model of DNA replication have been observed.[151]

While studies of this nature are extremely difficult to perform due to the low abundance of replicative intermediates that can be found in higher plant ctDNA, this approach has been successful in locating ctDNA replication origins in two algae after synchronization of the cultures by light-dark regimes. In *Euglena* the putative origin was found to be located within a polymorphic region on the plastome upstream from the 5' end of the 16S rRNA gene.[152,153] Sequence determination of this (A + T)-rich region revealed possibilities for a number of stem and loop conformations, which are believed to be a characteristic structural element of DNA replication origins.[154] In *Chlamydomonas,* two positions about 6.5 kbp apart (10 and 16.5 kbp upstream of the 5' end of a 16S rRNA gene, respectively) were observed to be origins of D-loop elongation.[155] One of the fragments harboring a D-loop initiation site on the *Chlamydomonas* plastome was shown to hybridize to one of several ctDNA fragments promoting autonomous replication of a yeast plasmid in *Chlamydomonas.*[156] Vallet and Rochaix[157] demonstrated, however, that this ARS and the origin of replication were 1.5 kbp apart. As in *Euglena,* the replication origin of *Chlamydomonas* ctDNA is A + T rich and shows potential for cruciform structures.[158,159] Chloroplast DNA fragments exhibiting autonomously replicating characteristics as well as putative origins (based on homology with algal sequences) have been identified in higher plants as well.[53,160-163] These sequences have structural features (A + T rich, putative stem and loop formations) similar to their algal and yeast counterparts, but their roles as true replication origins *in vivo* have not yet been determined.

Plastome deletion mutants that were derived from pollen cultures of wheat and barley have shown that at least 70% of the plastid genome is not necessary for its replication, since the truncated molecules appeared to be replicated. The area of the plastome common to all replication mutants examined is a portion of the large, single-copy region.[164-166]

G. ENZYMOLOGY OF DNA REPLICATION
1. DNA Polymerase

DNA polymerase activity in chloroplasts was first reported by Tewari and Wildman.[167] Since then a DNA polymerase activity from spinach chloroplasts has been extensively characterized,[168] and the enzyme from pea plastids has been purified to apparent homogeneity by McKown and Tewari[169] (Table 3). Both enzymes fall into the gamma class of DNA polymerases according to the criteria established by Weissbach et al.[93] for the classification of eukaryotic DNA polymerases. The chloroplast enzyme is remarkably insensitive to 2',3'-dideoxynucleotides, which is in variance to the behavior of animal gamma DNA polymerases.[95,96,170] Preliminary, unpublished results from our lab show that chloroplast DNA polymerase isolated from cultured soybean cells has similar properties to the enzymes from pea and spinach chloroplasts. The major difference between the enzymes from spinach[168] and pea[169] is the lack of preference for a poly(rA)·(dT)$_{12-18}$ template by the latter. The assay

TABLE 3
Chloroplast DNA Polymerases

Source	Pisum sativum[169]	Spinacea oleracea[168]
Molecular weight	105,000	90,000
Divalent cation	Mg^{2+}	Mn^{2+}
KCl optimum (mM)	120	100—120
Preferred template	Activated DNA[a]	Poly(rA)·dT$_{12-18}$[b]
Aphidicolin	Resistant	Resistant
ddTTP	ND	Resistant

Note: ND = not determined.

[a] Assays were performed in the presence of Mg^{2+}.
[b] Assays were performed in the presence of Mn^{2+}.

in this case was performed in the presence of Mg^{2+} rather than Mn^{2+}, under which conditions the enzyme from spinach[168] and soybean cells[225] fail to demonstrate a preference for the primed ribohomopolymer as well. Whether there is more than one DNA polymerase activity in the chloroplast, as suggested by a previous report,[171] has yet to be established. Since the enzyme from pea chloroplasts has been shown to catalyze "replication-like" synthesis when provided with accessory enzymes and a proper template,[172] this enzyme seems likely to be the replicative enzyme in chloroplasts.

2. Other Proteins Involved in DNA Replication

To date, there are no data available concerning the involvement of specific proteins in chloroplast DNA synthesis. However, in analogy to the requirements for the replication of prokaryotic and eukaryotic circular plasmid and virus DNAs,[74] and in light of the basic similarities between eukaryotic and prokaryotic replication mechanisms, one can predict that more or less the same enzymatic activities as described for nuclear DNA replication will be necessary to replicate the plastome.

Siedlecki et al.[173] reported the purification and characterization of a DNA topoisomerase I from spinach chloroplasts. The enzyme has a molecular weight of approximately 115,000 and is of the prokaryotic type. Lam and Chua[174] present evidence for the additional presence of a novobiocin-sensitive topoisomerase II, similar to *Escherichia coli* gyrase, in chloroplast extracts from pea. Although both enzymes seem to modulate the transcriptional activity of chloroplast genes *in vitro*,[174] their involvement in plastome replication was not investigated. It is, however, possible to think of roles for chloroplast topoisomerases similar to those detected in the SV40 DNA replication system.[89]

H. *IN VITRO* REPLICATION SYSTEMS

Attempts to examine the mechanisms of ctDNA replication *in vitro* have produced relatively limited success. *In vitro* incorporation of radiolabeled deoxynucleotides into the DNA of isolated chloroplasts has been reported by several groups.[175-179] Analysis of the products resulting from *in organello* DNA synthesis, when performed, was unable to distinguish if true replication had occured in contrast to artifactual "fill-in" synthesis. *In vitro* replication assays with chloroplast extracts have been reported for higher plants[172,176,179,180] as well as for *Chlamydomonas*.[159] Use of chloroplast extracts to promote DNA synthesis on chimeric plasmid templates containing cloned ctDNA fragments from maize demonstrated that only a small portion of the maize chloroplast genome was available for use as templates, and site specific initiation could not be demonstrated.[176] Similar experiments with chloroplast extracts from *Marchantia polymorpha*[179] exhibited preferential synthesis on several plasmid templates containing ctDNA fragments, but it was unclear whether replication or repair-type synthesis was taking place. De Haas et al.[180] demonstrated site-specific initiation on one of two cloned *Petunia* chloroplast ARS. Perhaps the most convincing studies to date have been conducted with a partially purified DNA polymerase fraction from pea chloroplasts using cloned maize plastome fragments as templates,[172] and with cell free extracts from *Chlamydomonas*[159] using chimeric plasmid templates that contain one of the D-loop initiation sites. In both systems, site-specific initiation was demonstrated by clearly showing preferential use of plasmid templates containing a certain segment of the chloroplast genome. The region of the maize plastome that appeared to act as a replication origin *in vitro* was 67% homologous to the *Chlamydomonas* ctDNA fragment that was identified to carry an origin of replication. Both are located within the ribosomal protein L16 coding region on the respective plastome.[172,181]

I. ROLE OF PLASTID MEMBRANES

Chloroplast DNA is located in distinct areas, termed nucleoids, in the chloroplast. In *Beta vulgaris* it has been estimated that 4 to 8 copies of chloroplast DNA constitute a nucleoid.[182] Electron microscope autoradiography of spinach chloroplast sections derived from leaf disks previously radiolabeled with thymidine indicate that the chloroplast DNA is associated with the thylakoid system. Grana lamellae in particular showed a high density of silver grains in addition to the low electron density areas believed to contain the DNA fibrils.[183] An extension of these studies with isolated thylakoid vesicles confirmed the association of chloroplast DNA with grana lamellae.[184] In wheat, barley, and oat, on the other hand, the plastid genome seems to be bound to the chloroplast envelope.[149,185,186] It has been suggested that the attachment of chloroplast DNA to the membrane might be important for the segregation of DNA molecules to the daughter compartments upon division of the organelle.[183,187] Though there is no direct evidence for this suggestion, recent electron and light microscopic observations on developing etioplasts from *Phaseolus vulgaris* support an involvement of thylakoids in the partitioning of chloroplast DNA nucleoids[188] in the organelle.

IV. MITOCHONDRIAL DNA

A. GENOME STRUCTURE

In contrast to the small (14 to 18 kbp) mammalian mtDNA, higher plant chondriomes are large and heterogeneous in size.[189-191] Based on restriction enzyme analysis and/or reassociation kinetics, values between 200 and 2500 kbp have been reported for various species, and large differences in mtDNA sizes are found even among the members of a taxonomic family. Chondriomes of *Cucurbitaceae,* for example, vary eightfold in size.[192] As shown by reassociation kinetics, the plant chondriome for the most part is not composed of highly reiterated sequences.[191,193] Where present, highly repetitive sequences do not represent more than 10% of the entire chondriome.[192,194]

Structural investigations of the mitochondrial genome of higher plants have revealed a complicated pattern of DNA forms and sizes. Electron microscopic observations are consistent with the existence of predominantly linear molecules of heterogeneous lengths, as well as minor circular species of varying size classes. In addition, complex restriction patterns with varying fragment stoichiometries are commonly observed that make it difficult to determine the *in vivo* size and shape of the higher plant chondriome.[191] A hypothesis consistent with the observed phenomena is the coexistence of various DNA species that arise by inter- and intramolecular recombinational events. The establishment of complete restriction maps for the chondriomes of *Brassica campestris,*[193] *B. hirta,*[196] *Zea mays,*[195] and spinach[133] has lent support to this hypothesis. In turnip and spinach, a tripartite genome consisting of a circular master molecule and two smaller subgenomic circular DNA species have been found. The latter arise from the master molecule by intragenomic recombination, promoted by two short direct repeat sequences on its DNA.[133,193] In maize, the chondriome appears to be multipartite, generated by recombination events at several distinct sites,[195] while the mtDNA from *B. hirta* consists of a single circular species.[196] In addition to these high molecular weight DNA species, plant mitochondria contain a variety of small, apparently independently replicating DNA and RNA components.[189-191]

While mammalian mtDNA encodes only a limited number of genes and exhibits an extremely economical gene arrangement, plant mitochondrial DNA, due to its larger size, has theoretically a much larger coding capacity. However, protein synthesis performed *in organello* revealed that the number of polypeptides synthesized by isolated plant mitochondria is within the range of those in animal and fungal mitochondria, and a large part of the plant chondriome coding capacity is as yet unaccounted for.[197,198] The genes so far identified on the plant chondriome, with few exceptions, are the same as those from other sources and include subunits of cytochrome oxidase, ATPase, and other electron transport components. In addition, genes for the mitochondrion-specific translation apparatus, namely ribosomal RNAs and a presumably full complement of transfer RNAs, are encoded by mtDNA.[199] Like chloroplast proteins, the majority of the mitochondrial proteins are nuclear encoded, cytoplasmically synthesized as precursors, and imported into the organelle post-translationally.[200] It is assumed that, at least in the case of animal mitochondria, all gene products necessary for the replication of the chondriome are nuclear encoded. In this context, the recent findings by Clayton's group that two mitochondrial replication proteins are associated with nuclear-encoded RNA moieties[201,202] is of special interest, since it implies transport of RNA across the mitochondrial membranes, which so far had been thought not to occur.

B. CHONDRIOME COPY NUMBERS

While there is quite an array of publications regarding plastome copy numbers during development and in different tissue types of the plant, comparable information on mitochondrial DNA is almost nonexistent. Lamppa and Bendich[203] followed the mtDNA content in *Pisum sativum* during development. In contrast to the large increase in ctDNA levels observed during leaf tissue development, the copy numbers of the chondriome decrease slightly from 410 copies per cell in embryos to around 220 to 260 copies in leaves. The amount of mtDNA per cell stayed remarkably constant in young and mature leaves, as well as in young, dark- and light-grown leaves. In roots, low copy numbers of both plastome and chondriome (130 per cell) were found. Within the *Cucurbitaceae,* similar mitochondrial genome copy numbers (100 to 300 per cell) were estimated,[192,204] corresponding to a few to less than one genome copy number per organelle as determined by electron microscopic studies on serially sectioned tissues.

C. REPLICATION MECHANISM

In contrast to the progress made on the mechanism of DNA replication in mammalian mitochondria, nothing is known about this process in higher plants. In mammals, mitochondrial DNA replication starts at a noncoding area, the heavy strand replication origin, with the formation of a D-loop that is extended unidirectionally. Only after the heavy strand has been replicated approximately 60% does light strand replication proceed from a separate origin far away on the circular genome and initiated on a single stranded template.[205]

D. ENZYMOLOGY OF DNA REPLICATION

It has been shown that priming at the heavy strand origin of mammalian mitochondrial DNA synthesis is provided by RNA transcripts initiated from the light strand promoter.[206] The primary transcripts are subsequently cleaved at short, conserved sequences on the template strand[207] by an RNA processing activity that contains a small, nuclear-encoded RNA species required for its activity.[202,208] Priming on the single-stranded L-strand origin, on the other hand, is mediated by a primase activity.[209] The enzyme, though associated with DNA polymerase gamma, could be separated from the latter by glycerol gradient centrifugation.[210] The activity of the enzyme appears to be dependent on associated RNA species, one of which has been identified as the cytoplasmic 5.8S ribosomal RNA.[201] No enzyme activity providing a primer for plant mitochondrial DNA synthesis has been reported.

Mitochondrial topoisomerase I has been described from human leukemia cells[211] and *Xenopus laevis* oocytes.[212] The human enzyme was shown to be sensitive to ATP.[213] Indirect evidence supports the participation of a DNA topoisomerase II in the replication of mammalian mitochondrial DNA.[214,215] Echeverria et al.[216] found a topoisomerase I in wheat embryo mitochondria. The enzyme has a molecular weight of 110,000 and is not affected by ATP. Interestingly, the enzyme displays a feature of eukaryotic topoisomerases I, namely, the ability to relax positively supercoiled DNA, and by several biochemical criteria is indistinguishable from an enzyme isolated from wheat germ cytosol.[90,91] An involvement of the enzyme from wheat germ mitochondria in mtDNA replication was not demonstrated.

As already mentioned above, the enzyme responsible for the replication of the animal chondriome is DNA polymerase gamma. While the enzyme has been isolated and characterized from a variety of eukaryotic sources,[94] reports on plant mitochondrial DNA polymerase are scarce. Castroviejo et al.[217] and Christophe et al.[218] isolated and characterized a DNA polymerase from purified wheat germ mitochondria that is resitant to *N*-ethylmaleimide and prefers the template poly(dA·dT) in the presence of Mg^{2+}. The enzyme is resistant to aphidicolin but sensitive to dideoxyTTP. Though wheat embryo mitochondrial DNA polymerase has some properties similar to DNA polymerase gamma from animal sources, the authors feel that it is sufficiently different to exclude it from this class of polymerases.[218] A mitochondrial DNA polymerase isolated from cultured soybean cells was indistinguishable by a variety of biochemical criteria from its chloroplast counterpart and can be classified as a gamma type enzyme.[225] *In vivo* studies with aphidicolin, an inhibitor of replicative nuclear DNA synthesis,[109] and ethidium bromide, which has been shown to affect organellar DNA synthesis,[219-221] demonstrated that a DNA polymerase other than DNA polymerase alpha was involved in organelle DNA replication in cultured rice cells[109] and isolated wheat embryo mitochondria.[222]

E. *IN VITRO* REPLICATION SYSTEMS

In organello DNA synthesis systems have been described for purified mitochondria from maize seedlings[223,224] and wheat embryos.[222] Though all three systems incorporated radiolabeled precursors into DNase-sensitive, acid-precipitable material, none supported specific initiation at presumed origins of replication. Preliminary results in one of the maize systems, however, suggest that site-speciific initiation in a manner comparable to heavy-strand initiation in mammalian mtDNA may have taken place.[224] Upon denaturing gel electrophoresis of labeled reaction products, small, discrete DNA species were found that could possibly represent D-loop DNA fragments, but were not characterized further.

V. CONCLUDING REMARKS

Some aspects of the rather broad topic "DNA synthesis in plant cells" have been reviewed here in depth, while others have been only briefly mentioned or left out altogether. While in part reflecting the authors' biases, this distinction has sometimes been necessitated by the dearth of information in some areas. Research concerned with the molecular basis of nuclear and organellar DNA replication in higher plants began comparatively recently, so that our knowledge at present lags far behind what is known about these processes in animal cells. Furthermore, there are at present no good virus model systems for studies of DNA synthesis in plants, analogous to SV40, adenovirus, and polyomavirus in animals. Although the general mechanisms of nuclear DNA replication appear to be similar in plants and animals, the fact that to date there is no evidence in plants for high molecular weight complexes of replication enzymes suggests that there may be fundamental differences between the two processes. The present lack of data concerning plant mitochondiral DNA synthesis does not allow any comparison to its animal counterpart. It seems, however, unlikely that the large plant chondriome of a few hundred kbp should be replicated in the same manner as the small mammalian mtDNA. Last, but not least, the sequence of the genetic entity unique to plants, the plastome, has recently been determined for two species. Despite the progress made on the elucidation of its coding potential, replication of chloroplast DNA has proven difficult to study, and our current knowledge of this process is quite sparse. The ongoing efforts of developing suitable vectors to modify plant nuclear and organellar DNA via foreign DNA transfer will undoubtedly lead to a better understanding of the molecular mechanisms and cellular controls governing DNA replication in the plant cell.

REFERENCES

1. **Bennett, M. D. and Smith, J. B.,** Nuclear DNA amounts in Angiosperms, *Philos. Trans. R. Soc. Lond.* B, 274, 227, 1976.
2. **Leutwiler, L. S., Hough-Evans, B. R., and Meyerowitz, E. M.,** The DNA of *Abrabidopsis thaliana, Mol. Gen. Genet.*, 194, 14, 1984.
3. **Van't Hof, J. and Sparrow, A. H.,** A relationship between DNA content, nuclear volume and minimum mitotic cycle time, *Proc. Natl. Acad. Sci. U.S.A.*, 49, 897, 1963.
4. **Van't Hof, J.,** Relationship between mitotic cycle duration, S period duration and the average rate of DNA synthesis in the root meristem cells of several plants, *Exp. Cell Res.*, 39, 48, 1965.
5. **Francis, D., Kidd, A. D., and Bennett, M. D.,** DNA replication in relation to C values, in *The Cell Division Cycle in Plants,* Bryant, J. A. and Francis, D., Eds., Cambridge University Press, Cambridge, 1985, 61.
6. **Moore, P. D.,** Nuclear DNA content as a guide to plant growth rate, *Nature (London)*, 318, 412, 1985.
7. **Walbot, V. and Golberg, R. B.,** Plant genome organization and its relationship to classical plant genetics, in *Nucleic Acids in Plants,* Vol. 1, Hall, T. C. and Davies, J. W., Eds., CRC Press, Boca Raton, FL, 1979, 3.
8. **Flavell, R.,** The molecular characterization and organization of plant chromosomal DNA sequences, *Annu. Rev. Plant Physiol.*, 31, 569, 1980.
9. **Goldberg, R. B., Hoschek, G., Kamalay, J. C., and Timberlake, W. C.,** Sequence complexity of nuclear and polysomal RNA in leaves of the tobacco plant, *Cell,* 14, 123, 1978.
10. **Kiper, M., Bartels, D., Herzfeld, F., and Richter, G.,** The expression of a plant genome in hnRNA and mRNA, *Nucleic Acids Res.*, 6, 1961, 1979.
11. **Nagl, W.,** DNA endoreduplication and polyteny understood as evolutionary strategies, *Nature (London)*, 261, 614, 1976.
12. **Evans, L. S. and Van't Hof, J.,** Is polyploidy necessary for tissue differentiation in higher plants?, *Am. J. Bot.*, 62, 1060, 1975.
13. **Libbenga, K. R. and Torrey, J. G.,** Hormone-induced endoreduplication prior to mitosis in cultured pea root cortex cells, *Am. J. Bot.*, 690, 293, 1973.
14. **Nagl, W., Phol, J., and Radler, A.,** The DNA endoreduplication cycle, in *The Cell Division Cycle in Plants,* Bryant, J. A. and Francis, D., Eds., Cambridge University Press, Cambridge, 1985, 217.
15. **Walbot, V. and Cullis, C. A.,** Rapid genome changes in higher plants, *Annu. Rev. Plant Physiol.*, 36, 367, 1985.
16. **Spiker, S.,** Plant chromatin structure, *Annu. Rev. Plant Physiol.*, 36, 235, 1985.
17. **Myrray, M. G. and Kennard, W. C.,** Altered chromatin conformation of the higher plant gene phaseolin, *Biochemistry,* 23, 4225, 1984.
18. **Spiker, S. and Isenberg, I.,** Cross-complexing patterns of plant histones, *Biochemistry,* 16, 1819, 1977.
19. **Spiker, S. and Isenberg, I.,** Evolutionary conservation of histone-histone binding sties: evidence from interkingdom complex formation, *Cold Spring Harbor Symp. Quant. Biol.*, 42, 157, 1978.

20. **Liberati-Langenbuch, J., Wilhelm, M. L., Gigot, C., and Wilhelm, F. X.,** Plant and animal histones are completely intercangeable in the nucleosome core, *Biochem. Biophys. Res. Commun.,* 94, 1161, 1980.
21. **Einck, L., Dibble, R., Frado, L. L. Y., and Woodcock, C. L. F.,** Nucleosomes as antigens. Characterization of determinants and cross-reactivity, *Exp. Cell Res.,* 139, 101, 1982.
22. **Hadlaczky, G.,** Structure of metaphase chromosomes of plants, *Int. Rev. Cytol.,* 94, 57, 1985.
23. **Howard, A. and Pelc, S. R.,** Synthesis of deoxyribonucleic acid in normal and irradiated cells and its relation to chromosome breakage, *Heredity* (Suppl.), 6, 216, 1953.
24. **Hotta, Y. and Stern, H. S.,** Inducibility of thymidine kinase by thymidine as a function of interphase stage, *J. Cell. Biol.,* 25, 99, 1965.
25. **Constabel, F., Kurz, W. G. W., Chatson, K. B., and Kirkpatrick, J. W.,** Partial synchrony in soybean cell suspension cultures induced by ethylene, *Exp. Cell Res.,* 105, 263, 1977.
26. **Eriksson, T.,** Partial synchronization of cell division in suspension cultures of *Haplopappus gracilis, Physiol. Plant.,* 19, 900, 1966.
27. **Jouanneau, J. P.,** Controle par les cytokinines de la synchronization des mitoses dans les cellules de tabac, *Exp. Cell. Res.,* 67, 329, 1971.
28. **King, P. J., Cox, B. J., Fowler, M. W., and Street, H. E.,** Metabolic events in synchronized cell cultures of *Acer pseudoplatanus* L., *Planta,* 117, 109, 1974.
29. **Chu, Y. and Lark, K. G.,** Cell-cycle parameters of soybean (*Glycine max* L.) cells growing in suspension culture: suitability of the system for genetic studies, *Planta,* 132, 259, 1976.
30. **Nishi, A., Kato, K., Takahashi, M., and Yoshida, R.,** Partial synchronization of carrot cell culture by auxin deprivation, *Physiol. Plant.,* 39, 9, 1977.
31. **Mi, C. C., Wang, A. S., and Phillips, R. L.,** Partial synchronization of maize cells in liquid suspension culture, *Maize Genet. Coop. News Lett.,* 56, 142, 1982.
32. **Nagata, T., Okada, K., and Takebe, I.,** Mitotic protoplasts and their infection with tobacco mosaic virus RNA encapsulated in liposomes, *Plant Cell Rep.,* 1, 250, 1982.
33. **Matthews, B. F.,** Isolation of mitotic chromosomes from partially synchronized carrot *(D. carota)* cell suspension cultures, *Plant Science Lett.,* 31, 165, 1983.
34. **Sala, F., Galli, M. G., Nielsen, E., Magnien, E., Devreux, M., Pedrali-Noy, G., and Spadari, S.,** Synchronization of nuclear DNA synthesis in cultured *Daucus carota* L. cells by aphidicolin, *FEBS Lett.,* 153, 204, 1983.
35. **Meyer, P., Walgenbach, E., Bussmann, K., Hombrecher, G., and Saedler, H.,** Synchronized tobacco protoplasts are efficiently transformed by DNA, *Mol. Gen. Genet.,* 201, 513, 1985.
36. **Heinhorst, S., Cannon, G. C., and Weissbach, A.,** Plastid and nuclear DNA synthesis are not coupled in suspension cells of *Nicotiana tabacum, Plant Mol. Biol.,* 4, 3, 1985.
37. **Heinhorst, S., Cannon, G. C., and Weissbach, A.,** Chloroplast DNA synthesis during the cell cycle in cultured cells of *Nicotiana tabacum:* inhibition by hydroxyurea and nalidixic acid, *Arch. Biochem. Biophys.,* 239, 475, 1985.
38. **Blaschke, J. R., Forche, E., and Neumann, K.-H.,** Investigations on the cell cycle of haploid and diploid tissue cultures of *Datura innoxia* Mill. and its synchronization, *Planta,* 144, 7, 1978.
39. **Szabados, L., Hadlaczky, G., and Dudits, D.,** Uptake of isolated plant chromosomes by plant protoplasts, *Planta,* 151, 141, 1981.
40. **Kovacs, C. and Van't Hof, J.,** Synchronization of a proliferative population in a cultured plant tissue, *J. Cell Biol.,* 47, 536, 1970.
41. **Galli, M. G. and Sala, F.,** Aphidicolin as synchronizing agent in root tip meristems of *Haplopappus gracilis, Plant Cell Rep.,* 2, 156, 1983.
42. **Sala, F., Galli, M. G., Pedrali-Noy, G., and Spadari, S.,** Synchronization of plant cells in culture and in meristems by aphidicolin, in *Methods in Enzymology,* Vol. 118, Weissbach, A., and Weissbach, H., Eds., Academic Press, Orlando, 1986, 87.
43. **Sala, F., Sala, C., Galli, M. G., Nielsen, E., Pedrali-Noy, G., and Spadari, S.,** Inactivation of aphidicolin by plant cells, *Plant Cell Rep.,* 2, 265, 1983.
44. **Huberman, J. A. and Riggs, A. D.,** Autoradiography of chromosomal DNA fibers from Chinese hamster cells, *Proc. Natl. Acad. Sci. U.S.A.,* 55, 599, 1966.
45. **Huberman, J. A. and Riggs, A. D.,** On the mechanisms of DNA replication in mammalian chromosomes, *J. Mol. Biol.,* 32, 327, 1968.
46. **Van't Hof, J.,** DNA fiber replication of chromosomes of pea root cells terminating in S, *Exp. Cell Res.,* 99, 47, 1976.
47. **Van't Hof, J.,** Replicon size and rate of fork movement in early S of higher plant cells *(Pisum sativum), Exp. Cell Res.,* 103, 395, 1976.
48. **Van't Hof, J. and Bjerknes, C. A.,** Similar replicon properties of higher plant cells with different S periods and genome sizes, *Exp. Cell Res.,* 136, 461, 1981.
49. **Van't Hof, J., Kuniyuki, A., and Bjerknes, C. A.,** The size and number of replicon families of chromosomal DNA of *Arabidopsis thaliana, Chromosoma,* 68, 269, 1978.

50. **Van't Hof, J.,** DNA fiber replication in chromosomes of a higher plant *(Pisum sativum), Exp. Cell Res.,* 93, 95, 1975.

51. **Murray, M. G., Cuellar, R. E., and Thompson, W. F.,** DNA sequence organization in the pea genome, *Biochemistry,* 17, 5781, 1978.

52. **Van't Hof, J.,** Control points within the cell cycle, in *The Cell Division Cycle in Plants,* Bryant, J. A. and Francis, D., Eds., Cambridge University Press, Cambridge, 1985, 1.

53. **Uchimiya, H., Ohtani, T., Ohgawara, T., Harada, H., Sugita, M., and Sugiura, M.,** Molecular cloning of tobacco chromosomal and chloroplast DNA segments capable of replication in yeast, *Mol. Gen. Genet.,* 192, 1, 1983.

54. **Andre, D., Jacquemin, J. M., and Masson, P.,** Isolation and characterization of DNA sequences from *Triticum aestivum* which function as origins of replication in *Saccharomyces cerevisiae, Plant Cell Rep.,* 2, 175, 1983.

55. **Ohtani, T., Kiyokawa, S., Ohgawara, T., Harada, H., and Uchimiya, H.,** Nucleotide sequences and stability of a *Nicotiana* nuclear DNA segment possessing autonomously replicating ability in yeast, *Plant Mol. Biol.,* 5, 35, 1985.

56. **Van't Hof, J., Lamm, S. S., and Bjerknes, C. A.,** Detection of replication initiation by a replicon family in DNA of synchronized pea *(Pisum sativum)* root cells using benzoylated naphthoylated DEAE-cellulose chromatography, *Plant Mol. Biol.,* 9, 77, 1987.

57. **Van't Hof, J., Hernandez, P., Bjerknes, C. A., and Lamm, S. S.,** Location of the replication origin in the 9-kb repeat size class of rDNA in pea *(Pisum sativum), Plant Mol. Biol.,* 9, 87, 1987.

58. **Henson, P.,** The presence of single-stranded regions in mammalian DNA, *J. Mol. Biol.,* 119, 487, 1978.

59. **Garcia-Herdugo, G., Gonzalez-Fernandez, A., and Lopez-Saez, J. F.,** DNA replication in the presence of protein synthesis inhibitors in higher plant cells, *Exp. Cell Res.,* 104, 1, 1976.

60. **Schvartzman, J. B. and Van't Hof, J.,** In the higher plant *Pisum sativum* maturation of nascent DNA is blocked by cycloheximide, but only after 4—8 replicons are joined, *Nucleic Acids Res.,* 10, 6196, 1982.

61. **Schvartzman, J. B., Chenet, B., Bjerknes, C., and Van't Hof, J.,** Nascent replicons are synchronously joined at the end of S phase or during G2 phase in peas, *Biochim. Biophys. Acta,* 653, 185, 1981.

62. **Van't Hof, J.,** Pea *(Pisum sativum)* cells arrested in G2 have nascent DNA with breaks between replicons and replication clusters, *Exp. Cell Res.,* 129, 231, 1980.

63. **Bryant, J. A.,** Enzymology of nuclear DNA replication in plants, *CRC Crit. Rev. Plant Sci.,* 3, 169, 1985.

64. **Li, J. J. and Kelly, T. J.,** Simian virus 40 DNA replication *in vitro, Proc. Natl. Acad. Sci. U.S.A.,* 81, 6973, 1984.

65. **Tjian, R.,** The binding site on SV40 for a T antigen-related protein, *Cell,* 13, 165, 1978.

66. **Myers, M. R. and Tjian, R.,** Construction and analysis of simian virus 40 origins defective in tumor antigen binding and DNA replication, *Proc. Natl. Acad. Sci. U.S.A.,* 77, 6491, 1980.

67. **Stahl, H., Droege, P., and Knippers, R.,** DNA helicase activity of SV40 tumor antigen, *EMBO J.,* 5, 1939, 1986.

68. **Dean, F. B., Bullock, P., Murakami, Y., Wobbe, C. R., Weissbach, L., and Hurwitz, J.,** Simian virus 40 (SV40) DNA replication: SV40 large T antigen unwinds DNA containing the SV40 origin of replication, *Proc. Natl. Acad. Sci. U.S.A.,* 84, 16, 1987.

69. **Dodson, M., Dean, F. B., Bullock, P., Echols, H., and Hurwitz, J.,** Unwinding of duplex DNA from the SV40 origin of replication by T antigen, *Science,* 238, 964, 1987.

70. **Hotta, Y. and Stern, H.,** DNA unwinding protein from meiotic cells of *Lilium, Biochemistry,* 17, 1872, 1978.

71. **Chase, J. W. and Williams, K. R.,** Single stranded DNA binding proteins required for DNA replication, *Annu. Rev. Biochem.,* 55, 103, 1986.

72. **Hotta, Y. and Stern, H.,** A DNA binding protein in meiotic cells of *Lilium, Dev. Biol.,* 26, 87, 1971.

73. **Hotta, Y. and Stern, H.,** The effect of dephosphorylation on the properties of a helix-destabilizing protein from meiotic cells and its partial reversal by a protein kinase, *Eur. J. Biochem.,* 95, 31, 1979.

74. **Kornberg, A.,** *DNA Replication,* W. H. Freeman, San Francisco, 1980.

75. **Conaway, R. C. and Lehman, I. R.,** A DNA primase activity associated with DNA polymerase alpha from *Drosophila melanogaster* embryos, *Proc. Natl. Acad. Sci. U.S.A.,* 79, 2523, 1982.

76. **Shioda, M., Nelson, E. M., Bayne, M. L., and Benbow, R. M.,** DNA primase activity associated with DNA polymerase alpha from *Xenopus laevis* ovaries, *Proc. Natl. Acad. Sci. U.S.A.,* 79, 7209, 1982.

77. **Plevani, P., Badaracco, G., Augl, C., and Chang, L. M. S.,** DNA polymerase I and DNA primase complex in yeast, *J. Biol. Chem.,* 259, 7532, 1984.

78. **Wong, S. W., Paborsky, L. R., Fisher, P. A., Wang, T. S.-F., and Korn, D.,** Structural and enzymological characterization of immunoaffinity-purified DNA polymerase alpha DNA primase complex from KB cells, *J. Biol. Chem.,* 261, 7958, 1986.

79. **Vishvanatha, J. K., Coughlin, S. A., Wesolowski-Owen, M., and Baril, E. F.,** A multiprotein form of DNA polymerase alpha from HeLa cells, *J. Biol. Chem.,* 261, 6619, 1986.

80. **Ottiger, H., Frei, P., Haessig, M., and Huebscher, U.,** Mammalian DNA polymerase alpha: a replication competent holoenzyme form from calf thymus, *Nucleic Acids Res.,* 15, 4789, 1987.
81. **Huebscher, U.,** The mammalian primase is part of a high molecular weight DNA polymerase alpha polypeptide, *EMBO J.,* 2, 133, 1983.
82. **Hay, R. T., Hendrickson, E. A., and DePamphilis, M. L.,** Sequence specificity for the initiation of RNA-primed simian virus 40 DNA synthesis *in vivo, J. Mol. Biol.,* 175, 131, 1984.
83. **Anderson, S. and DePamphilis, M. L.,** Metabolism of Okazaki fragments during simian virus 40 DNA replication, *J. Biol. Chem.,* 254, 11495, 1979.
84. **Graveline, J., Tarrago-Litvak, L., Castroviejo, M., and Litvak, S.,** DNA primase activity from wheat embryos, *Plant Mol. Biol.,* 3, 207, 1984.
85. **Marchesi, M. L., Villaggi, F., Spadari, S., Pedrali-Noy, G., and Sala, F.,** Properties of a DNA primase from rice cells, *Mutat. Res.,* 181, 93, 1987.
86. **Gellert, M.,** DNA topiosomerases, *Annu. Rev. Biochem.,* 50, 879, 1981.
87. **Uemura, T. and Yanagida, M.,** Isolation of type I and II DNA topoisomerase mutants from fission yeast: single and double mutants show different phenotypes in cell growth and chromatin organization, *EMBO J.,* 3, 1737, 1984.
88. **Holm, C., Goto, T., Wang, J. C., and Botstein, D.,** DNA topoisomerase II is required at the time of mitosis in yeast, *Cell,* 41, 553, 1985.
89. **Yang, L., Wold, M. S., Li, J. J., Kelly, T. J., and Liu, L. F.,** Roles of DNA topoisomerases in simian virus 40 DNA replication *in vitro, Proc. Natl. Acad. Sci. U.S.A.,* 84, 950, 1987.
90. **Dynan, W. S., Jendrisak, J. J., Hager, D. A., and Burgess, R. R.,** Purification and characterization of wheat germ DNA topoisomerase I (nicking-closing enzyme), *J. Biol. Chem.,* 256, 5860, 1981.
91. **Been, M. D., Burgess, R. D., and Champoux, J. J.,** DNA strand breakage by wheat germ type I topoisomerase, *Biochim. Biophys. Acta,* 782, 304, 1984.
92. **Fukuta, H., Ohgami, K., and Fukusawa, H.,** Isolation and characterization of DNA topoisomerase II from cauliflower inflorescences, *Plant Mol. Biol.,* 6, 137, 1986.
93. **Weissbach, A., Baltimore, D., Bollum, F., Gallo, R., and Korn, D.,** Nomenclature of eukaryotic DNA polymerases, *Eur. J. Biochem.,* 59, 1, 1975.
94. **Weissbach, A.,** Cellular and viral-induced eukaryotic polymerases, in *The Enzymes,* Vol. 14, Boyer, P.D., Ed., Academic Press, Orlando, 1981, 67.
95. **Huebscher, U., Kuenzle, C. C., and Spadari, S.,** Functional roles of DNA polymerases beta and gamma, *Proc. Natl. Acad. Sci. U.S.A.,* 76, 2316, 1979.
96. **Zimmermann, W., Chen, S. M., Bolden, A., and Weissbach, A.,** Mitochondrial DNA replication does not involve DNA polymerase alpha, *J. Biol. Chem.,* 255, 11847, 1980.
97. **Bensch, K. G., Tanaka, S., Hu, S.-Z., Wang, T. S.-F., and Korn, D.,** Intracellular localization of human DNA polymerase alpha with monoclonal antibodies, *J. Biol. Chem.,* 257, 8391, 1982.
98. **DePamphilis, M. L. and Wassarman, P. M.,** Replication of eukaryotic chromosomes: a close-up of the replication fork, *Annu. Rev. Biochem.,* 49, 627, 1980.
99. **Miller, M. R., Ulrich, R. G., Wang, T. S.-F., and Korn, D.,** Monoclonal antibodies against human DNA polymerase alpha inhibit DNA replication in permeabilized human cells, *J. Biol. Chem.,* 260, 134, 1985.
100. **Byrnes, J. J., Downey, K. M., Black, V. L., and So, A. G.,** A new mammalian DNA polymerase with 3' to 5' exonuclease activity: DNA polymerase delta, *Biochemistry,* 15, 2817, 1976.
101. **Crute, J. J., Wahl, A. F., and Bambara, R. A.,** Purification and characterization of two new high molecular weight forms of DNA polymerase delta, *Biochemistry,* 25, 26, 1986.
102. **Amileni, A., Sala, F., Cella, R., and Spadari, S.,** The major DNA polymerase in cultured plant cells: partial purification and correlation with cell multiplication, *Planta,* 146, 521, 1979.
103. **Sala, F., Parisi, B., Burroni, D., Amileni, A. R., Pedrali-Noy, and Spadari, S.,** Specific and reversible inhibition by aphidicolin of the alpha-like DNA polymerase of plant cells, *FEBS Lett.,* 117, 93, 1980.
104. **Misumi, M. and Weissbach, A.,** The isolation and characterization of DNA polymerase alpha from spinach, *J. Biol. Chem.,* 257, 2323, 1982.
105. **Castroviejo, M., Tharaud, D., Tarrago-Litvak, L., and Litvak, S.,** Multiple deoxyribonucleic acid polymerases from quiescent wheat embryos, *Biochem. J.,* 181, 183, 1979.
106. **Castroviejo, M., Fournier, M., Gatius, M., Gandar, J. C., Labouesse, B., and Litvak, S.,** Tryptophanyl-tRNA synthetase is found closely associated with and stimulates DNA polymerase alpha-like activity from wheat embryos, *Biochem. Biophys. Res. Comm.,* 107, 294, 1982.
107. **Litvak, S. and Castroviejo, M.,** DNA polymerases from plant cells, *Mutat. Res.,* 181, 81, 1987.
108. **Ward, C. and Weissbach, A.,** Isolation and characterization of DNA polymerase alpha from spinach, in *Methods in Enzymology,* Vol. 118, Weissbach, A. and Weissbach, H., Eds., Academic Press, New York, 1986, 97.
109. **Sala, F., Galli, M. G., Levi, M., Burroni, D., Parisi, B., Pedrali-Noy, G., and Spadari, S.,** Functional roles of the plant alpha-like and gamma-like DNA polymerases, *FEBS Lett.,* 124, 112, 1981.

110. **Ikegami, S., Taguchi, T., Ohashi, M., Oguro, M., Nagano, H., and Mano, Y.**, Aphidicolin prevents mitotic cell division by interfering with the activity of DNA polymerase alpha, *Nature (London)*, 275, 458, 1978.

111. **Spadari, S., Sala, F., and Pedrali-Noy, G.**, Aphidicolin: a specific inhibitor of nuclear DNA replication in eukaryotes, *Trends Biochem. Sci.*, 7, 29, 1982.

112. **D'Alesandro, M. M., Jaskot, R. H., and Dunham, V. L.**, Soluble and chromatin-bound DNA polymerases in developing soybean, *Biochem. Biophys. Res. Commun.*, 94, 233, 1980.

113. **Chivers, H. and Bryant, J. A.**, Molecular weights of the major DNA polymerases in a higher plant, *Pisum sativum* L. (pea), *Biochem. Biophys. Res. Commun.*, 110, 632, 1983.

114. **Sawai, Y., Sugano, N., and Tsukada, K.**, Ribonuclease H activity in cultured plant cells, *Biochim. Biophys. Acta*, 518, 181, 1978.

115. **Tsukada, K. and Nishi, A.**, Polynucleotide ligase from cultured plant cells, *J. Biochem.*, 70, 541, 1971.

116. **Kessler, B.**, Isolation, characterization and distribution of a DNA ligase from higher plants, *Biochim. Biophys. Acta*, 240, 496, 1971.

117. **Howell, S. H. and Stern, H.**, The appearance of DNA breakage and repair activities in the synchronous meiotic cycle of *Lilium*, *J. Mol. Biol.*, 55, 357, 1971.

118. **Cannon, G. C., Heinhorst, S., and Weissbach, A.**, Organellar DNA synthesis in permeabilized soybean cells, *Plant Mol. Biol.*, 7, 331, 1986.

119. **Roman, R., Caboche, M., and Lark, K. G.**, Replication of DNA by nuclei isolated from soybean suspension cultures, *Plant Physiol.*, 66, 726, 1980.

120. **Caboche, M. and Lark, K. G.**, Preferential replication of repeated DNA sequences in nuclei isolated from soybean cells grown in suspension culture, *Proc. Natl. Acad. Sci. U.S.A.*, 78, 1731, 1981.

121. **Palmer, J. D.**, Comparative organization of chloroplast genomes, *Annu. Rev. Genet.*, 19, 325, 1985.

122. **Palmer, J. D. and Thompson, W. F.**, Chloroplast DNA rearrangements are more frequent when a large inverted repeat is lost, *Cell*, 29, 537, 1982.

123. **Bohnert, H. J., Crouse, E. J., and Schmitt, J. M.**, Organization and expression of plastid genomes, in *Encyclopedia of Plant Physiology*, New Series, Vol. 14B, Parthier, B. and Boulter, D., Eds., Springer Verlag, Berlin, 1982, 475.

124. **Crouse, E. J., Schmitt, J. M., and Bohnert, H. J.**, Chloroplast and cyanobacterial genomes, genes and RNAs: a compilation, *Plant Mol. Biol. Rep.*, 3, 43, 1985.

125. **Ohyama, K., Fukuzawa, H., Kohchi, T., Shirai, H., Sano, T., Sano, S., Umesono, K., Shiki, Y., Takeuchi, M., Chang, Z., Aota, A., Inokuchi, H., and Ozeki, H.**, Chloroplast gene organization deduced from complete sequence of liverwort *Marchantia polymorpha* chloroplast DNA, *Nature (London)*, 322, 572, 1986.

126. **Shinozaki, K., Ohme, M., Tanaka, M., Wakasugi, T., Hayashida, N., Matsubayashi, T., Zaita, N., Chunwongse, J., Obokata, J., Yamaguchi-Shinozaki, K., Ohto, C., Torazawa, K., Meng, B. Y., Sugita, M., Deno, H., Kamogashira, T., Yamada, K., Kusuda, J., Takaiwa, F., Kato, A., Tohdoh, N., Shimada, H., and Sugiura, M.**, The complete nucleotide sequence of the tobacco chloroplast genome: its organization and expression, *EMBO J.*, 5, 2043, 1986.

127. **Stern, D. B. and Lonsdale, D. M.**, Mitochondrial and chloroplast genomes of maize have a 12-kilobase DNA sequence in common, *Nature (London)*, 299, 698, 1982.

128. **Lonsdale, D. M., Hodge, T. P., Howe, C. J., and Stern, D. B.**, Maize mitochondrial DNA contains a sequence homologous to the ribulose-1,5-bisphosphate carboxylase large subunit gene of chloroplast DNA, *Cell*, 34, 1007, 1983.

129. **Scott, N. S. and Timmis, J. N.**, Homologies between nuclear and plastid DNA in spinach, *Theor. Appl. Genet.*, 67, 279, 1984.

130. **Timmis, J. N. and Scott, N. S.**, Sequence homology between spinach nuclear and chloroplast genomes, *Nature (London)*, 305, 65, 1983.

131. **Stern, D. B. and Palmer, J. D.**, Extensive and widespread homologies between mitochondrial DNA and chloroplast DNA in plants, *Proc. Natl. Acad. Sci. U.S.A.*, 81, 1946, 1984.

132. **Whisson, D. L. and Scott, N. S.**, Nuclear and mitochondrial DNA have sequence homology with a chloroplast gene, *Plant Mol. Biol.*, 4, 267, 1985.

133. **Stern, D. B. and Palmer, J. D.**, Tripartite mitochondrial genome of spinach: physical structure, mitochondrial gene mapping, and locations of transposed chloroplast DNA sequences, *Nucleic Acids Res.*, 14, 5651, 1986.

134. **Kemble, R. J., Mans, R. J., Gabay-Laughnan, S., and Laughnan, J.**, Sequences homologous to episomal mitochondrial DNAs in the maize nuclear genome, *Nature (London)*, 304, 744, 1983.

135. **Carlson, J. E., Erickson, L. R., and Kemble, R. J.**, Cross hybridization between organelle RNAs and mitochondrial and chloroplast genomes in *Brassica*, *Curr. Gent.*, 11, 161, 1986.

136. **Lamppa, G. K. and Bendich, A. J.**, Changes in chloroplast DNA levels during development of pea *(Pisum sativum)*, *Plant Physiol.*, 64, 126, 1979.

137. **Boffey, S. A. and Leech, R. M.,** Chloroplast DNA levels and the control of chloroplast division in light grown wheat leaves, *Plant Physiol.,* 69, 1387, 1982.
138. **Tymms, M. J., Scott, N. S., and Possingham, J. V.,** DNA content of *Beta vulgaris* chloroplasts during leaf cell expansion, *Plant Physiol.,* 71, 785, 1983.
139. **Scott, N. S. and Possingham, J. V.,** Changes in chloroplast DNA levels during growth of spinach leaves, *J. Exp. Bot.,* 34, 1756, 1983.
140. **Cannon, G. C., Heinhorst, S., Siedlecki, J., and Weissbach, A.,** Chloroplast DNA synthesis in light and dark grown cultured *Nicotiana tabacum* cells as determined by molecular hybridization, *Plant Cell Rep.,* 4, 41, 1985.
141. **Lawrence, M. E. and Possingham, J. V.,** Microspectrofluorometric measurement of chloroplast DNA in dividing and expanding leaf cells of *Spinacia oleracea, Plant Physiol.,* 81, 708, 1986.
142. **Bendich, A. J.,** Why do chloroplasts and mitochondria contain so many copies of their genome?, *Bioassays,* 6, 279, 1987.
143. **Tymms, M. J., Scott, N. S., and Possingham, J. V.,** Chloroplast and nuclear DNA content of cultured spinach leaf discs, *J. Exp. Bot.,* 33, 831, 1982.
144. **Ellis, R. J.,** Chloroplast protein synthesis: principles and problems, in *Subcellular Biochemistry,* Vol. 9, Roodyn, D. B., Ed., Plenum Press, New York, 1983, 237.
145. **Rose, R. J., Cran, D. G., and Possingham, J. V.,** Changes in DNA synthesis during cell growth and chloroplast replication in greening spinach leaf discs, *J. Cell Sci.,* 17, 27, 1975.
146. **Leonard, J. M. and Rose, R. J.,** Sensitivity of the chloroplast division cycle to chloramphenicol and cycloheximide in cultured spinach leaves, *Plant Sci. Lett.,* 14, 159, 1979.
147. **Walbot, V. and Coe, E. H.,** Nuclear gene *iojap* conditions a programmed change to ribosome-less plastids in *Zea mays, Proc. Natl. Acad. Sci. U.S.A.,* 76, 2760, 1979.
148. **Herrmann, R. G. and Feierabend, J.,** The presence of DNA in ribosome-deficient plastids of heat-bleached rye leaves, *Eur. J. Biochem.,* 104, 603, 1980.
149. **Scott, N. S., Cain, P., and Possingham, J. V.,** Plastid DNA levels in albino and green leaves of the "albostrians" mutant of *Hordeum vulgare, Z. Pflanzenphysiol.,* 108, 187, 1982.
150. **Kolodner, R. D. and Tewari, K. K.,** Presence of displacement loops in the covalently closed circular chloroplast deoxyribonucleic acid from higher plants, *J. Biol. Chem.,* 250, 8840, 1975.
151. **Kolodner, R. D. and Tewari, K. K.,** Chloroplast DNA from higher plants replicates by both the Cairns and the rolling circle mechanism, *Nature (London),* 256, 708, 1975.
152. **Koller, B. and Delius, H.,** Origin of replication in chloroplast DNA of *Euglena gracilis* located close to the region of variable size, *EMBO J.,* 1, 995, 1982.
153. **Ravel-Chapuis, P., Heizmann, P., and Nigon, V.,** Electron microscopic localization of the replication origin of *Euglena gracilis* chloroplast DNA, *Nature (London),* 300, 78, 1982.
154. **Schlunegger, B. and Stutz, E.,** The *Euglena gracilis* chloroplast genome: structural features of a DNA region possibly carrying the single origin of replication, *Curr. Genet.,* 8, 629, 1984.
155. **Waddell, J., Wang, X.-M., and Wu, M.,** Electron microscopic localization of the chloroplast DNA replicative origin in *Chlamydomonas reinhardii, Nucleic Acids Res.,* 12, 3843, 1984.
156. **Rochaix, J. D., van Dillewijn, J., and Rahire, M.,** Construction and characterization of autonomously replicating plasmids in the green unicellular alga *Chlamydomonas reinhardii, Cell,* 36, 925, 1984.
157. **Vallet, J. M. and Rochaix, J. D.,** Chloroplast origins of DNA replication are distinct from chloroplast ARS sequences in two green algae, *Curr. Genet.,* 9, 321, 1985.
158. **Wang, X.-M., Chang, C. H., Waddell, J., and Wu, M.,** Cloning and delimiting one chloroplast DNA replicative origin of *Chlamydomonas, Nucleic Acids Res.,* 12, 3857, 1984.
159. **Wu, M., Lou, J. K., Chang, D. Y., Chang, C. H., and Nie, Z. Q.,** Structure and function of a chloroplast DNA replication origin of *Chlamydomonas reinhardtii, Proc. Natl. Acad. Sci. U.S.A.,* 83, 6761, 1986.
160. **Ohtani, T., Uchimiya, H., Kato, A., Harada, H., Sugita, M., and Sugiura, M.,** Location and nucleotide sequence of a tobacco chloroplast DNA segment capable of replication in yeast, *Mol. Gen. Genet.,* 195, 1, 1984.
161. **Overbeeke, N., Haring, M. A., Nijkamp, H. J. J., and Kool, A. J.,** Cloning of *Petunia hybrida* chloroplast DNA sequences capable of autonomous replication in yeast, *Plant Mol. Biol.,* 3, 235, 1984.
162. **De Haas, J. M., Boot, K. J. M., Haring, M. A., Kool, A. J., and Nijkamp, H. J. J.,** A *Petunia hybrida* chloroplast DNA region, close to one of the inverted repeats, shows sequence homology with the *Euglena gracilis* chloroplast DNA region that carries the putative replication origin, *Mol. Gen. Genet.,* 202, 48, 1986.
163. **Briat, J. F. and Dalmon, J.,** *Spinacia oleracea* chloroplast DNA sequence homology with ARS and ARC elements within the inverted repeat, upstream of the rDNA operon, *Nucleic Acids Res.,* 14, 8223, 1986.
164. **Day, A. and Ellis, T. H. N.,** Chloroplast DNA deletions associated with wheat plants regenerated from pollen: possible basis for maternal inheritance of chloroplasts, *Cell,* 39, 359, 1984.
165. **Day, A. and Ellis, T. H. N.,** Deleted forms of plastid DNA in albino plants from cereal anther culture, *Curr. Genet.,* 9, 671, 1985.

166. **Ellis, T. H. N. and Day, A.**, A hairpin plastid genome in barley, *EMBO J.*, 5, 2769, 1986.
167. **Tewari, K. K. and Wildman, S. G.**, DNA polymerase in isolated tobacco chloroplasts and nature of the polymerized product, *Proc. Natl. Acad. Sci. U.S.A.*, 58, 689, 1967.
168. **Sala, F., Amileni, A. R., Parisi, B., and Spadari, S.**, A gamma-like DNA polymerase in spinach chloroplasts, *Eur. J. Biochem.*, 112, 211, 1980.
169. **McKown, R. L. and Tewari, K. K.**, Purification and properties of a pea chloroplast DNA polymerase, *Proc. Natl. Acad. Sci. U.S.A.*, 81, 2354, 1984.
170. **Wernette, C. M. and Kaguni, L. S.**, A mitochondrial DNA polymerase from embryos of *Drosophila melanogaster, J. Biol. Chem.*, 261, 14764, 1986.
171. **Tewari, K. K., Kolodner, R. D., and Dobkin, W.**, Replication of circular chloroplast DNA, in *Genetics and Biogenesis of Chloroplasts and Mitochondria*, Buecher, T., et al., Eds., Elsevier/North Holland Biomedical Press, Amsterdam, 1976, 379.
172. **Gold, B., Carrillo, N., Tewari, K. K., and Bogorad, L.**, Nucleotide seuqence of a preferred maize chloroplast genome template for *in vitro* DNA synthesis, *Proc. Natl. Acad. Sci. U.S.A.*, 84, 194, 1987.
173. **Siedlecki, J., Zimmermann, W., and Weissbach, A.**, Characterization of a prokaryotic topoisomerase I activity in chloroplast extracts from spinach, *Nucleic Acids Res.*, 11, 1523, 1983.
174. **Lam, E. and Chua, N.-H.**, Chloroplast DNA gyrase in *in vitro* regulation of transcription by template topology and novobiocin, *Plant Mol. Biol.*, 8, 415, 1987.
175. **Scott, N. S., Shah, V. C., and Smillie, R. M.**, Synthesis of chloroplast DNA in isolated chloroplasts, *J. Cell Biol.*, 38, 151, 1968.
176. **Zimmermann, W. and Weissbach, A.**, Deoxyribonucleic acid synthesis in isolated chloroplasts and chloroplast extracts of maize, *Biochemistry*, 21, 3334, 1982.
177. **Mills, W. R. and Baumgartner, B. J.**, Light-driven DNA biosynthesis in isolated pea chloroplasts, *FEBS Lett.*, 163, 124, 1983.
178. **Overbeeke, N., de Waard, J. H., and Kool, A. J.**, Characterization of *in vitro* DNA synthesis in an isolated chloroplast system of *Petunia hybrida*, in *Proteins Involved in DNA Replciation*, Huebscher, U. and Spadari, S., Eds., Plenum Press, New York, 1984, 107.
179. **Tanaka, A., Yamano, Y., Fukuzawa, H., Ohyama, K., and Komano, T.**, *In vitro* DNA synthesis by chloroplasts isolated from *Marchantia polymorpha* L. cell suspension cultures, *Agric. Biol. Chem.*, 48, 1239, 1984.
180. **De Haas, J. M., Kool, A. J., Overbeeke, N., van Brug, W., and Nijkamp, H. J. J.**, Characterization of DNA synthesis and chloroplast DNA replication initiation in a *Petunia hybrida* chloroplast lysate system, *Curr. Genet.*, 12, 377, 1987.
181. **Lou, J. K., Wu, M., Chang, C. H., and Cuticchia, A. J.**, Localization of a r-protein gene within the chloroplast DNA replication origin of *Chlamydomonas, Curr. Genet.*, 11, 537, 1987.
182. **Herrmann, R. G. and Kowallik, K. V.**, Multiple amounts of DNA related to the size of chloroplasts. II. Comparison of electron microscopic and autoradiographic data, *Protoplasma*, 69, 365, 1970.
183. **Rose, R. J. and Possingham, J. V.**, The localization of [³H]thymidine incorporation in the DNA of replicating spinach chloroplasts by electron microscope autoradiography, *J. Cell Sci.*, 20, 341, 1976.
184. **Rose, R. J.**, The association of chloroplast DNA with photosynthetic membrane vesicles from spinach chloroplasts, *J. Cell Sci.*, 36, 169, 1979.
185. **Sellden, G. and Leech, R. M.**, Localization of DNA in mature and young wheat chloroplasts using the fluorescent probe 4'-6-diamidino-2-phenylindole, *Plant Physiol.*, 68, 731, 1981.
186. **Hashimoto, H.**, Changes in distribution of nucleoids in developing and dividing chloroplasts and etioplasts of *Avena sativa, Protoplasma*, 127, 119, 1985.
187. **Rose, R. J., Cran, D. G., and Possingham, J. V.**, Distribution of DNA in dividing spinach chloroplasts, *Nature (London)*, 251, 641, 1974.
188. **Lindbeck, A. G. C., Rose, R. J., Lawrence, M. E., and Possingham, J. V.**, The role of chloroplast membranes in the location of chloroplast DNA during the greening of *Phaseolus vulgaris* etioplasts, *Protoplasma*, 139, 92, 1987.
189. **Leaver, C. J. and Gray, M. W.**, Mitochondrial genome organization and expression, *Annu. Rev. Plant Physiol.*, 33, 373, 1982.
190. **Sederoff, R. R.**, Structural variation in mitochondrial DNA, in *Advances in Genetics*, Vol. 22, Scandalios, J. G., Ed., Academic Press, New York, 1984, 1.
191. **Pring, D. R. and Lonsdale, D. M.**, Molecular biology of higher plant mitochondrial DNA, *Int. Rev. Cytol.*, 97, 1, 1985.
192. **Ward, B. L., Anderson, R. S., and Bendich, A. J.**, The mitochondrial genome is large and variable in a family of plants *(Cucurbitaceae), Cell*, 25, 793, 1981.
193. **Palmer, J. D. and Shields, C. R.**, Tripartite structure of the *Brassica campestris* mitochondrial genome, *Nature (London)*, 307, 437, 1984.
194. **Kolodner, R. D. and Tewari, K. K.**, Physicochemical characterization of mitochondrial DNA from pea leaves, *Proc. Natl. Acad. Sci. U.S.A.*, 69, 1830, 1972.

195. **Lonsdale, D. M., Hodge, T. P., and Fauron, C. M.-R.,** The physical map and organization of the mitochondrial genome from the fertile cytoplasm of maize, *Nucleic Acids Res.,* 12, 9249, 1984.
196. **Palmer, J. D. and Herbon, L. A.,** Unicircular structure of the *Brassica hirta* mitochondrial genome, *Curr. Genet.,* 11, 565, 1987.
197. **Hack, E. and Leaver, C. J.,** The alpha subunit of the maize F_1-ATPase is synthesized in the mitochondrion, *EMBO J.,* 2, 1783, 1983.
198. **Leaver, C. J., Forde, B. G., Dixon, L. K., and Fox, T. D.,** Mitochondrial genes and cytoplasmically inherited variation in higher plants, in *Mitochondrial Genes,* Slonimski, P., Borst, P., and Attardi, G., Eds., Cold Spring Harbor Press, Cold Spring Harbor, NY, 1982, 457.
199. **Eckenrode, V. K. and Levings, C. S., III,** Maize mitochondrial genes, *In Vitro Cell. Dev. Biol.,* 22, 169, 1986.
200. **Neupert, W. and Schatz, G.,** How proteins are transported into mitochondria, *Trends Biochem. Sci.,* 6, 1, 1981.
201. **Wong, T. W. and Clayton, D. A.,** DNA primase of human mitochondria is associated with structural RNA that is essential for enzymatic activity, *Cell,* 45, 817, 1986.
202. **Chang, D. D. and Clayton, D. A.,** A mammalian mitochondrial RNA processing activity contains nucleus-encoded RNA, *Science,* 235, 1178, 1987.
203. **Lamppa, G. K. and Bendich, A. J.,** Changes in mitochondrial DNA levels during development of pea *(Pisum sativum* L.), *Planta,* 162, 463, 1984.
204. **Bendich, A. J. and Gauriloff, L. P.,** Morphometric analysis of cucurbit mitochondria: the relationship between chondriome volume and DNA content, *Protoplasma,* 119, 1, 1984.
205. **Clayton, D. A.,** Replication of animal mitochondria, *Cell,* 28, 693, 1982.
206. **Chang, D. D., Hauswirth, W. W., and Clayton, D. A.,** Replication priming and transcription initiate from precisely the same site in mouse mitochondrial DNA, *EMBO J.,* 4, 1559, 1985.
207. **Chang, D. D. and Clayton, D. A.,** Priming of human mitochondrial DNA replication occurs at the light-strand promoter, *Proc. Natl. Acad. Sci. U.S.A.,* 82, 351, 1985.
208. **Chang, D. D. and Clayton, D. A.,** A novel endoribonuclease cleaves at a priming site of mouse mitochondrial DNA replication, *EMBO J.,* 6, 409, 1987.
209. **Wong, T. W. and Clayton, D. A.,** *In vitro* replication of human mitochondrial DNA: accurate initiation at the origin of light-strand synthesis, *Cell,* 42, 951, 1985.
210. **Wong, T. W. and Clayton, D. A.,** Isolation and characterization of a DNA primase from human mitochondria, *J. Biol. Chem.,* 260, 11530, 1985.
211. **Castora, F. J., Lazarus, G. M., and Kunes, D. L.,** The presence of two mitochondrial topoisomerases in human acute leukemia cells, *Biochem. Biophys. Res. Commun.,* 130, 854, 1985.
212. **Brun, G., Vannier, P., Scovassi, I., and Callen, J. C.,** DNA topoisomerase I from mitochondria of *Xenopus laevis* oocytes, *Eur. J. Biochem.,* 118, 407, 1981.
213. **Castora, F. J. and Kelly, W. G.,** ATP inhibits nuclear and mitochondrial type I topoisomerases from human leukemia cells, *Proc. Natl. Acad. Sci. U.S.A.,* 83, 1680, 1986.
214. **Castora, F. J., Vissering, F., and Simpson, M. V.,** The effect of bacterial DNA gyrase inhibitors on DNA synthesis in mammalian mitochondria, *Biochem. Biophys. Acta,* 740, 417, 1983.
215. **Callen, J. C., Tourte, M., Dennebouy, N., and Mounolou, J. C.,** Changes in D-loop frequency and superhelicity among the mitochondrial DNA molecules in relation to organelle biogenesis in oocytes of *Xenopus laevis, Exp. Cell Res.,* 143, 115, 1983.
216. **Echeverria, M., Martin, M. T., Ricard, B., and Litvak, S.,** A DNA topoisomerase type I from wheat embryo mitochondria, *Plant Mol. Biol.,* 6, 417, 1986.
217. **Castroviejo, M., Tharaud, D., Mocquot, B., and Litvak, S.,** Factors affecting the onset of deoxyribonucleic acid synthesis during wheat embryo germination, *Biochem. J.,* 181, 193, 1979.
218. **Christophe, L., Tarrago-Litvak, L., Castroviejo, M., and Litvak, S.,** Mitochondrial DNA polymerase from wheat embryos, *Plant Sci. Lett.,* 21, 181, 1981.
219. **Mahler, H. R. and Bastos, R. N.,** Coupling between mitochondrial mutation and energy transduction, *Proc. Natl. Acad. Sci. USA,* 71, 2241, 1974.
220. **Flechtner, V. R. and Sager, R.,** Ethidium bromide induced selective and reversible loss of chloroplast DNA, *Nature New Biol.,* 241, 277, 1973.
221. **Gillham, N. W., Boynton, J. E., and Harris, E. H.,** Specific elimination of mitochondrial DNA from *Chlamydomonas* by intercalating dyes, *Curr. Genet.,* 12, 41, 1987.
222. **Ricard, B., Echeverria, M., Chrisophe, L., and Litvak, S.,** DNA synthesis in isolated mitochondria and mitochondrial extracts from wheat embryos, *Plant Mol. Biol.,* 2, 167, 1983.
223. **Bedinger, P. and Walbot, V.,** DNA synthesis in purified maize mitochondria, *Curr. Genet.,* 10, 631, 1986.
224. **Carlson, J. E., Brown, G. L., and Kemble, R. J.,** *In organello* mitochondrial DNA and RNA synthesis in fertile and cytoplasmic male sterile *Zea mays* L., *Curr. Genet.,* 11, 151, 1986.
225. **Heinhorst, S., Cannon, G. C., and Weissbach, A.,** unpublished observations.

Chapter 7

CHROMATIN STRUCTURE OF PLANT GENES

Robert J. Ferl and Anna-Lisa Paul

TABLE OF CONTENTS

I. INTRODUCTION

The study of chromatin structure in plants has moved rapidly forward only in the last few years. In fact, as recently as 1985, Steven Spiker prefaced his review of the subject[1] by answering, "Yes, no, and we don't know" to the question, "Is plant chromatin just like animal chromatin?" That "don't know" comment pretty well summed up the situation, especially with respect to those aspects of chromatin structure that are now known to be directly involved in gene regulation. While many descriptions of the basic structure of bulk plant chromatin were available at that time, detailed research into the specific chromatin structure of well-characterized plant genes was only just beginning. Today there are a good number of published articles on chromatin structure and plant gene regulation. Much of that research was fueled by direct interests in plant gene regulation.

In many cases studies on the chromatin structure of plant genes (particularly the changes in chromatin structure presumed to accompany the regulated expression of a gene) were designed primarily to understand basic aspects of gene regulation, not chromatin structure per se. Yet when considered on the whole, these sets of experiments do provide fundamental insights into the structure of plant chromatin.

The goal of this review is to present a compendium of the current state of published knowledge of plant chromatin structure. We have, wherever possible, drawn upon the animal literature for a base of background information, but our review of the non-plant literature is not meant to be complete. Much like the research itself, we have also focused on those aspects of plant chromatin structure most profoundly or directly associated with transcriptional regulation.

II. THE NUCLEOSOME ARRAY AND ITS RELATION TO GENE REGULATION

A. BACKGROUND

The genetic material of the eukaryotic cell is a complex organization of DNA and protein. An association of protein facilitates the packaging of DNA into an ordered structure suitable for rapid replication and transcription as the needs of the cell dictate. Chromatin is the end product of such packaging, a linear condensation of about 10,000-fold from naked DNA to a metaphase chromosome. This condensation proceeds in three general steps termed "first", "second", and "higher" order structure.

First order chromatin structure is the nucleosome array. "Beads on a string" has been the most common analogy used to describe the appearance of this structure as seen with electron microscopy,[2] but a more apt description coined by Eissenberg et al.[3] is "string wrapping around beads". In either case, the "beads" (nucleosomes) are octamers of histone proteins together with the DNA to which they are bound and the "string" is the DNA that links the nucleosomes forming a 10 nm fiber.[3] The amount of DNA in the nucleosome, ~146 base pairs (bp), is fairly constant between species or tissues. The length of the linker DNA is more variable, ranging from 20 to 200 bp.

Most of the regulation associated with chromatin is in first order structure but there are cases where second and even higher orders of condensation are involved in regulation, such as the inactivation of one of the X chromosomes in mammalian females.[4] There are often regions of permanently condensed chromatin seen in plants but it has generally not been linked with any particular chromosome or family of genes. More often, regions of condensed chromatin are related to the amount of repetitive DNA chracteristic of that particular plant species.[5] A possible exception has been found in maize. When seedlings are exposed to anaerobic stress, an increase in the degree of chromatin condensation is observed with electron microscopy.[6] Anaerobic stress in maize is correlated with the reduction in levels of general transcription and induction of a limited set of anaerobic stress proteins.[7]

Although changes in second-order structure are not usually associated with transcriptional activation, there are some aspects that may be of importance to regulation. It is certain that the histone H1 is crucial to the stability of second-order chromatin.[8] In addition, there is evidence to suggest H1 may be bound to transcriptionally active chromatin differently than it is in inactive chromatin.[9] This would imply that gene-specific differences in second-order structure may exist, but little evidence has been accumulated in this regard.

Transcriptional activation is often reflected in the structure of first-order chromatin, hence the phrases "active" and "inactive" chromatin or nucleosomes. There are two major categories of structural changes observed in active chromatin with regard to the nucleosome array. The first concerns irregularities in the histones and other proteins associated with the nucleosome, and the second concerns changes in the linker length or the distribution of nucleosomes over the gene.

The nucleosome core particles are uniform structures and virtually indistinguishable regardless of source,[10-15] but histone variants have been associated with transcriptional activity in a variety of systems. Examples of hyperacetylated histones in active chromatin can be seen in *Physarum*[16] and rat liver.[17] Ubiquinated histones have also been associated with both active[18,19] and inactive[20] chromatin. In addition, the non-histone "high mobility group" proteins HMG14 and HMG17 have been found complexed with the nucleosomes along transcriptionally active genes in animals[21-23] and have been employed in the purification of active nucleosomes.[24]

The length of DNA associated with the core particle may vary somewhat between organisms but the differences observed may be a function of the experimental procedure rather than the source. Significant variation is, however, seen in the spacing of nucleosomes. The amount of DNA that links the nucleosomes in the first-order array can vary with species, tissues, and stage of development.[3,25]

Changes in first-order chromatin structure are, for the most part, visualized with the aid of two endonucleases: micrococcal nuclease and DNase I. Micrococcal nuclease preferentially cleaves linker DNA so that a mild digestion will result in "ladders" of DNA in units of nucleosome repeat length (Figure 1). Changes in the repeat length are usually not an indicator of transcriptional activation, but the correlation has been made on occasion.

DNase I has been used extensively in the study of chromatin structure. Since it will cut any region of unprotected DNA, DNase I is a useful tool in discerning regions of chromatin which are less tightly associated with proteins than the surrounding chromatin. The chromatin within carefully isolated nuclei is assumed to be very close to its natural configuration *in vivo*. Thus, digestions of nuclei with DNase I should accurately reflect the condition of the chromatin *in vivo*.

There are two types of phenomena that are frequently observed after digestion of nuclei with DNase I: (1) a general increase in sensitivity of active chromatin in comparison to inactive chromatin, and (2) hypersensitivity which describes regions of increased sensitivity which are superimposed on areas of general sensitivity. DNase I digestion is not limited to cut linker DNA, suggesting regions of general sensitivity must be of a different configuration than a typical nucleosome array. What is usually envisioned is a relaxation of the array to create open regions of DNA which are better able to accommodate the transcriptional machinery. A relaxation of the chromatin without actual loss of nucleosomes may be accomplished by alterations in histone-DNA binding characteristics that afford DNase I better access to the DNA. General DNase I sensitivity is usually a hallmark of transcriptional readiness.

Hypersensitive sites are small regions (typically 50 to 200 bp) of chromatin that are thought to be nucleosome free.[26] Nonhistone proteins are often found associated with hypersensitive sites and may function in keeping the sites free of nucleosomes or in some other way participate in transcriptional regulation. There have been several cases where such proteins have been directly implicated as transcription factors.[27-29]

micrococcal nuclease

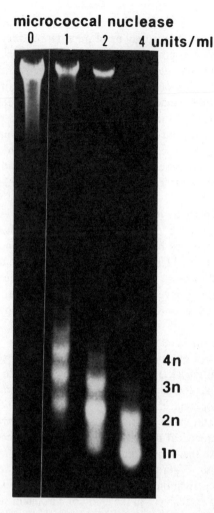

FIGURE 1. An example of nucleosome ladders from maize. Isolated nuclei were incubated with micrococcal nuclease, then the DNA was isolated and resolved on a 1.3% agarose gel. Since micrococcal nuclease digests the linker DNA between nucleosomes, all bands are multiples of the size of monomeric nucleosome DNA (n). In this case n = ~160 bp.

There is a wealth of information correlating DNase I sensitivity and hypersensitivity to transcriptional activity in animal systems (see References 3, 30, 31 for reviews). Plants have received far less attention in this regard but recent studies have shown plant and animal systems to be very similar.

B. CHROMATIN STRUCTURE OF PLANT GENES
1. General Nuclease Sensitivity
a. Reassociation Kinetics

Reassociation kinetics was one of the first means used to evaluate the effects of DNase I on actively transcribing plant chromatin.[32] Nuclei were isolated from wheat embryos, then divided into two sets; one set was digested with DNase I while the other acted as a control. The DNAs were then compared with respect to the speed with which they would drive reassociation reactions with cDNA. DNase I-digested DNA was found to reassociate more slowly than the more intact DNA from the control nuclei. To evaluate the effects of DNase I between transcriptionally active and inactive genes, comparisons were made between types of cDNA, total poly(A)RNA from unimbibed embryos for inactive genes and polysomal poly(A)RNA from imbibed embryos for active genes. The source of cDNA had no effect on the reassociation kinetics with DNA from either control or DNase I digested nuclei. In explanation, Spiker et al.[32] suggest that the chromatin of the dormant embryo is stored in an open, "potentially transcribable" state such that chromatin does not undergo restructuring at the onset of transcription.

b. Seed Storage Proteins

The regulation of the seed-storage-protein gene *phaseolin* was examined by Murray and Kennard.[33] They compared the chromatin of leaf tissue, which does not express the gene, to chromatin from cotyledon, where the gene is active. Nuclei were isolated from both tissues and challenged with DNase I. The chromatin of cotyledons where *phaseolin* is active showed an increase in general sensitivity around the 5' flanking region and the adjacent 1 kb of the coding region of the *phaseolin* gene.

Legumin, which encodes a seed storage protein in peas, is also actively transcribed in cotyledons and inactive in leaves. Isolated nuclei digested with DNase I showed that the chromatin of the *legumin* gene of actively transcribing cotyledon tissue had a greater sensitivity to digestion than did chromatin isolated from leaf tissue. In this case the coding and 3' flanking region of the gene was five times more sensitive in cotyledons than in leaves.[34]

c. Tissue-Specific Differences in Structure

Steinmüller et al.[35] examined a number of genes from barley for differences in chromatin structure among tissues and under different environmental conditions. Three genes showed tissue-specific differences in their sensitivity to DNase I: the seed storage protein hordin, the light-harvesting chlorophyll a/b protein (LHCP), and the 15-kDa polypeptide. Tissues in which these were expressed showed an increase in DNase I sensitivity over tissues in which they are inactive.[35] Evidence to suggest that chromatin structure functions in the regulation of a gene that can recognize an external stimulus was seen with the 15 kDa protein gene. The 15 kDa protein gene from dark grown leaves became less sensitive to DNase I digestion as the leaves were exposed to light.[35] This change in chromatin structure parallels the cessation of transcription of the 15 kDa gene when leaves are illuminated.

2. DNase I Hypersensitivity
a. Pea Ribosomal Genes

The large number of ribosomal RNA genes often found in plants suggested to Kaufman et al.[36] that a modification of chromatin structure associated with these genes may help to distinguish transcriptionally active copies from inactive ones. In the Alaska cultivar of pea there are two major length classes of rDNA (long, L and short, S). Both length variants are induced by light and contain DNase I-hypersensitive sites. There are 12 hypersensitive sites in the L variant: 9 residing within the subrepeat array of the non-coding spacer regions between the rRNA genes and 3 near the start of the 18S gene. Induction (exposure of dark-grown plants to light) does not change the position or intensity of these sites. The S variant contains both constitutive and inducible hypersensitive sites. The constitutive sites lie within or just 3' to the subrepeat region. Three other sites are evident only after light induction; their positions correspond to the hypersensitive sites closest to the start of the 18S gene in the L variant. The change in S variant chromatin structure with light induction may be correlated with the increased rate of rRNA synthesis in light grown plants.[36]

b. Barley Hordin and PCR

Steinmüller and Apel[37] isolated chromatin (rather than nuclei) from etiolated barley leaves and used DNase I digestions to characterize the genes encoding NADPH-protochlorophyllide oxidoreductase (PCR), which is actively transcribed in leaves, and *hordin*, a seed storage protein gene. PCR was generally more sensitive to DNase I digestion than *hordin* and showed evidence of hypersensitive sites (although these sites were not mapped by the authors). *Hordin*, which is inactive in leaves, showed no indications of hypersensitivity.

c. Maize Shrunken

The 5′ region of the maize *shrunken* gene has also been found to contain DNase I-hypersensitive sites.[38] Digestion of nuclei from endosperm (where the gene is active) resulted in five hypersensitive sites. Two were located far upstream at −1050 and −2150, one closer to the start of transcription at −550 and two within the first intron at +150 and +550. The −550 site is adjacent to a pair of direct repeats which has sequence homology to enhancer elements and contains five copies of the trinucleotide GTG that is potentially associated with protein binding domains. The site at −1050 was missing in tissues where *shrunken* is only weakly expressed, suggesting this region may play a role in *shrunken* regulation.

d. Maize Adh

The chromatin structure of the maize *alcohol dehydrogenase* genes has been examined by a number of researchers. Maize ADH is encoded for two anaerobically inducible genes, *Adh1* and *Adh2*,[39] making this system particularly useful for comparisons of active and inactive genes.

Paul et al.[40] found both constitutive and inducible hypersensitive sites in the 5′ region of *Adh1* (Figure 2). The constitutive sites range furthest upstream between positions −160 to −700. Anaerobic induction caused an increase in sensitivity in the constitutive sites and generated two new hypersensitive sites. The inducible sites are centered around positions −140 and −40 and define an "anaerobic response region". This region encompasses TATAA at −30 and CAAT at −90.

The importance of the 5′ flanking region of *Adh1* defined by these hypersensitive sites has been demonstrated by a number of deletion and transient expression studies. Howard et al.[41] showed the 5′ region alone (up to −1096) of the *Adh1-S* allele was sufficient to induce the expression of a *cat* (chloramphenicol acetyltransferase) reporter gene in maize protoplasts under anaerobic stress. Walker et al.[42] expanded transient expression experiments in protoplasts with hybrid promoters and linker scanning mutations. The results delineated a region between positions −140 and −99 in the *Adh1* promoter that they refer to as the "anaerobic response element" (ARE). The functional significance of the ARE region was further confirmed in studies by Lee et al.[43] and Ellis et al.[44] This region is essential for the anaerobic response, is sufficient in itself to promote transcription, and correlates directly with hypersensitive sites.[40]

There is almost no homology between the 5′ regions of *Adh1* and *Adh2*, although the coding regions are virtually identical.[46] Ashraf et al.[47] examined the chromatin structure of the 5′ region of *Adh2* and used this to compare the divergent upstream regions of the two *Adh* genes.

Maize *Adh2* contains three constitutive hypersensitive sites in the 5′ region, centered around −55, −305, and −455.[47] All sites are presented regardless of the transcriptional state of the gene but anaerobic induction elicits two responses: an increase in the sensitivity of each of the sites and an extension in the 3′ direction of the −55 site. The extension of the −55 site uncovers the TATAA box.

e. T-DNA

Crown gall disease initiated by *Agrobacterium tumefaciens* naturally transforms dicotyledonous plants as part of its pathogenic life cycle. A portion of a tumor-inducing (Ti)

0.1 ug/ml
DNase I
C   U   I

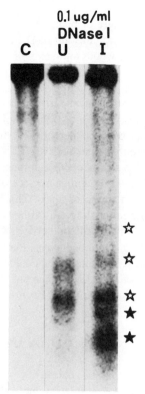

FIGURE 2.   An example of DNase I hypersensitive sites in a
plant gene. Nuclei were isolated from maize tissues that were
either actively expressing *Adh1* (I-induced) or were not expressing the gene (U-uninduced) and incubated with DNase I.
The purified DNA was restricted to produce an end common
to all fragments generated by DNase I. The resulting fragments
were resolved on a 2% agarose gel, transferred to a supporting
membrane and hybridized with a probe homologous to the common restricted end. Maize *Adh1* shows several hypersensitive
sites denoted here by stars. Open stars are common to both
active (I) and inactive (U) chromatin, whereas solid stars are
unique to the chromatin of cells where *Adh1* activity has been
induced by anaerobic stress.

plasmid is transferred to the host genome to initiate production of compounds essential to
the infecting bacterium. Octopine is one such compound, and it was this class of Ti plasmid
that was examined with regard to chromatin structure.

The portion of the Ti plasmid that is transferred is usually divided into left and right
regions. The tumor inducing and octopine synthesis genes are on the left (T-left) while other
genes not necessary for tumorigenesis are coded for on the right (T-right). It was supposed
that the essential tumorigenic genes of T-left would be associated with relaxed chromatin
typical of transcriptionally active genes, while the chromatin of T-right would be in a more
tightly configured structure. DNase I digestion showed very little difference between T-left
and T-right chromatin, although both integrated sequences were generally more sensitive to
DNase I digestion than the surrounding host DNA. Micrococcal nuclease digestion gave
evidence that both T-left and T-right were associated with nucleosomes but the patterns were
diffuse and superimposed on a smear.[48] The most likely explanation is that all T-DNA
sequences incorporated into the host genome are in a state of transcriptional readiness in
which the nucleosomes are in a more open configuration.

C. PERSPECTIVES

The fundamental chromatin structure of plant genes is, in fact, quite similar to that of animal genes in many respects. The organization of DNase I-hypersensitive sites into constitutive and inducible regions seen in both animals[45,49-51] and plants[36-38,40,47] suggests two related models for the role of chromatin in gene regulation. First, the constitutive regions may function in recognizing signals from the environment by binding factors that trigger transcription. This is supported by the increased sensitivity of constitutive sites during activation of single-copy plant genes,[37,38,40,47] as if the act of the signal factor binding creates an even more exposed region of chromatin. Alternatively, the DNase-sensitive conformation of the chromatin around the constitutive sites may itself act as a guide to direct or facilitate access to the induced regions which then receive the transcription signal directly.

In a review of the literature, it would seem on the face of it that plant chromatin research has lagged behind similar work from animal systems. This has certainly been influenced by the "natural" barriers to chromatin research found in plants. The cell wall makes generating intact nuclei a challenge. There is limited availability of tissue culture lines, making it difficult to obtain a source of cells which are homogeneous. Plants also seem to be fraught with endogenous nucleases and proteases compared to animal systems, and strong counteractive measures have sometimes been necessary (Reference 36, for example).

In spite of these obstacles, the elucidation of plant chromatin structure is now proceeding rapidly.

III. SPECIFIC PROTEIN-DNA INTERACTIONS

A. BACKGROUND

The characterization of *trans*-acting regulatory molecules represents a situation where the study of chromatin structure directly overlaps basic studies in gene transcription. "Chromatin", by its very definition, includes the rather disperse group of molecules known simply as non-histone chromosomal proteins. *Trans*-acting regulatory factors certainly fit into that group. And as a class of regulatory protein molecules, these factors must interact directly with the DNA to accomplish their function. The gene-specific interactions of these factors together with the more general associations of the nucleosomes are responsible for the nuclease hypersensitive sites discussed above. In fact, the spatial disruptions caused by changes in nucleosome distributions can be viewed as the mechanism by which access to the DNA is provided for the regulatory factors.[3,31]

Recent advances in both *in vivo* and *in vitro* experimental protocols have allowed the characterization of several plant gene regulatory factors. By far the major innovation for the *in vitro* study of nuclear extracts has been the application of the gel retardation assay.[52] The assay is a sensitive indicator of the specific binding of proteins to DNA fragments and has been successfully used in the analysis of several plant gene systems. Combined with binding competition assays and DNase I footprinting, the binding sites of several plant regulatory proteins have been precisely determined. In animal systems, the *in vivo* detection of protein-DNA interactions (Reference 53, for example) has relied on DMS footprinting and genomic sequencing.[54] There is only one example in the literature of the application of this technology to plants.

Currently the bulk of the published work on plant chromatin structure related to gene activity is on well-characterized gene systems. Thus, the binding of regulatory factors can (at least in some cases) be correlated with prior data on nuclease hypersensitive sites or identified *cis*-acting DNA sequences.

B. *TRANS*-ACTING FACTORS IN PLANTS

1. *In Vitro* Detection of Protein-DNA Interactions

a. Maize Zeins

Maier et al.[55] have used the gel retardation assay and *in vitro* DNase I footprinting to identify a protein factor that binds to sequences 5' to a zein gene of maize. The zein gene family encodes proteins accumulated and stored during the development of the maize endosperm. When a restriction fragment corresponding to positions -250 to -360 was used in a filter binding assay, it was found to be capable of binding a protein in a crude nuclear extract from developing maize endosperms. Within that fragment lies a 15 bp sequence that had been discovered by sequence inspection to be 5' to all zein genes. When oligonucleotides of that consensus sequence were used in gel retardation assays, binding to nuclear protein was again observed. DNase I footprinting revealed that the only binding activity observed occurred directly over that 15 bp consensus sequence.

Because the zeins are transcribed only in the developing endosperm, and because other endosperm-specific genes also possess the 15 bp consensus sequence, Maier et al.[55] appear to have initially characterized a tissue-specific *trans*-acting regulatory factor. However, extracts from other tissues were not tested for binding activity.

b. Soybean Lectin

In a similar context, the soybean lectin storage protein gene is transcribed only in the developing embryo. Jofuku et al.[56] have examined the 5' flanking sequences of the lectin gene for DNA elements that serve as binding sites for *trans*-acting factors and have closely correlated that data with transcriptional assays. The lectin gene is transcribed only in the developing embryos, not in the leaves, stems, or roots. When nuclear extracts were tested for their ability to bind to DNA from the 5' end of the lectin gene, binding activity could be detected using gel retardation assays in extracts from embryos, but not leaves, stems, or roots. Furthermore, the presence of the binding activity in the embryo extracts was stage specific. The maximum amount of binding activity was observed in extracts from embryos 40 days after flowering—precisely the time of maximal lectin gene transcription.

The approximate location of the binding site of the factors was determined by using various deletion fragments of the promoter region as substrate in gel-retardation binding assays. Only the region from -214 to -44 was capable of binding to nuclear proteins. The precise binding site was localized by DNase I-protection footprinting to two regions, region I from -184 to -173 and region II from -165 to -126. From the footprint data alone, it is unclear whether the two regions represent the bipartite binding site of a single protein factor, or the separate binding sites of independent factors. However, Jofuku et al.[56] were able to further characterize the binding activity by fractionating the nuclear extract on SDS-polyacrylamide gels. After electrophoresis, the proteins were transferred to nitrocellulose supports and reacted with labeled lectin gene DNA. The labeled DNA bound to a 60 kDa protein that was present in embryos at high levels only at 40 days after flowering. The 60 kDa protein was not present in stem nuclear extracts, nor was it able to bind DNA from the leghemoglobin gene (which is not coordinately expressed with lectin). Therefore it would appear that this 60 kDa protein represents a truly tissue-specific, stage-specific *trans*-acting regulatory factor of the lectin gene.

c. Wheat Histones

Plant and animal histone genes possess a conserved hexameric DNA sequence in their 5' flanking regions that have been hypothesized to play a role in the transcriptional regulation of histone genes. Mikami et al.[57] have used synthetic oligonucleotides representing this conserved consensus in gel retardation experiments utilizing wheat germ and seedling nuclear extracts. Specific binding of nuclear protein(s) was observed in both the seedling and germ extracts. Further experiments utilizing methylation interference reactions indicated that several G residues within and adjacent to the hexamer are intimately involved in the binding of the protein to the sequence element.

d. Ribulose Bisphosphate Carboxylase

In what represents the most comprehensive study of the protein interactions with a plant promoter published to date, Green et al.[58] have defined several interactions within the promoter of the pea rbcS3A (ribulose bisphosphate carboxylase) gene. Utilizing gel retardation assays, proteins were identified that bound to regions of the promoter previously shown to be directly involved in the light-induced activation of the gene. These regions of DNA are referred to as LREs, for light responsive elements. Without these *cis*-acting sequences, the rbcS3A will not respond to light as a transcriptional induction signal.

DNase I footprinting and methylation interference experiments showed that factor(s) bind to four separate regions in the promoter. Two of the regions, located next to each other between approximately -100 and -160, correspond directly with the most prominent LREs. Two additional binding sites were located further upstream (-210 to -260) and correlate to additional, redundant LREs that can be activated when the two more proximal LREs are deleted. The binding of nuclear factors to the LREs could be detected in extracts prepared either in the presence of light or in the dark. However, the character of the binding activity may be slightly different in the two types of extracts.

2. In Vivo Detection of Trans-Acting Factors

Adaptations of genomic sequencing[54] have allowed the *in vivo* detection of protein-DNA interactions in several animal gene systems and in one plant gene, maize *Adh1*.[59] Dimethyl sulfate (DMS) can be employed as a direct chemical probe for protein binding sites. Intact, living plant cells can be effectively treated with DMS without any manipulation that might disturb the chromatin structure or specific protein-DNA interactions. DMS rapidly penetrates plant cell walls and membranes to modify guanine (G) residues in genomic DNA *in situ*. Close protein-DNA contacts can be detected by DMS since proximity of amino acid chains within the major groove will either inhibit (protect) or enhance the ability of the DMS molecule to contact and modify the G residue.

The maize *Adh1* gene, which is inducible by anaerobic stress, was scanned for altered DMS reactivities that would indicate the binding of regulatory proteins.[59] In both aerobic and anaerobic cells, factors were detected bound to sequence elements in the area from -110 to -140. This region corresponds directly with the region of the promoter that earlier DNase I studies showed to be resistant to nuclease digestion regardless of the transcriptional state of the gene (Reference 40, discussed above). The constant presence of a factor or factors bound to this area near the induced hypersensitive region could well be responsible for the relative protection of that area from DNase I digestion.

The key feature of these protein interactions is that they change binding characteristics upon the onset of transcriptional induction. The pattern of DMS interactions changes as the cells become anaerobic. A likely explanation would be a change in the conformation of the regulatory protein. Furthermore, these binding sites correspond to the anaerobic response elements (AREs) that Walker et al.[42] have identified as being required for anaerobic induction in transient transformation assays.

The anaerobic induction of *Adh1* gene activity is further accompanied by the subsequent binding of factors at -90 to -100 and at -175 to -185. DMS footprints are seen over these sites only in cells that have been induced, and the sites appear to be vacant until anaerobic conditions occur.

C. PERSPECTIVES

In all of the systems, plant or animal, that have been studied *in vitro*, it is likely that only a few of what must be many *trans*-acting transcription factors have been characterized thus far. An inherent attribute of the *in vitro* assays is that they will primarily detect the most abundant protein factors present in the extracts. Therefore, other minor proteins that may play a key role in the regulation of the gene discussed above might well have yet escaped detection. *In vivo* DMS experiments may detect more factors, as it has the potential to detect every factor stably bound to the DNA. However, it is limited in that only G residue interactions are detectable. It is likely that the best overall approach will combine *in vivo* and *in vitro* methods in order to detect as many of the members of the regulated transcription complex as possible, and to confirm that the proteins analyzed *in vitro* are truly associated with the target DNA *in vivo*.

The study of DNA-binding proteins that perform *trans*-acting regulatory functions is currently a very active pursuit among plant molecular biologists. The methods for detecting protein-DNA interactions *in vitro* are simple and accurate. It is clear then that the examples cited in the previous section of this review represent only the initial observations from plant gene systems, and that many more examples will soon exist.

From the work published to date, there are no evident redundancies between the animal and plant *trans*-acting factors. None of the plant footprints or binding sites match, for example, the SP1 binding site.[60] No clear consensus binding sites can be derived among the plant binding sites so far presented. So it appeares that plant gene systems will make a unique contribution to the general knowledge of gene regulation and the details of chromatin structure that accomplish regulated gene transcription.

ACKNOWLEDGMENTS

The authors recognize the support of the United States Department of Agriculture Competitive Grant 86-CRCR-1-1997 during the writing of this paper for experiments on maize chromatin structure.

REFERENCES

1. **Spiker, S.,** Plant chromatin structure, *Annu. Rev. Plant Physiol.,* 36, 235, 1985.
2. **Olins, A. L. and Olins, D. E.,** Spheroid chromatin units (ν bodies), *Science,* 183, 330, 1974.
3. **Eissenberg, J. C., Cartwright, I. L., Thomas, G. H., and Elgin, S. C. R.,** Selected topics in chromatin structure, *Annu. Rev. Genet.,* 19, 485, 1985.
4. **Lyon, M. F.,** X-chromosome inactivation and developmental patterns in mammals, *Biol. Rev.,* 47, 1, 1972.
5. **Nagl, W.,** Nuclear ultrastructure: condensed chromatin in plants is species specific (karyotypical) but not tissue specific (functional), *Protoplasma,* 100, 53, 1979.
6. **Aldrich, H. C., Ferl, R. J., Hils, M. H., and Akin, D. E.,** Ultrastructural correlates of anaerobic stress in corn roots, *Tissue Cell,* 17, 341, 1985.
7. **Sachs, M., Freeling, M., and Okimoto, R.,** Anaerobic proteins of maize, *Cell,* 20, 761, 1980.
8. **Thoma, F., Koller, T., and Klug, A.,** An involvement of histone H1 in the organization of the nucleosome and the salt-dependent superstructure of chromatin, *J. Cell Biol.,* 83, 403, 1979.
9. **Weintraub, H.,** Histone H1 dependent chromatin superstructures and the supression of gene activity, *Cell,* 38, 17,1984.
10. **Lutz, C. and Nagl, W.,** A reliable method for preparation and electron microscopic visualization of nucleosomes in higher plants, *Planta,* 149, 408, 1980.
11. **Moreno, M. L., Sogo, J. M., and de la Torre, C.,** Higher plant nucleosomes. A micromethod for isolating and dispersing chromatin fibers, *New Phytol.,* 81, 681, 1978.
12. **McGhee, J. D. and Engle, J. D.,** Subunit structure of chromatin is the same in plants and animals, *Nature (London),* 254, 449, 1975.
13. **Greimers, R. and Deltour, R.,** Organization of transcribed and non-transcribed chromatin in isolated chromatin of *Zea maize* root cells, *Eur. J. Cell Biol.,* 23, 303, 1981.

14. **Richmond, T. J., Finch, J. T., Rushton, B., Rhodes, D., and Klug, A.,** Structure of the nucleosome core particle at 7A resolution, *Nature (London),* 311, 532, 1984.

15. **McGhee, J. D. and Felsenfeld, G.,** Nucleosome structure, *Annu. Rev. Biochem.,* 49, 1115, 1980.

16. **Johnson, E. M., Sterner, R., and Allfrey, V. G.,** Altered nucleosomes of active nucleolar chromatin contain accessible histone H3 in its hyperacetylated forms, *J. Biol. Chem.,* 262, 6943, 1987.

17. **Allegra, P., Sterner, R., Clayton, D. F., and Allfrey, V. G.,** Affinity chromatographic purification of nucleosomes containing transcriptionally active DNA sequences, *J. Mol. Biol.,* 196, 397, 1987.

18. **Levinger, L. and Varshavsky, A.,** Selective arrangement of ubiquinated and D1 protein containing nucleosomes within the *Drosophila* genome, *Cell,* 28, 375, 1982.

19. **Anderson, M. W., Ballal, N. R., Goldknopf, I. L., and Busch, H.,** Protein A24 lyase activity in nucleoi of thioacetamide-treated rat liver releases histone 2A and ubiquitin from conjugated protein A24, *Biochemistry,* 20, 1100, 1981.

20. **Matsui, S., Seon, B. K., and Sandberg, A. A.,** Disappearance of a structural chromosomal protein A4 in mitosis: implications for molecular basis of chromatin condensation, *Proc. Natl. Acad. Sci. U.S.A.,* 76, 6386, 1979.

21. **Weisbrod, S. and Weintraub, H.,** Isolation of a subclass of nuclear proteins responsible for conferring a DNase-I-sensitive structure on globin chromatin, *Proc. Natl. Acad. Sci. U.S.A.,* 76, 630, 1979.

22. **Weisbrod, S., Groudine, M., and Weintraub, H.,** Interactions of HMG 14 and 17 with actively transcribed genes, *Cell,* 19, 289, 1980.

23. **Sandeen, G., Wood, W. I., and Felsenfeld, G.,** The interaction of high mobility proteins HMG14 and 17 with nucleosomes, *Nucleic Acids Res.,* 8, 3757, 1981.

24. **Weisbrod, S. and Weintraub, H.,** Isolation of actively transcribed nucleosomes using immobilized HMG 14 and 17 and an analysis of alpha-globin chromatin, *Cell,* 23, 391, 1981.

25. **Yaniv, M. and Cereghini, S.,** Structure of transcriptionally active chromatin, *CRC Crit. Rev. Biochem.,* 21, 1, 1986.

26. **Elgin, S. C. R.,** Anatomy of hypersensitive sites, *Nature (London),* 309, 213, 1984.

27. **Wu, C.,** Two protein binding sites in chromatin implicated in the activation of heat-shock genes, *Nature (London),* 309, 229, 1984.

28. **Davidson, B. L., Elgy, J. M., Mulvihill, E. R., and Chambon, P.,** Formation of stable preinitiation complexes between eukaryotic class B transcription factors and promoter sequences, *Nature (London),* 301, 680, 1983.

29. **Dynan, W. S. and Tjian, R.,** The promoter specific transcription factor SP1 binds to upstream sequences in the SV40 early promoter, *Cell,* 35, 79, 1983.

30. **Weisbrod, S.,** Active chromatin, *Nature (London),* 297, 289, 1982.

31. **Thomas, G. H., Seigfried, E., and Elgin, S. C. R.,** DNase I hypersensitive sites: a structural feature of chromatin associated with gene expression, in *Chromosomal Proteins and Gene Expression,* Reeck, G., Goodwin, G., and Puigdomench, P., Eds., Plenum Press, New York, 1985, 77.

32. **Spiker, S., Murray, M. G., and Thompson, W. F.,** DNase I sensitivity of transcriptionally active genes in intact nuclei and isolated chromatin of plants, *Proc. Natl. Acad. Sci. U.S.A.,* 80, 815, 1983.

33. **Murray, M. A. and Kennard, W. C.,** Altered chromatin structure of higher plant gene phaseolin, *Biochemistry,* 23, 4225, 1984.

34. **Sawyer, R. M., Boulter, D., and Gatehouse, J. A.,** Nuclease sensitivity of storage-protein genes in isolated nuclei of pea seeds, *Planta,* 171, 254, 1987.

35. **Steinmüller, K., Batchaun, A., and Apel, K.,** Tissue-specific and light dependent changes of chromatin organization in barley *(Hordeum vulgare), J. Biochem.,* 158, 519, 1986.

36. **Kaufman, L. S., Watson, J. C., and Thompson, W. F.,** Light regulated changes in DNase I hypersensitive sites in the rRNA genes of *Pisum sativum, Proc. Natl. Acad. Sci. U.S.A.,* 84, 1550, 1987.

37. **Steinmüller, K. and Apel, K.,** A simple and efficient procedure for isolating plant chromatin which is suitable for studies of DNase I sensitive domains and hypersensitive sites, *Plant Mol. Biol.,* 7, 87, 1986.

38. **Wurtzel, E. T., Burr, F. A., and Burr, B.,** DNase I hypersensitivity and expression of the *Shrunken-1* gene of maize, *Plant Mol. Biol.,* 8, 251, 1987.

39. **Freeling, M.,** Simultaneous induction by anaerobiosis or 2, 4-D of multiple enzymes specified by 2 unlinked genes: differential *Adh1, Adh2* expression in maize, *Mol. Gen. Genet.,* 127, 215, 1973.

40. **Paul, A-L., Vasil, V., Vasil, I. K., and Ferl, R. J.,** Constitutive and anaerobically induced DNase I hypersensitive sites in the 5' region of the maize *Adh1* gene, *Proc. Natl. Acad. Sci. U.S.A.,* 84, 799, 1987.

41. **Howard, E. A., Walker, J. C., Dennis, E. S., and Peacock, W. J.,** Regulated expression of an alcohol dehydrogenase 1 chimeric gene introduced into maize protoplasts, *Planta,* 170, 535, 1987.

42. **Walker, J. C., Howard, E. A., Dennis, E. S., and Peacock, W. J.,** DNA sequences required for anaerobic expression of an alcohol dehydrogenase 1 chimeric gene, *Proc. Natl. Acad. Sci. U.S.A.,* 84, 6624, 1987.

43. **Lee, L., Fenoll, C., and Bennetzen, J. L.,** Construction and homologous expression of a maize *Adh1* based *NCOI* cassette vector, *Plant Physiol.,* 85, 327, 1987.

44. **Ellis, J. G., Llewellyn, D. J., Dennis, E. S., and Peacock, W. J.,** Maize *Adh*1 promoter sequences control anaerobic regulation: addition of upstream promoter elements from constitutive genes is necessary for expression in tobacco, *EMBO J.,* 6, 11, 1987.

45. **Becker, P., Renkawitz, R., and Schultz, G.,** Tissue specific DNase I hypersensitive sites in the 5' flanking sequences of the tryptophan oxygenase and the tryptophan aminotransferase genes, *EMBO J.,* 3, 2015, 1984.

46. **Dennis, E. S., Sachs, M. M., Gerlach, W. L., Finnegan, E. J., and Peacock, W.,** Alcohol dehydrogenase 2 (*Adh2*) gene of maize, *Nucleic Acids Res.,* 13, 727, 1985.

47. **Ashraf, M., Vasil, V., Vasil, I. K., and Ferl, R. J.,** Chromatin structure of the 5' promoter region of the maize *Adh2* gene and its role in gene regulation, *Mol. Gen. Genet.,* 208, 185, 1987.

48. **Coats, D., Taliercio, E. W., and Gelvin, S. B.,** Chromatin structure of integrated T-DNA in crown gall tumors, *Plant Mol. Biol.,* 8, 159, 1987.

49. **Nelson, J. A. and Groudine, M.,** Transcriptional regulation of human cytomegalovirus major intermediate-early gene is associated with induction of DNase I hypersensitive sites, *Mol. Cell. Biol.,* 6, 452, 1986.

50. **Fritton, H. P., Igo-Kemenes, T., Nowock, J., Stretch-Jurk, U., Thiesen, M., and Sippel, A. E.,** Alternative sets of DNase I hypersensitive sites characterize the various functional states of the chicken lysozyme gene, *Nature (London),* 311, 163, 1984.

51. **Burch, J. B. E. and Weintraub, H.,** Temporal order of chromatin structural changes associated with the activation of the major chicken vitellogenin gene, *Cell,* 33, 65, 1983.

52. **Fried, M. and Crothers, D. M.,** Equilibria and kinetics of lac repressor-operator interactions by polyacrylamide gel electrophoresis, *Nucleic Acids Res.,* 9, 6505, 1981.

53. **Church, G. M., Ephrussi, A., Gilbert, W., and Tonegawa, S.,** Cell-type-specific contacts to immunoglobulin enhancers in nuclei, *Nature (London),* 798, 801, 1985.

54. **Church, G. M. and Gilbert, W.,** Genomic sequencing, *Proc. Natl. Acad. Sci. U.S.A.,* 81, 1991, 1984.

55. **Maier, U.-G., Brown, J. W. S., Toloczyki, C., and Feiz, G.,** Binding of a nuclear factor to a consensus sequence in the 5' flanking region of zein genes from maize, *EMBO J.,* 17, 22, 1987.

56. **Jofuku, K. D., Okamuro, J., and Goldberg, R. B.,** Interaction of an embryo DNA binding protein with a soybean lectin gene upstream region, *Nature (London),* 328, 734, 1987.

57. **Mikami, K., Tabata, T., Kawata, T., Nakayama, T., and Iwabuchi, M.,** Nuclear protein(s) binding to the conserved DNA hexameric sequence postulated to regulate transcription of wheat histone genes, *FEBS Lett.,* 223, 273, 1987.

58. **Green, P. J., Kay, S. A., and Chua, N.-H.,** Sequence-specific interactions of a pea nuclear factor with light-responsive elements upstream of the rbcS-3A gene, *EMBO J.,* 6, 2543, 1987.

59. **Ferl, R. J. and Nick, H. S.,** *In vivo* detection of regulatory factor binding sites in the 5' flanking region of maize *Adh1, J. Biol. Chem.,* 262, 7947, 1987.

60. **Dynan, W. and Tjian, R.,** The promoter-specific transcription factor Sp1 binds to upstream sequences in the SV40 early promoter, *Cell,* 35, 79, 1983.

Chapter 8

PLOIDY MANIPULATIONS IN THE POTATO

S. J. Peloquin, Georgia L. Yerk, and Joanna E. Werner

TABLE OF CONTENTS

I. INTRODUCTION

The potato, *Solanum tuberosum* (2n = 4x = 48), is the best organism in which to manipulate whole sets of chromosomes. One can obtain haploids (sporophytes with the gametic chromosome number) from either unfertilized eggs or immature male gametophytes. Also, several meiotic variants have been identified which result in male and female gametophytes and gametes with the unreduced chromosome number (2n gametes).[1] Further, it is possible to go from 48 to 24 to 12 chromosome plants either through two cycles of parthenogenesis or by obtaining haploids from male gametophytes via anther culture. The ability to readily obtain sporophytes with the gametophytic chromosome number and gametophytes with the sporophytic chromosome number is the basis for the ease of manipulating sets of chromosomes in potatoes.

These ploidy manipulations are worthwhile in their own right. However, even more important and interesting is the recognition that they are the essential ingredients of: (1) new breeding methods which make use of haploids and 2n gametes to obtain maximum heterozygosity and epistasis in tetraploids, 4x, from 4x-2x and 2x-2x crosses, (2) effective and efficient transfer of germplasm from wild 2x species to 4x cultivars where haploids put the germplasm in usable form and 2n gametes transmit it to the 4x cultivated forms, (3) the origin and evolution of polyploids by sexual polyploidization in that 2n gametes provide the opportunity for both increasing the chromosome number and introgression from lower to higher ploidy levels, and (4) mapping genes in relation to the centromere by half-tetrad analysis.

The main cultivated potato of North America, South America, and Europe is a tetraploid with four similar sets of chromosomes. It is fortunate to have many wild and cultivated relatives. The tuber-bearing Solanums are represented by about 150 species.[2] These species extend from the U.S. through Chile, but are most abundant in the Andean regions of Peru and Bolivia. They grow from sea level to altitudes of more than 4000 m and in many different ecological conditions. More important, they are a rich source of diverse germplasm to be used in improving the tetraploid cultivated potato. Finally, the species represent a polyploid series with chromosome numbers of 24, 36, 48, 60, and 72 — with about 70% of the species having 24 chromosomes.

The genetics of the 4x cultivated potato is complicated by the presence of 4 sets of chromosomes, particularly in comparison to organisms such as corn which have only 2 sets of chromosomes.[3] A simple comparison using coins and dice illustrates the point. The chance of obtaining 2 heads on flipping 2 coins is 1/4 (1/2 × 1/2). This is the same chance one has of obtaining a homozygous recessive genotype *aa* in corn following selfing of a heterozygous *A/a* individual, where heads represents the dominant allele *A* and tails the recessive *a*, and one coin represents the egg and the other the sperm (1/2 the eggs and sperms carry *A* and 1/2 are *a*). The potato with 4 sets of chromosomes has 4 alleles at each locus and an *AAaa* individual on selfing produces the gametes 1*AA*: 4*Aa*: 1*aa*, assuming random chromosome segregation. Since a die has 6 sides we can let an ace represent the *aa* gamete. The chance of getting an ace on shaking a die is 1/6, the same as that of obtaining an *aa* gamete. With one die representing the male gamete and the other the female gamete, the chance of obtaining an *aaaa* plant is 1/6 × 1/6 = 1/36, the same as the chance of obtaining 2 aces on shaking the dice. The difference between 1/4 and 1/36 is large, but if we consider the chance of obtaining a plant homozygous recessive for three loci from the heterozygote it would be 1/64 in corn and 1/46,656 in potatoes. Thus, the ability to manipulate potatoes from 4x to 2x through haploidization simplifies the genetics enormously — it is flipping coins rather than rolling dice.

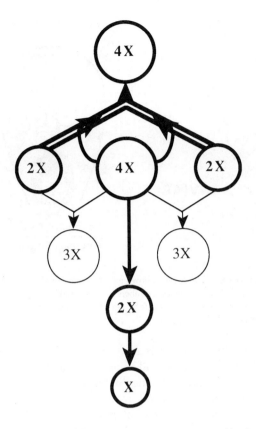

FIGURE 1.   Major ploidy manipulations in potatoes. Numbers represent ploidy levels. Circle on left of an arrow indicates female parent, circle on right indicates male parent, and circle with one line indicates haploidization. ⸗ 2n gamete; —— n gamete; —— less frequent.

II. PLOIDY MANIPULATIONS

The main ploidy manipulations discussed are illustrated in Figure 1. The 4x (tetraploid) can be reduced to 2x and then x through either two successive cycles of haploidization or by deriving, via anther culture, haploids (2n = 2x = 24) from developing microspores of a 4x and then haploids (2n = x = 12) from a 2x plant. If a 4x plant is crossed to a 2x plant that produces 2n pollen, mainly 4x plants are recovered. Similarly, if a 2x plant that produces 2n eggs is crossed to a 4x plant, mainly 4x progeny are recovered. These are both examples of unilateral sexual polyploidization. If the 2x plant producing 2n eggs is crossed with a 2x plant forming 2n pollen, 4x progeny are obtained (bilateral sexual polyploidization).[4] There are other ploidy manipulations involving 3x, 5x, and 6x plants, but they are of lesser importance.[5]

The product of functioning of an *n* gamete from a 4x and an *n* gamete from a 2x is a triploid embryo (3x). However, 3x plants are rarely recovered, since the associated endosperms do not develop normally and the developing seeds usually abort. An explanation of the basis of "triploid block" is provided by the endosperm balance number (EBN) hypothesis.[6] This hypothesis assumes that normal endosperm development occurs when EBN in the endosperm is a balance of 2 EBN from the female to 1 EBN from the male. Deviations from this 2:1 ratio result in abnormal endosperm development and seed failure. Since the endosperm in the developing seed of a 2x plant originates from a cell that contains 2 polar

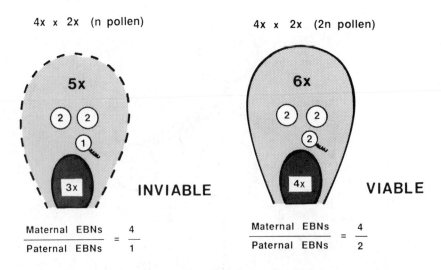

FIGURE 2.   Outcome of 4x × 2x crosses following functioning of either n or 2n gamete according to Endosperm Balance Number hypothesis. Numbers in circles indicate contribution of polar nuclei to endosperm, circle with tail indicates contribution of male nucleus. Lightly shaded areas = endosperms; darker shaded areas = zygotes.

nuclei each contributing 1 EBN and one male gamete nucleus with one EBN, the 2:1 ratio is maintained in 2x × 2x crosses. Figure 2 illustrates the situation in the endosperm following a 4x × 2x cross. The 2 polar nuclei from the 4x female each contribute 2 EBN and the 2x male contributes 1 EBN resulting in a 4:1 ratio and abnormal endosperm development. However, if in the 4x × 2x cross, the 2x plant produces 2n pollen, viable seeds containing 4x embryos are obtained. The 2n pollen doubles the EBN contribution of the male gamete nucleus to the endosperm. Thus, normal endosperm development occurs with 4 EBN from the female and 2 EBN from the male. EBN and 2n gametes are intimately associated with the ability to obtain many 4x progeny from 4x × 2x and 2x × 2x crosses. This association is vital to breeding methods involving these crosses, germplasm transfer from the 2x to 4x chromosome level, and in the origin of tetraploids from 2x populations and the introgression of 2x into 4x species in nature.

III. HAPLOIDIZATION

Twenty-eight haploid plants from seven clones of the common potato were obtained in the original trials.[7] They were found among seedling progenies of interploid crosses, Tuberosum (2n = 4x = 48) × Phureja (2n = 2x = 24), designed to make detection of haploids easy, through use of an appropriate marker gene. The success of the initial trials was encouraging in terms of the vigor and fertility of the haploids. It was evident that if the full potential of the haploid approach was to be exploited, large numbers of haploids from a wide range of 4x potatoes would be essential. Thus, it was important to find methods of increasing haploid frequency.

The results related to increasing haploid frequency have been rewarding in three areas: (1) Decapitation of the seed-parent resulted in a 10- to 15-fold increase in fruits per pollination and a similar increase in haploid frequency. This technique consists of cutting off the upper portion (inflorescence plus three or four leaves) of the plant at the time the first flower opens and placing the cut stem in a water-filled container in an air-conditioned greenhouse. (2) Selection of the seed-parent can also result in about a 10-fold increase in haploid frequency. (3) Most important and surprising was the finding that the ''pollinator'' (source of pollen) had a very significant effect on haploid frequency.[8] The use of superior ''pollinators'' resulted

TABLE 1
Polyploidizing Methods and Their Effects on Some Relevant Progeny
Characteristics in Potatoes

Progeny characteristics	Polyploidizing method		
	Sexual polyploidization	Somatic hybridization	Somatic doubling
Polyploidy	+	+	+
Heterosis	+	+	−
Genetic variability	+	−	−

in a 5- to 15-fold increase in haploid frequencies over a wide range of seed-parents. Haploid frequencies in the range of 30 to 100 per 100 fruit have been obtained consistently from combinations of elite seed-parents and elite "pollinators". Thus, thousands of haploids representing hundreds of seed parents have been obtained with modest effort in several countries.[9-12]

Attempts to obtain haploids of 4x potatoes from immature male gametophytes through anther culture have met with very limited success.[13] Further improvement in anther culture techniques may help, but it is easy to obtain haploids from 4x × 2x interploid crosses through the use of superior "pollinators" with appropriate marker genes, so that this will continue to be the major method of obtaining haploids.

The potato is unique in that the chromosome number has been reduced from 48 to 24 to 12 through two cycles of haploid extraction. Haploids (2n = x = 12) of 2n = 2x = 24 clones have been obtained from both 2x × 2x crosses and anther culture.[14,15] More than 500, 12-chromosome haploids were obtained from over 2 million seeds from 2x × 2x crosses.[16] As expected, the 12-chromosome plants were very unthrifty and most failed to produce tubers, so they were maintained in stem tip cultures. Induced chromosome doubling of five, 12 chromosome clones produced homozygous 24 and 48 chromosome plants. These plants were modestly improved in vigor and tuberization, but were highly sterile. Anther culture has been used extensively in attempts to obtain 12 chromosome plants from 24 chromosome clones. Although a few 12 chromosome haploids were obtained, 99% of the regenerated plants had 24 or 48 chromosomes originating from 2n gametes, sporophytic tissue, and/or doubled monoploid cells.[17] Haploids (2n = x = 12) have also been obtained from 2x wild *Solanum* species, particularly *S. verrucosum,* through anther culture.[12]

IV. 2N GAMETES

It is interesting and instructive to compare the three major polyploidizing methods, sexual polyploidization, somatic hybridization, and somatic doubling, in relation to their effect on relevant progeny characteristics (Table 1). Sexual polyploidization provides for increased chromosome number, heterosis, and genetic variability. However, somatic hybridization does not generate genetic variability, and somatic doubling only increases the chromosome number. Intercrossing, absence of crossing-over, and first division restitution in both sexes will be the nuclear equivalent of somatic hybridization of two individuals.

Inherited variations in the meiotic process that result in 2n gametes provide the opportunity for manipulating whole sets of chromosomes to increase the ploidy level following either 4x-2x or 2x-2x matings. Such variations affecting microsporogenesis result in 2n pollen, and those affecting megasporogenesis lead to 2n eggs. Thus, one can obtain 4x progeny from either 4x × 2x crosses following functioning of 2n pollen, or from 2x × 4x matings if 2n eggs are formed, and from 2x × 2x hybridization if both 2n pollen and 2n eggs function in the respective parents.

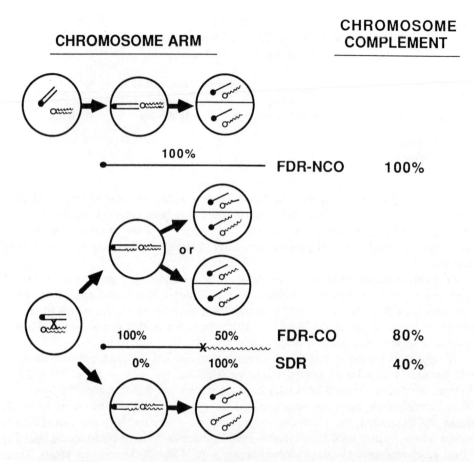

FIGURE 3. Percent parental heterozygosity transmitted to offpsring by 2n gametes for either a chromosome arm or the entire 2x chromosome complement if 2n gametes are formed by FDR-NCO (first division restitution without crossing-over), FDR-CO (first division restitution with crossing-over), or SDR (second division restitution).

There are six distinct possible modes of 2n gamete formation: (1) premeiotic doubling, (2) first division restitution (FDR), (3) chromosome replication during meiotic interphase, (4) second division restitution (SDR), (5) post meiotic doubling, and (6) apospory — the development of the gametophyte from a somatic cell. The two modes of 2n gamete formation in potatoes are FDR and SDR.

The significance of meiotic mutants that result in 2n gametes resides in their genetic consequences as illustrated in Figure 3. With FDR all loci from the centromere to the first crossover, that are heterozygous in the parent, will be heterozygous in the gametes, and one half the loci between the first and second crossovers, that are heterozygous in the parent, will be heterozygous in the gametes. One can roughly estimate the total amount of heterozygosity transmitted from parent to offspring by FDR by estimating arm ratios and chiasmata for each of 12 bivalents, and assuming each chromosome has similar amounts of genetic information. For example, for a chromosome with a 1:0 arm ratio, 75% of the heterozygosity is transmitted; with a 1:1 arm ratio and a crossover in the middle of one arm, 88% of the heterozygosity is transmitted. Using this method it is estimated that about 80% of the heterozygosity is transmitted from parent to offspring with FDR. In addition, a large proportion of the epistasis is also transmitted from parent to progeny. FDR with crossing-over is the predominant mode of 2n pollen formation in potatoes.[18]

The genetic consequences are different from FDR 2n gametes in which no crossing-over occurs. The lack of crossing-over results in the intact genotype of the parent being incorporated into each 2n gamete. Significantly, 100% of the heterozygosity and epistasis of the parent can be transmitted to all progeny.

The genetic consequences with SDR are extremely different. All loci from the centromere to the first crossover will be homozygous in the gametes, and all loci between the first and second crossovers, that are heterozygous in the parent, will be heterozygous in the gametes. For a chromosome with a 1:0 arm ratio, 50% of the heterozygosity will be transmitted and for one with a 1:1 arm ratio, 25%. Following the same method used to calculate the percentage of heterozygosity transmitted with FDR, less than 40% of the parental heterozygosity will be transmitted from parent to progeny with SDR. SDR is the most prevalent mode of 2n egg formation in potatoes.

There are several features of meiotic variations resulting in 2n gametes which should be noted: (1) almost all of those reported are controlled by a single locus; (2) they are invariably recessive; (3) sometimes the character usually associated with a particular genotype fails to appear — penetrance is not always 100%; (4) the frequency of meiocytes that express the parental genotype can vary from 1 to 100% — this variable expressivity is the rule, not the exception; and (5) expression is significantly modified by genetic, environmental, and developmental factors.

A. 2N POLLEN

The existence of male gametophytes with the sporophytic chromosome number has been known for some time.[19] However, its enormous potential in ploidy manipulations was not perceived, perhaps because it was viewed as a rare event. The elucidation of the potential of 2n pollen in ploidy manipulations had its origin in research results related to evolution and breeding methods. One such result was the experimental demonstration in *S. chacoense* that tetraploid progeny could result from 4x × 2x crosses if the 2x plant produces 2n pollen, and from 2x selfing if it produces both 2n pollen and 2n eggs.[20] Thus, from an evolutionary standpoint 4x progeny could arise from 2x plants by a sexual process, and introgression from the 2x to 4x level in nature could occur through functioning of 2n pollen.

The first meiotic mutants that result in 2n pollen were found while testing 4x-2x crosses as a breeding method. Twenty, 4x cultivars were crossed with more than 100, 2x Phureja-haploid Tuberosum hybrids.[21] The progeny were mainly 4x, very vigorous, and the mean tuber yield of many 4x families exceeded that of the cultivar parent. Several 2x parents were exceptional in that many rather than few seeds/fruit were obtained when they were used as pollen parents, and more than 99% of the progeny were 4x; this suggests the possibility of a systematic method of 2n pollen formation. Among exceptional 2x clones there were two types based on characteristics of the progeny. The first type gave the higher yielding, more uniform progeny.

Cytological analyses provided the clues to explain the breeding results. In the normal sequence of microsporogenesis the first meiotic division is not followed by cytokinesis, the two spindles in the second division are oriented such that their poles define a tetrahedron, and cytokinesis results in a tetrad of four n microspores (Figure 4). In the first type of exceptional clones the first division is normal, but in some microsporocytes second division spindles are parallel, and following cytokinesis two, 2n microspores are formed.[22] The correlation between the frequency of 2n pollen and parallel spindles was high. The parallel orientation of second division spindles is inherited as a simple recessive, *ps*.[23]

The results of parallel spindles is genetically equivalent to FDR. Thus, the high yields and uniformity of the 4x progeny from 4x × 2x crosses could be explained on the basis of the cytological mechanism underlying their production. FDR 2n gametes, as indicated earlier, transmit approximately 80% of the parental heterozygosity and a large proportion of the epistasis to the progeny and thus generate uniform and high-yielding families from 4x × 2x crosses.

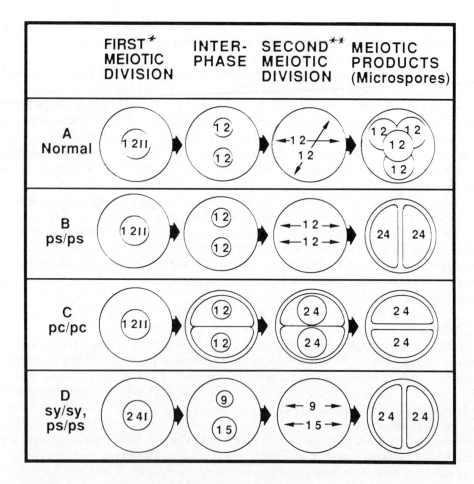

FIGURE 4. Cytological consequences of normal microsporogenesis (A) and microsporogenesis in three meiotic mutants that result in 2n pollen: parallel spindles (B), premature cytokinesis (C), and combination of parallel spindles with synaptic mutant (D). Smaller circles represent nuclei. *12II = 12 bivalents; 24I = 24 univalents. **Thin arrows indicate meiotic spindles.

The second type of exceptional 2x clones gave 4x progeny that were not uniform and lower yielding. Cytological analysis revealed a different meiotic variation. The first division in many sporocytes is followed by cytokinesis; the second division does not occur, except that the chromatids fall apart, and a dyad of two, 2n microspores is formed.[22] This variation, premature cytokinesis (*pc*) is also simply inherited.[23] Premature cytokinesis is genetically equivalent to SDR, and transmits about one half of the heterozygosity of FDR.

A third mechanism of 2n pollen formation combines a synaptic mutant with parallel spindles. These synaptic mutants are characterized by poor pairing and/or reduced chiasma frequencies in microsporogenesis. The synaptic variant, *sy-3*, was found in Phureja-haploid hybrids.[24] It is characterized, in particular clones, by lack of chiasma and 24 univalents at the first meiotic division. Univalents are distributed at random in anaphase I, for example, 15 to one pole and 9 to the other. This behavior usually leads to complete male sterility, since the chance of the right 12 chromosomes going to one pole is 1/4,096. Fortunately, if clones with *sy-3* are also homozygous for parallel spindles, a high frequency of functional 2n pollen is produced. The unequal distribution of chromosomes in the first division is rectified by parallel spindles in the second division, which results in two sets of 12 chromosomes in each 2n spore of the dyad. The genetic significance of this combination of meiotic mutants is that the intact genotype of the parent is incorporated into each 2n male gamete.

Two other synaptic variants, *sy-2* and *sy-4*, have been found, one in *S. commersonii*, the other in haploids of the cultivar Atzimba.[25,26] Bivalents are almost always lacking in *sy-2*, but a small and variable number of bivalents occur in clones with *sy-4*.

With the cytogenetics and genetics of 2n pollen formation elucidated, information concerning the prevalence of 2n pollen among both wild and cultivated taxons provided the final bit of evidence necessary in the realization of the enormous potential 2n pollen has for ploidy manipulations. A summary of the work in this area is available.[5,18] These data overwhelmingly support the concept that the occurrence of 2n pollen is widespread among the taxons. In addition, the allele conditioning the formation of 2n pollen is present in high frequencies in many of the species examined.[27]

The predominant mechanism of 2n pollen formation in all the materials examined has been parallel spindles. In addition to being present in wild species, parallel spindles occur in high frequencies in the cultivated taxons, Phureja (2x), Stenotomum (2x), Andigena (4x), and Tuberosum (4x).[18,28]

The abundance of 2n pollen, its simple inheritance and ease of identification, as well as its ability to transmit heterozygosity (intralocus interactions) and epistasis (interlocus interactions), have made it a powerful tool for manipulating ploidy levels.

B. 2N EGGS

There are two steps in evaluating clones for 2n egg production: (1) identifying clones that produce 2n eggs and (2) determining the mode, SDR or FDR, and mechanism of 2n egg formation. The occurrence of 2n eggs can be established on the basis of obtaining 4x progeny following 2x × 4x crosses.[21] This interploid cross results in almost exclusively 4x progeny, since there is a severe "triploid block". The 4x endosperm (2 maternal EBN:2 paternal EBN) associated with triploid embryos develop abnormally and the seeds abort.[6] The number of seeds per fruit following 2x × 4x crosses is a measure of the frequency of 2n eggs. Seeds per fruit varied from 1 to about 100.[29]

The mode of 2n egg formation can be most effectively determined cytologically by microscopic observations of megasporogenesis and megagametogenesis, and genetically through use of half-tetrad analysis. An important advance for studying the cytology of 2n egg formation was the development of a stain-clearing technique for whole ovules.[30] It is very much faster and significantly more effective in distinguishing n from 2n eggs than the traditional embedding-sectioning method. Through application of this new method, cytological analysis of 2n egg formation in a large number of 2x clones representing haploids of Tuberosum, haploid species hybrids, and species has been possible. In several clones 60% of the ovules contained 2n eggs.[29]

Five mechanisms of 2n egg formation have been identified in diploid potatoes through cytological observations of stained-cleared whole ovules.[29,31] Normal megasporogenesis consists of two meiotic divisions, each followed by cytokinesis to give four n megaspores (Figure 5). The five mechanisms that can result in 2n eggs are (1) omission of the second meiotic division, (2) failure of cytokinesis after the second meiotic division followed by nuclear fusion, (3) abnormal chromosome movement in the second division with formation of restitution nuclei, (4) synaptic variant with mainly univalents at the first division, followed by restitution and a normal second division, and (5) omission of the first division and a mitotic-like second division. Variations 1, 2, and 3 are SDR mechanisms, and 4 and 5 are FDR mechanisms. Cytological analysis also indicated that within a clone, 2n eggs can be formed by more than one mechanism such that mixtures of FDR and SDR 2n eggs can occur.

Half-tetrad analysis provides a method of mapping a gene in relation to the centromere and of genetically determining the mode of 2n gamete formation.[32] The procedure for determining the mode of 2n egg formation involves crossing a 2x clone heterozygous for a

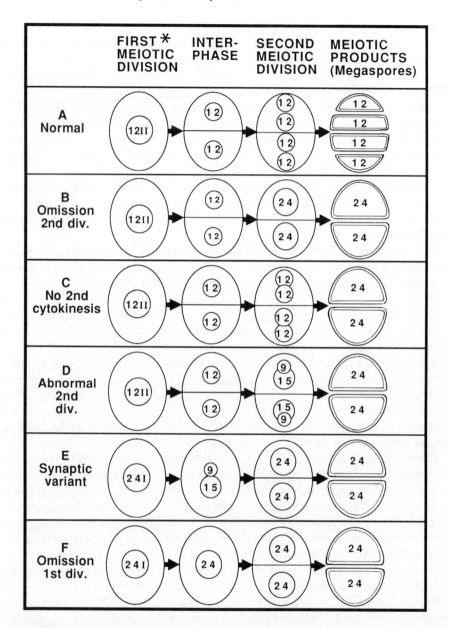

FIGURE 5.   Cytological consequences of normal megasporogenesis (A) and megasporogenesis in five meiotic variants that result in 2n eggs (B,C,D,E, and F). Circles represent nuclei. *12II = 12 bivalents; 24I = 24 univalents.

marker gene with a monoallelic 4x clone (*Y/y* × *yyyy*), and determining the frequency of *yyyy* progeny. For example, if *yyyy* results in white and *Y-* in yellow tuber flesh and the *Y* locus is 18 map units from the centromere, then if 2n eggs are formed by SDR, 32% of the 4x progeny will have white tuber flesh, while 9% are expected to have white flesh if FDR is the mode of formation. The genetic results with half-tetrad analysis demonstrate that a large majority of 2x clones form 2n eggs by SDR, a few by FDR, and a few by both SDR and FDR.[33] The recent finding of marker genes within a few map units of the centromere will greatly improve half-tetrad analysis as a method of determining the mode of 2n egg formation.[34]

Meiotic variations that result in 2n eggs are required for the ploidy manipulations essential for breeding schemes based on obtaining 4x progeny from 2x × 4x and 2x × 2x crosses. They are also an integral part of the ploidy manipulations which have contributed to the evolution of the 4x potatoes by allowing for the origin of tetraploids from diploid populations.

REFERENCES

1. **Peloquin, S. J.,** Meiotic mutants in potato breeding, *Stadler Genet. Symp.,* 14, 1982, 99.
2. **Huaman, Z. and Ross, R. W.,** Updated listing of potato species names, abbreviations, and taxonomic status, *Am. Potato J.,* 62, 629, 1985.
3. **Hougas, R. W. and Peloquin, S. J.,** The potential of potato haploids in breeding and genetic research, *Am. Potato J.,* 35, 701, 1958.
4. **Mendiburu, A. O., Peloquin, S. J., and Mok, D. W. S.,** Potato breeding with haploids and 2n gametes, in *Haploids in Higher Plants,* Kasha, K., Ed., University of Guelph, Guelph, Canada, 1974, 249.
5. **den Nijs, T. P. N. and Peloquin, S. J.,** 2nd gametes in potato species and their function in sexual polyploidization, *Euphytica,* 26, 585, 1977.
6. **Johnston, S. A., den Nijs, T. P. N., Peloquin, S. J., and Hanneman, R. E., Jr.,** The significance of genic balance to endosperm development in interspecific crosses, *Theor. Appl. Genet.,* 57,5, 1980.
7. **Hougas, R. W., Peloquin, S. J., and Ross, R. W.,** Haploids of the common potato, *J. Hered.,* 49, 103, 1958.
8. **Hougas, R. W., Peloquin, S. J., and Gabert, A. C.,** Effect of seed-parent and pollinator on frequency of haploids in *Solanum tuberosum, Crop Sci.,* 4, 593, 1964.
9. **Bender, K.,** Uber die Erzeugung und Entstehing dihaploider Pflanzen bei *Solanum tuberosum, Z. Pflanzenzuecht.,* 50, 141, 1963.
10. **Frandsen, N. O.,** Haploidproduktion ans einem Kartoffelzuchtmaterial mit intersiven Wildarteinkreuzing, *Zuchter,* 37, 120, 1967.
11. **Hermsen, J. G. T. and Verdenius, J.,** Selection from *Solanum tuberosum* Group Phureja of genotypes combining high frequency haploid induction with homozygosity for embryo-spot, *Euphytica,* 22, 244, 1973.
12. **Irikura, Y.,** Cytogenetic studies on haploid plants of tuber-bearing Solanums. I. Induction of haploid plants of tuber-bearing Solanums, *Res. Bull. Hokkaido Natl. Agric. Exp. Stn.,* 112, 1, 1975.
13. **Johansson, L.,** Improved methods for induction of embryogenesis in anther cultures of *Solanum tuberosum, Potato Res.,* 29, 179, 1986.
14. **van Breukeln, E. W. N., Ramanna, M. S., and Hermsen, J. G. T.,** Monohaploids (2n = x = 12) from autotetraploid *Solanum tuberosum* (2n = 4x = 48) through two successive cycles of female parthenogenesis, *Euphytica,* 24, 567, 1975.
15. **Foroughi-Nehr, B., Wilson, H. M., Mix, G., and Gaul, H.,** Monohaploid plants from anthers of a dihaploid genotype of *Solanum tuberosum* L., *Euphytica,* 26, 361, 1977.
16. **Uijtewaal, B. A., Huigen, D. J., and Hermsen, J. G. T.,** Production of potato monohaploids (2n = x = 12) through prickle pollination, *Theor. Appl. Genet.,* 73, 751, 1987.
17. **Uhrig, H.,** Genetic selection and liquid medium conditions improve the yield of androgenetic plants from diploid potatoes, *Theor. Appl. Genet.,* 71, 455, 1985.
18. **Watanabe, K.,** Occurrence, frequency, cytology, and genetics of 2n pollen; and sexual polyploidization in tuber-bearing Solanums, Ph.D. thesis, University of Wisconsin, Madison, 1988.
19. **Bleier, H.,** Untersuchungen uber die sterilitat der kartoffel, *Arch. Pflanzenbau,* 5, 545, 1931.
20. **Marks, G. E.,** The origin and significance of intraspecific polyploidy: experimental evidence from *Solanum chacoense, Evolution,* 20, 552, 1966.
21. **Hanneman, R. E., Jr. and Peloquin, S. J.,** Ploidy levels of progeny from diploid-tetraploid crosses in the potato, *Am. Potato J.,* 45, 255, 1968.
22. **Mok, D. W. S. and Peloquin, S. J.,** Three mechanisms of 2n pollen formation in diploid potatoes, *Can. J. Genet. Cytol.,* 17, 217, 1975.
23. **Mok, D. W. S. and Peloquin, S. J.,** The inheritance of three mechanisms of diplandroid (2n pollen) formation in diploid potatoes, *Heredity,* 35, 295, 1975.
24. **Okwuagwu, C. O. and Peloquin, S. J.,** A method of transferring the intact parental genotype to the offspring via meiotic mutants, *Am. Potato J.,* 58, 512, 1981.
25. **Johnston, S. A., Rhude, R. W., Ehlenfeldt, M. K., and Hanneman, R. E., Jr.,** Inheritance and microsporogenesis of a synaptic mutant *(sy-2)* from *Solanum commersonii* Dun., *Can. J. Genet. Cytol.,* 28, 520, 1986.

26. **Iwanaga, M.,** Discovery of a synaptic mutant in potato haploids and its usefulness for potato breeding, *Theor. Appl. Genet.,* 68, 87, 1984.
27. **Camadro, E. L. and Peloquin, S. J.,** The occurrence and frequency of 2n pollen in three diploid Solanums from northwest Argentina, *Theor. Appl. Genet.,* 56, 11, 1980.
28. **Iwanaga, M. and Peloquin, S. J.,** Origin and evolution of cultivated tetraploid potatoes via 2n gametes, *Theor. Appl. Genet.,* 61, 161, 1982.
29. **Werner, J. E. and Peloquin, S. J.,** Frequency and mechanisms of 2n eggs formation in haploid Tuberosum-wild species F₁ hybrids, *Am. Potato J.,* 64, 641, 1987.
30. **Stelly, D. M., Peloquin, S. J., Palmer, R. G., and Crane, C. F.,** Mayer's hemalum-methyl salicylate: a stain-clearing technique for observations within whole ovules, *Stain. Technol.,* 59, 155, 1984.
31. **Stelly, D. M. and Peloquin, S. J.,** Formation of 2n megagametophytes in diploid tuber-bearing Solanums, *Am. J. Bot.,* 73, 1351, 1986.
32. **Mendiburu, A. O. and Peloquin, S. J.,** Gene-centromere mapping by 4x-2x matings in potatoes, *Theor. Appl. Genet.,* 54, 177, 1979.
33. **Stelly, D. M. and Peloquin, S. J.,** Diploid female gametophyte formation in 24-chromosome potatoes: genetic evidence for the prevalence of the second meiotic division restitution mode, *Can. J. Genet. Cytol.,* 28, 101, 1986.
34. **Douches, D. S. and Quiros, C. F.,** Use of 4x-2x crosses to determine gene-centromere map distances of isozyme loci in *Solanum* species, *Genome,* 29, 519,1987.

Chapter 9

THE CHLOROPLAST GENOME AND REGULATION OF ITS EXPRESSION

Peta C. Bonham-Smith and Don P. Bourque

TABLE OF CONTENTS

I. INTRODUCTION

Scientists are currently orchestrating and performing the Herculean task of determining the sequence of the entire human genome. One might imagine that publication of the complete nucleotide sequence of chloroplast DNA[1,2] would, in short order, lead to a solution of all the major biological questions regarding the functions of chloroplast DNA and regulation of chloroplast gene expression. This is a naive conclusion — the battle has just begun. In this review, we highlight what has been learned from knowing the complete sequence of two chloroplast genomes and integrate this information with the current state of knowledge regarding regulation of chloroplast gene expression. We hope to encourage continued research by discussing some of the remaining opportunities for discovery in chloroplast molecular biology.

Numerous previous reviews of chloroplast genome structure and function exist.[3-6] This review presents a summary of significant recent discoveries, rather than a comprehensive literature compilation. In particular, attention is focused on the subject of regulation of chloroplast gene expression. Potential regulatory mechanisms are discussed, with emphasis on the ribosomal protein genes as an example of a group of genes coding for proteins with related functional roles, i.e., involved in chloroplast protein synthesis.

II. FEATURES OF CHLOROPLAST DNA AND ITS GENES

Chloroplasts are found in all members of the kingdom *Plantae* and some members of the kingdom *Protista*. Actively dividing meristematic cells of higher plants contain small, undifferentiated organelles called proplastids which, depending upon the fate of the cell, are capable of differentiating into amyloplasts, chromoplasts, or chloroplasts.[7-10] Plastid molecular biology research has concentrated on chloroplasts, the organelles responsible for photosynthesis.

A. INVERTED REPEATS AND rRNA OPERONS

Chloroplasts contain one or more copies of a DNA genome which, in the majority of higher plants, is double-stranded, circular, and organized into two single copy regions of different lengths, separated by two identical repeat regions (the size of which varies) whose orientations are inverted, respectively (Figure 1).[11] Exceptions to this general structure include some legumes, such as *Vicia faba*[12] (broad bean) and *Pisum sativum*[13,14] (pea), and the red alga *Griffithsia pacifica,*[15] which contain only one copy of the region corresponding to the inverted repeat. In the legumes this single region is split almost in half, with each half being located separately on the genome.[16,17] Another exception to the general chloroplast structure is found in the green alga *Euglena gracilis,* where different strains can contain one to five tandemly arranged copies.[18]

In the majority of plant species, each repeated segment carries genes coding for a variable number of proteins and one operon for ribosomal RNAs. The rRNAs are arranged, as are their homologous genes in *Escherichia coli* and the blue-green alga *Anacystis nidulans,*[19] in the conserved order: [5']-16S rRNA-spacer (containing tRNA^{Ile} and tRNA^{Ala})-23S rRNA-[4.5S rRNA]-spacer-5S rRNA-[3'].[5] Unlike bacteria, chloroplast ribosomes of most higher plants contain a 4.5S rRNA with 67% (tobacco)[20] and 71% (maize)[21] homology to the 3'-end of eubacterial 23S rRNA. This RNA species results from processing of the chloroplast 23S rRNA gene transcript.[22-24] Instead of a 4.5S rRNA, the ribosomes of *Chlamydomonas reinhardii* contain two rRNA species, 7S and 3S, which are homologous (7S, between 65 and 67%, and 3S, between 60 and 74%) to the 5' terminal region of prokaryotic and other chloroplast 23S rRNAs.[25] In *Chlamydomonas,* the rRNA genes are arranged [5']16S-7S-3S-23S[3'].

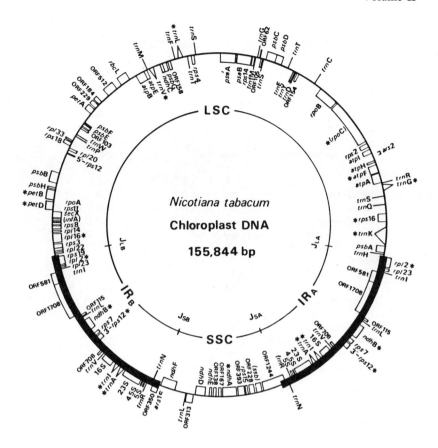

FIGURE 1.   Circular map of the tobacco chloroplast genome. Genes shown on the outside of the map are coded on the A strand and genes on the inside of the map are on the B strand. An asterisk indicates an intron containing gene. (From Shinozaki, K., et al., *Plant Mol. Biol. Rep.*, 4, 112, 1986. With permission.)

In higher plants, introns occur in the tRNA^{Ile} and tRNA^{Ala} genes;[5] however, in *Chlamydomonas reinhardii* both of these genes are uninterrupted.[26] In contrast to higher plants, the 23S rRNA gene of the two algae, *C. reinhardii*[27] and *Chlorella ellipsoidea*,[28] possess introns. Furthermore, the rRNA genes of *C. ellipsoidea* are located in two operons. Operon 1 contains the 16S rRNA and tRNA^{Ile} genes while operon 2 has the tRNA^{Ala}-23S rRNA-5S rRNA segment. Due to an inversion of operon 2, it is transcribed in the opposite direction to that of operon 1.[28] Polarity reversal of inverted repeats,[29] low rates of nucleotide substitution within them[30] and frequent rearrangements[17,31] of the chloroplast genome have been observed. Thus, several variations of chloroplast RNA operon and gene structure can be identified among different plant species.

B. HETEROGENEITY OF CHLOROPLAST GENOMES

Much experimental evidence supports the idea that the population of chloroplast DNA molecules within a single chloroplast, a cell, or even within an entire plant is homogeneous.[31] However, the inheritance of unique T1 RNase digestion fragments of tobacco chloroplast 16S rRNA,[32] the relative orientation of single copy and inverted repeat regions in bean,[29] the different copy numbers of specific restriction fragment length polymorphisms in individual rice plants,[33] and heteroplasmicity in *Chlamydomonas*[34,35] provide evidence that plastid genome heterogeneity exists within an individual plant. This situation is consistent with that found for plant mitochondrial genomes.[36-38]

TABLE 1
Comparison of Chloroplast DNA Genes

Gene classification		Tobacco[1,39]	Marchantia[2,40]
rRNA		4	4
tRNA		30	32
Total RNA genes		34	36

Protein function	Gene		
ATPase	atp[a]	A,B,E,F,H,I	A,B,E,F,H,I
Photosystem I	psa	A,B,C	A,B
Photosystem II	psb	A,B,C,D,E,F,G	A,B,C,D,E,F,G
Cytochrome b/f	pet	A,B,D	A,B,D
RNA polymerase	rpo	A,B	A,B,C1,C2
NADH dehydrogenase	ndh	A,B,C,D,E,F	1,2,3,4,4L,5,6
RuBisCo	rbc	L	L
Ribosomal proteins			
Small subunit	rps	12	11
Large subunit	rpl	7	8
	sec	X	X
Initiation factor	inf	A	A
Membrane transport protein	mbp		X,Y
Fe-S ferredoxin	frx		A,B,C
Total protein genes[b]		48	54
URFs[c]		43	At least 28
Intron-containing genes		15	12

[a] Gene nomenclature according to chloroplast convention.[41]
[b] Total number of identified protein-coding genes.
[c] Unidentified reading frame.

C. CODING FUNCTIONS OF CHLOROPLAST DNA

In 1986, the complete chloroplast DNA nucleotide sequences of an angiosperm, *Nicotiana tabacum,*[1] and a liverwort, *Marchantia polymorpha,*[2] were published. The tobacco sequence is 155,844 bp, while that of *Marchantia* is 121,024 bp. The size difference between these genomes is due to the 15 kbp greater length of each inverted repeat region of the tobacco chloroplast DNA. In general, the coding functions of these genomes for RNA and proteins have been highly conserved during evolution (Table 1). Most of these chloroplast genes have been shown to be transcribed.[31,42-44] Four rRNA species, 23S, 16S, 5S, and 4.5S and thirty tRNA genes are common to both the tobacco and *Marchantia* genome. In addition, *Marchantia* chloroplast DNA contains tRNA[Arg] (CCG) and pseudo-tRNA[Pro] (GGG) genes.[2] Among the unidentified reading frames (URFs), many sequences are conserved between tobacco and *Marchantia,* suggesting that they may indeed encode functional polypeptides which are unique to the chloroplast.

Plastid metabolism requires the products of hundreds of genes, of which the plastid genome only contains a small fraction. The majority of chloroplast proteins are nuclear-encoded polypeptides which are translated by cytoplasmic 80S ribosomes and subsequently transported into the chloroplast.[45] However, the chloroplast genome does code for both soluble stromal and thylakoid membrane proteins. The genes for stromal proteins (three putative RNA polymerase subunits, 19 ribosomal proteins, and the translation initiation factor If-1) together with the tRNA and rRNA genes all code for macromolecules involved in expression of the chloroplast genome. In the case of the tobacco chloroplast translation system, a number of nuclear-gene-encoded proteins must also be essential, including 35 or more additional ribosomal proteins.[46] The large subunit of the CO_2-fixing enzyme ribulose-

1,5-bisphosphate carboxylase/oxygenase (RuBisCo) is encoded by the rbcL gene of chloroplast DNA in all plants,[5] while the small subunit of RuBisCo (rbcS) is nuclear-encoded, with a few exceptions in cyanelles and *Rhodophyceae* where it is chloroplast-encoded.[47,48] Accumulation of the small subunit in the chloroplast, but not its synthesis in the cytoplasm, appears to be regulated by a proteolytic mechanism in the chloroplast.[49] Many of the polypeptides which are assembled into four major thylakoid membrane complexes (photosystems I and II, the cytochrome b_6/f complex, and the chloroplast ATPase complex) are encoded by the chloroplast genome[1,2] while the remaining polypeptides and the proteins of the fifth major thylakoid membrane complex (the light-harvesting chlorophyll protein complex) appear to be entirely nuclear-encoded.[1,2,50]

The tobacco,[40] *Marchantia*,[51] rice, sugar beet, and broad bean[52] chloroplast genomes contain sequences which translate into polypeptides homologous to six subunits (seven in *Marchantia*) of human mitochondrial NADH dehydrogenase. Northern blot hybridization has demonstrated that all six (seven) genes (ndh A-F or ndh 1—6 in *Marchantia*) are actively transcribed in the chloroplasts of plants grown under light or dark conditions.[40,51] Thus, an NADH-oxidizing electron-transfer chain similar to that reported in *Chlamydomonas*[53-55] may exist in higher plants.

D. INTRACELLULAR GENE TRANSFER AND EVOLUTION

The presence of the NADH dehydrogenase genes on the chloroplast genome may be the result of a gene transfer from mitochondria to chloroplasts, the reverse of chloroplast to mitochondria gene transfer for which there is suggestive evidence.[56,57] Transfer of DNA from the chloroplast to the nucleus during evolution may have occurred in the case of the tufA gene. A complete gene for the protein synthesis elongation factor Tu has been sequenced in the chloroplast genomes of the green algae *Chlamydomonas reinhardii*[58] and *Euglena gracilis*.[59] In higher plants, the gene is thought to be nuclear, suggesting that a chloroplast to nucleus gene transfer occurred early during higher plant evolution. Although the direction of transfer is unknown, examples of homologous chloroplast gene sequences which are present in spinach in both the nucleus and the chloroplast have been reported.[60,61]

There is no evidence that RNA from nuclear genes is transported into and expressed in chloroplasts. However, there is one report of transport of a nuclear-encoded 138-nucleotide RNA, which is essential for RNase MRP (mitochondrial RNA processing) activity, into mammalian mitochondria.[62] RNase MRP is a site-specific, single-stranded endonuclease involved in RNA primer metabolism during mammalian mitochondrial DNA replication.

The marked sequence homology between chloroplast genes and prokaryotic genes, as well as similar operon arrangements, give considerable evidence for a bacterial origin of chloroplasts. For example, two atp gene clusters in chloroplasts are also found in the cyanobacterium *Synechococcus* 6301.[63] Significant homologies also occur between the ribosomal protein gene clusters of chloroplasts and the str, S10, and spc ribosomal protein operons of *E. coli*.[1,2,64] Among various angiosperms, vascular plants, and land plants, the relative loci of genes on the chloroplast genome are often highly conserved and exemplified by that of the tobacco and *Marchantia* genomes.

E. CHLOROPLAST GENE INTRONS AND *TRANS*-SPLICING

Although the chloroplast genome has numerous prokaryotic features, a comparison of tobacco[1] and *Marchantia*[2] chloroplast DNA nucleotide sequences shows 15 and 18 genes, respectively, to contain introns — a characteristic eukaryotic feature. Of the 18 intron-containing genes in *Marchantia*, three are URFs of unknown function, but the remaining 15 correspond to the intron-containing genes of tobacco. In addition, six tRNA genes in each species contain introns.

Chloroplast introns can be classified into three groups. Group I introns, found in the tRNA[Leu] (UAA) genes of broad bean[65] and maize,[66] can be folded in a similar manner to the self-splicing intron of the *Tetrahymena* rRNA precursor.[67] The introns found in the genes for the 32-kDa protein of photosystem II[68] and those in the 23S rRNA[69] in *C. reinhardii* can also be classified as group I introns. Furthermore, the introns of these genes contain open reading frames which could code for maturase-like proteins;[70] this possibility has not been examined. Group II introns, found in the tRNA[Ile] and tRNA[Ala] genes,[71-73] are capable of forming secondary structures similar to those found in yeast[74] and maize[66] mitochondrial cytochrome oxidase introns. Group III introns, found in a number of tobacco intron-containing genes,[75,76] are bounded by conserved sequences which resemble both those of *E. gracilis* chloroplast introns[77] and of the boundary sequences of intron-containing nuclear genes.[78] Group III intron excision may occur by a mechanism similar to that described for nuclear mRNA precursors.[78] Messenger RNAs from 14 of the 15 tobacco intron-containing genes are presumably processed by *cis*-splicing of the appropriate exons. The mechanisms for chloroplast intron splicing are unknown.

The *rps*12 gene, encoding the chloroplast 30S ribosomal subunit protein S12, adds yet another twist to the complexity of chloroplast transcript processing and mRNA maturation. This gene consists of three exons, two of which (exons 2 and 3) are located in the inverted repeat of the tobacco genome and are transcribed as identical RNA species. However, exon 1, located in the large single copy region, 29 kilobase pairs downstream from the nearest copy of exons 2 and 3, is transcribed as a second RNA species. The processing of these two transcripts to produce a mature and translatable mRNA requires *cis*-splicing of exons 2 and 3 and also the unusual process of *trans*-splicing of the exon 1 transcript to the exon 2- and 3-containing transcript.[79-81] A similar discontinuous arrangement of the coding sequence of *rps*12 has been reported in maize[82] and soybean,[83] whereas in *E. gracilis* the *rps*12 gene is continuous and lacks introns.[84] Transcripts of the *psa*A$_2$ gene (P700 chlorophyll *a*-apoprotein of the photosystem I reaction center) in *C. reinhardii* are also likely to require a *trans*-splicing event for mRNA maturation.[85]

Unlike maize and tobacco, the chloroplast genome of *Euglena* does not contain introns in tRNA genes, although introns comprise approximately 32 kb (20%) of the 145 kb *Euglena* chloroplast genome.[44] However, in contrast to the chloroplast-encoded *rbc*L and *psb*A genes of higher plants, the *Euglena rbc*L and *psb*A genes do possess introns.[86] Comparing diverse taxa, there appears to be no evolutionary trend which predicts the presence or absence of introns within a given chloroplast gene. Continued examination of RNA processing in chloroplasts may reveal additional novel features and may contribute to establishing the mechanisms of chloroplast intron splicing.

III. REGULATION OF EXPRESSION OF CHLOROPLAST GENES

A. TRANSCRIPTION

Major changes in chloroplast mRNA transcript levels result from light-induced processes.[87-93] For example, plants can be grown under light conditions chosen to favor the excitation of one photosystem over the other. Exposure to "red" light excites PS I and "yellow" light excites PS II. Pea plants show a light-induced differential expression of reaction center genes such that transcription of the *psa*A gene (PS I) is induced by red light and *psb*B, *psb*C, and *psb*D (PS II) gene transcription is induced by yellow light.[94] Accumulation of reaction center protein complexes in the thylakoid membrane reflects the selective effects of light quality on transcription of these genes.[94] Changes in the transcript levels of light-regulated plastid genes in mustard (*psb*A)[95] and pea (*psb*A and *rbc*L),[96] as well as light-regulated, nuclear-encoded polypeptide genes,[97-101] appear to be mediated by phytochrome or the phytochrome photoreceptor. Although "global" changes in chloroplast transcriptional activity, as a result of light-induction, have been observed, a more specific control of transcription of individual genes can also arise due to variations in strength of the promoter elements preceding these genes. In some cases, however, light-induction of plastid protein levels must be regulated post-transcriptionally, possibly at the translational level[102,103] since mRNA levels are relatively unaffected by light.

1. Promoter Sequences

In prokaryotes, two regions of conserved sequences, 5'TATAAT3' and 5'TTGACA3', are centered −10 and −35 nucleotides, respectively, upstream of transcription start sites. These consensus sequences are separated by an average of 17 nucleotides and have been identified as elements of promoter sequences, i.e., RNA polymerase recognition sites.[104] Similar −35 and −10 promoter sequences (called ctp1 and ctp2, respectively), separated by spacers of from 9 to 23 nucleotides, have been located upstream of the transcription start site in 60 chloroplast genes.[105] The identity of these sequences as functional promoters has been verified in 24 cases using the criteria that (1) they are protected by E. coli RNA polymerase against DNase digestion or (2) they are active in initiating gene expression in either a homologous or heterologous in vitro system (reviewed in Reference 105).

Analysis of deletions 5' to the initiation start sites in the psbA gene of mustard,[106] the tRNA^{Met}, rbcL, atpB, and psbA of spinach[107,108] and the rRNA operon of maize[109] shows the ctp1 and ctp2 sequences to be essential for in vitro transcription in chloroplast-derived systems. Point mutations within these regions result in greatly decreased promoter strength. For example, a site-specific mutation of the invariant 3'-terminal T of the ctp2 element, TATAGT, of the maize rRNA operon results in a 98% decrease in transcription when compared to control levels.[109]

In mustard, a third element, 5'TATATAA3', resembling the eukaryotic "TATA" box[106] is found between the ctp1 and ctp2 elements. By itself, this sequence is insufficient to support a high level of transcription in constructions which lack the ctp1 (TATACT) and ctp2 (TTGACA) elements.[106] This third "TATA" element has also been located between the ctp1 and ctp2 elements of chloroplast-encoded protein genes of a number of plants, but is lacking 5'-upstream from tRNA or rRNA genes in these plants.[105] Link[106] suggests that selection of the less efficient TATA-like element in the etioplast and a preferential RNA polymerase recognition of the more efficient ctp1 element in the chloroplast would allow for differential expression of the psbA gene in chloroplasts of light-grown and etioplasts of dark-grown mustard plants. This "promoter-switching" hypothesis is a simple and powerful mechanism to provide developmental regulation of the PS II (psbA) gene. The validity of such a mechanism deserves future experimental attention.

In E. coli,[110] variations within the −35 consensus sequence can be correlated with transcriptional efficiency of the resulting promoter. To determine if a similar correlation occurs with chloroplast promoters, synthetic DNA fragments of defined transcriptional start sites and upstream proximal residues of the rbcL, atpB, and psbA genes were fused to a trnM$_2$ promoter deficient mutant.[107] Gruissem and Zurawski[108] demonstrated, from in vitro transcription assays, that the corresponding ctp1 sequences from the atpB, psbA, and rbcL genes of spinach chloroplast DNA could be ordered TTGACA>TTGCTT>TTGCGC with respect to their promoter strengths for trnM$_2$ transcription. However, in vivo experiments by Russell and Bogorad[111] have shown that the ctp1 and ctp2 elements, respectively, of rsp4 (TAGATA and TAATAT), rbcL (TTGCGC and TACAAT), and atpB (TTGACA and TAGTAT) in maize all have equal promoter strength. The corresponding elements (TTGACA and TAGTAT) for the maize psaB gene are three- to fivefold more efficient. These reports (especially for the atpB and rbcL genes) suggest that additional factors, other than recognition by RNA polymerase and relative binding efficiency to promoter sequences, may have a role in determining the resulting promoter strength.

2. Chloroplast RNA Polymerase

In eukaryotes, different promoter sequence elements are recognized by the three major RNA polymerases [I (rRNA), II (mRNA), and III (tRNA)], depending upon the class of gene to be transcribed. In the chloroplast, a variant of the ctp1 and ctp2 promoter sequences are found 5' to the coding sequence of many genes. These genes might conceivably be transcribed by a unique RNA polymerase which recognizes the conserved nature of these sequences, but whose affinity for the different promoters (see above) is altered by an association with a molecule such as ppGpp[112,113] or macromolecular transcription factors. It has been reported that the addition of a chloroplast fraction, containing a 27.5 kDa polypeptide (the S factor), to a purified chloroplast RNA polymerase activity from maize selectively increases transcription of chloroplast genes present on chimeric plasmids.[114,115]

Two different RNA polymerase activities have been isolated from *Euglena gracilis* chloroplasts. One activity is tightly bound to chloroplast DNA[116,117] and transcribes rRNA genes.[117] The second activity is found in a soluble extract of chloroplasts and specifically transcribes tRNA and mRNA genes.[118,119] Differences in the specificities of these RNA polymerase activities could result from an absence, or presence, of the ctp1 and/or ctp2 sequences (or other unknown recognition sequences) upstream of the rRNA, tRNA, and mRNA genes.[120] At least one chloroplast RNA polymerase activity has been identified in spinach,[121] pea,[122] wheat,[123] mustard,[124] and tobacco.[125] The presence of *rpo*A, *rpo*B, and *rpo*C genes in tobacco[1] and *Marchantia*[2] chloroplast DNA is evidence for an *E. coli*-like chloroplast RNA polymerase. Inhibition of chloroplast RNA polymerase activities *in vitro* has been observed using antibodies to fusion polypeptides synthesized from plasmid constructions containing chloroplast DNA sequences homologous to *E. coli* RNA polymerase genes *rpo*A, *rpo*B, and *rpo*C.[126] This is substantial evidence for the existence of an *E. coli*-type RNA polymerase enzyme in chloroplasts, but further research is required to fully characterize this enzyme and resolve the issue of uniqueness or multiplicity of chloroplast RNA polymerases.

3. Transcription Termination

Attenuation and antitermination play a major role in regulation of transcription in prokaryotes,[127-129] but there is no evidence that these mechanisms regulate chloroplast gene expression. Postulated secondary structures in the 5'-leader region of spinach chloroplast 16S rRNA transcripts may allow transcriptional regulation by an attenuation mechanism.[130] Transcription termination in chloroplasts may involve secondary structural features of the transcript which are related to those known to be important in transcription termination in prokaryotes.[131] For instance, short inverted repeats, capable of forming stem-loop structures, occur just upstream of the transcription termination site on *rbc*L mRNA in spinach, maize, pea,[132] and tobacco.[133] Deletions of the larger 5'-inverted repeat segment of the suggested stem-loop structure of the tobacco *rbc*L gene results in loss of *in vitro* termination control at this position.[133] However, the inverted repeats might also act as RNA processing sites or as protective structures against nucleolytic degradation, as suggested for similar structures in bacteria[134] and animal cells.[135]

The function of 3'-terminal repeats found in the RNA transcripts of some chloroplast genes was investigated using *in vitro* synthetic RNA products from plasmid constructs containing the spinach plastid *psb*A promoter region. This promoter preceded a polylinker suitable for the insertion of 3'-ends of plastid genes ligated to various downstream inverted repeat regions.[136] The resulting data demonstrated that most plastid 3'-inverted repeats are inefficient transcription terminators but that their presence greatly enhances the stability of the synthetic RNAs.[136] The degree of stability was dependent upon the specific inverted repeat sequences involved and RNAs lacking inverted repeats were rapidly degraded. These results led Stern and Gruissem[136] to suggest that 3' inverted repeats allow for a differential stability of mRNAs for several different components of the photosynthetic apparatus during light-induced development of active chloroplasts.

B. POST-TRANSCRIPTIONAL PROCESSING

Primary transcripts from the chloroplast genome are usually processed to shorter, functional tRNA,[137,138] mRNA,[79-81,139] and rRNA[23] products as a result of cleavage events which probably differ in detail for each RNA class. *In vitro* experiments have shown that both 5'- and 3'-endonucleolytic cleavages occur in chloroplast tRNA maturation.[137,138] Processing of the 5' end of spinach and tobacco tRNA^{Phe} precursors requires an RNase P-like activity.[140] A number of chloroplast genes whose primary transcripts are processed contain introns which can be classified into three groups (see Section I). The manner in which an intron is removed appears to depend upon the intron group. To date, very little is known about splicing mechanisms in chloroplasts. However, the introns in the *pet*B and *pet*D genes of tobacco are spliced out quickly and prior to any processing of the polycistronic message for *pet*D, *pet*B, *psb*B and *psb*H genes.[139] A similar distinction is found in the unusual splicing of the three exons comprising the *rps*12 mRNA in tobacco. The intron between exons 2 and 3, encoded on one transcript, is excised rapidly, possibly prior to the *trans*-splicing of exon 1 to the exon 2—3 transcript.[79-81]

A number of other chloroplast gene clusters, usually comprising genes from the same protein family, are also co-transcribed on polycistronic messages, e.g., the *rrna* operon,[23] the *psb* cluster,[139,141] a *trna* cluster,[142] the *pet* clusters,[143] and the ribosomal protein clusters.[64] Barkan et al.[144] suggest that the kinetics of RNA processing of these polycistronic messages may function *de facto* as a post-transcriptional mode of regulation of chloroplast gene expression.

C. TRANSLATION
1. Initiation

Chloroplast ribosomes possess numerous highly conserved features which are characteristic of *E. coli* and other prokaryotic ribosomes. Examples of these properties are rRNA nucleotide sequences,[145-148] sedimentation rates of rRNA, intact ribosomes and ribosomal subunits (reviewed in References 149 to 151), ribosome shape,[152] use of fMet as the N-terminal amino acid,[153,154] antibiotic sensitivities (with respect to streptomycin,[155-157] spectinomycin,[158,159] erythromycin, and kanamycin[160]), and a preference (98% of 175 plant chloroplast-encoded protein genes) for an AUG initiation codon (Table 2). Chloroplast genes also utilize GUG (in the *rps*19 gene of tobacco, spinach, and maize but not *Marchantia*) and UAA (in the *inf*A gene of tobacco) as initiation codons. These alternative initiation codons may function less efficiently than the AUG codon, since they are used less frequently to initiate translation in both prokaryotes[162,163] and chloroplasts. Compensatory structural features, permitting use of GUG and UUA codons, may involve important sequences which flank the initiation codon. In addition, other components of the initiation complex (e.g., initiation factors, 30S subunits, fMet-tRNA^{fmet}) may interact with chloroplast mRNA to permit proper alignment of initiation codons.

The Shine-Dalgarno (SD) sequence,[164] GGAGG, in the 5' noncoding region of prokaryotic mRNAs, is complementary to a CCUCC sequence (ASD) near the 3'-end of prokaryotic 16S rRNA (Figure 2). These two sequences form a base-paired complex prior to initiation of translation.[161,165-167] This base pairing provides a significant alignment factor permitting accurate translation initiation. A proposed stem-loop structure 5' to the SD-ASD alignment is thought to aid in this pairing (Figure 2). However, analysis of many prokaryotic sequences has identified several other conserved sequence elements involved in base pairing between mRNA and 16S rRNA during initiation. Some generalizations derived from these analyses are (1) a three base pair interaction between 16S rRNA and a GGA or GAG or a

TABLE 2

**Comparison of the Frequency of Bases in the
25 Positions 5′ and 3′ to the Translation Start
Site of Chloroplast and Prokaryotic Genes as
Determined from Genomic Sequences[a]**

	Chloroplast				Prokaryotic[e]			
(5′)	A	G	T	C	A	G	T	C
−25	66	23	69	15[b]	32	28	37	26
−24	65	32	65	11	33	26	35	29
−23	71	30	57	15	30	27	36	30
−22	61	20	69	23	31	23	37	32
−21	66	11	79	17	38	29	32	34
−20	70	20	67	16	56	15	25	27
−19	73	31	52	14	50	16	31	26
−18	62	32	55	24	29	20	39	35
−17	64	40	60	9	39	24	40	20
−16	71	36	60	7[c]	47	14	31	31
−15	71	27	63	14[d]	37	24	40	22
−14	82	26	57	10	42	22	32	27
−13	61	29	63	22	49	32	25	17
−12	75	38	42	20	45	46	22	10
−11	81	47	38	9	37	64	11	11
−10	72	59	35	9	42	65	8	8
−9	73	50	40	12	41	62	12	8
−8	83	38	49	5	43	47	22	11
−7	58	42	52	21	47	30	30	16
−6	75	36	48	16	40	27	42	14
−5	48	35	75	17	46	21	41	15
−4	60	17	63	35	50	19	36	18
−3	79	21	54	21	51	18	26	28
−2	48	12	76	39	36	15	47	25
−1	41	3	95	36	44	19	31	29
+1	170	4	1	0	116	6	1	0
+2	0	0	175	0	0	0	123	0
+3	1	172	2	0	0	123	0	0
+4	75	54	25	21	46	40	24	13
+5	40	21	35	79	35	18	15	55
+6	79	9	59	18	38	15	48	22
+7	87	20	21	47	61	28	10	24
+8	47	31	43	44	56	23	23	21
+9	77	17	47	22	36	23	41	23
+10	68	31	26	49[c]	49	19	36	19
+11	58	32	43	41	35	10	49	29
+12	64	20	71	19	47	15	39	22
+13	75	23	41	35	55	19	13	26
+14	49	23	61	41	52	12	35	24
+15	84	19	62	9	36	20	38	29
+16	49	41	50	35	49	27	21	26
+17	68	27	32	47	34	17	39	33
+18	56	31	67	20	26	27	41	29
+19	75	43	24	32	34	36	23	30
+20	52	25	43	49	35	20	35	33
+21	67	21	71	15	29	24	33	37

TABLE 2 (continued)
Comparison of the Frequency of Bases in the
25 Positions 5′ and 3′ to the Translation Start
Site of Chloroplast and Prokaryotic Genes as
Determined from Genomic Sequences[a]

	Chloroplast				Prokaryotic[e]			
(5′)	A	G	T	C	A	G	T	C
+22	53	41	38	42	33	37	22	31
+23	52	32	57	33	42	21	36	24
+24	63	14	81	16	27	22	40	34
+25	59	46	43	26	47	30	16	30
(3′)								

[a] Bonham-Smith and Bourque, unpublished results.
[b] 173 sequences analyzed in positions −17 to −25.
[c] 174 sequences analyzed in positions −16 and +10 to +25.
[d] 175 sequences (referenced in Table 3) analyzed in positions −15 to +9.
[e] 123 sequences. (From Ganoza, M. C., Kofoid, E. C., Martier, P., and Louis G. G., *Nucleic Acids Res.*, 15, 345, 1987. With permission.)

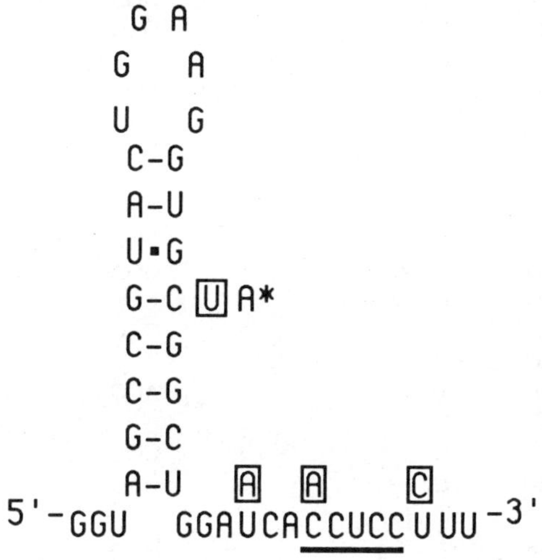

FIGURE 2.   The 3′ terminal sequence of the 16S rRNA from *Escherichia coli*,[145] tobacco,[146] maize,[147] *Marchantia*,[148] pea,[14] soybean,[168] spinach,[73] *Cyanophora paradoxa*,[169] *Spirodela oligorhiza*,[170] and *Euglena gracilis*.[171] The anti-Shine-Dalgarno[164] sequence is underlined. The boxed bases occur in *Euglena gracilis* and A* is found in *Spirodela oligorhiza*.

SD-like sequence is often sufficient for effective translation initiation;[172] (2) an A or U is the preferred nucleotide in every position between the SD sequence and the initiation codon;[173] and (3) the optimal upstream distance of the SD sequence from the initiation codon is 7 (±4) bases.[174]

The relationship of chloroplast mRNA translation initiation sequences to those of prokaryotes was examined in our laboratory by computer-assisted analysis[175] of 175 plant chloroplast gene sequences (Tables 2 and 3). In 70% of the sequences in Table 3, a related

TABLE 3
Distance of the Shine-Dalgarno Sequence from Chloroplast Protein Coding Gene ATG Start Codons

N[a]	Shine-Dalgarno sequence 5'	Gene 3'	Plant species	N[a]	Shine-Dalgarno sequence 5'	Gene 3'	Plant species
2	GGA	*pet*A	T,BB,W,S	20	GGA	*rps*18[b]	T
3	GGA	*psb*F	T	21	GGA	*psb*C	T,S,P,L,S6803
	GGA	*rbc*L	BB				
	GGA	*rps*2[b]	P	22	GGAG	*psb*A	Eg
					GGA	*rps*7[b]	M
4	GGA	*rbc*L	A		GAG	*rpo*A	P
	GGA	*rps*2[b]	W				
				23	GGA	*rpl*20[b]	Eg
5	GGAG	*rps*2[b]	T		GGAGG	*rps*8	T
	GGAGG	*rpl*33[b]	T				
	GGAG	*atp*I	L	24	AGG	*rpl*22[b]	M
	GGAG	*rbc*L	P				
				25	GGA	*rbc*L	Cm
6	GGAG	*rps*4[b]	T,S,L	26	GGA	*rps*11[b]	P
7	GGAG	*psb*E	T,O	28	GGA	*psb*E	L
	GGAGG	*psb*F	L,O		GGAG	*psb*A	C
	GGAGG	*psa*A	T				
	GGAGG	*psa*A$_1$	P,M	29	GGA	*psa*A	Eg
	GAGG	*atp*F	L		GAG	*psb*D	C
	GGAG	*atp*H	L				
	GAGG	*rpo*B	L	30	AGG	*ndh*4	L
	GGAGG	*rbc*L	T,B,M,PT,S,L				
	GGAG	*rps*8[b]	L	32	GGAG	*psb*A	A,L
	GGAG	*psa*A$_1$	C				
8	GGA	*psa*A	L	33	GGA	*psb*A	T,L
	GGAG	*psb*G	L,M		GAG	*frx*B	W,P
	GGAGG	*psb*D	T,S,P		GGAG	*psb*A	PT
	GGAG	*psb*D	L				
	GGA	*rpo*B	T	35	GAG	*atp*A	L
	GGAGG	*ndh*C	T				
	GGAGG	*ndh*E	T	38	AGG	*rps*12[b]	L
	GGA	*ndh*2	L				
	GGAGG	*ndh*3	L	39	GGA	*rpl*16[b]	So
	GGA	*ndh*4L	L		GAG	*rpo*A	T
	GAGG	*rps*16[b]	T				
	GGA	*rpl*14[b]	T	40	GGA	*ndh*F	T
					AGG	*rpl*16[b]	T
9	AGG	*psb*B	S6803				
	GGA	*rps*15[b]	T	41	GGA	*rps*12[b]	T
	GGAG	*rps*15[b]	L		AGG	*atp*A	T
	GGA	*rps*19[b]	L	45	GAG	*rpo*A	S
	GGAG	*mbp*X	L				
	GGAG	*frx*A	L	50	GGA	*rpl*20[b]	T
	GGA	*frx*B	L		AGG	*inf*A	S
	GGAGG	*sec*X[b]	P (L36)[c]				
				57	GGAGG	*atp*B	C
10	GAGG	*atp*H	T	59	GGA	*psb*A	Sn
	GGAG	*atp*C	P				
	GGA	*rps*14[b]	M	62	GGA	*rpo*A	L
	GGAGG	*rps*19[b]	M,S				
	GGA	*sec*X[b]	M (L36)[c]				

TABLE 3 (continued)
Distance of the Shine-Dalgarno Sequence from Chloroplast Protein Coding Gene ATG Start Codons

N[a]	Shine-Dalgarno sequence 5'	Gene 3'	Plant species	N[a]	Shine-Dalgarno sequence 5'	Gene 3'	Plant species
				63	GGA	atpB	T
				65	GGA	psbA	S
11	AGG	psbB	T,S,L				
	GGA	rpoC	L				
	GAGG	rpoC2	L	67	GGAG	rps7[b]	L
	GGA	rps14[b]	T				
	GAG	rps11[b]	Y,S	77	GGA	rpl16[b]	L
	GGAG	rps19[b]	T				
	GGAG	rpl22[b]	L	82	GGA	atpB	M
12	AGG	secX[b]	T[c]	90	GGA	atpB	R
	GAG	secX[b]	L[c]				
	GGAG	atpE	L	91	GAG	rps7[b]	T
	GAG	rpl20[b]	L				
	GAG	ndhD	T	99	GGA	rpl2[b]	T
					No Shine-Dalgarno sequence	atpB	B,P
						psbA	S6803
13	GAGG	atpH	W			psbG	T
	GAGG	atpI	T			ndh5	L
	GGA	rpl23[b]	T			rps12[b]	SB,Eg,M
	GGAG	infA	L			rps3[b]	M,L
	GGAG	psaA$_2$	C			rpl16[b]	M
						rps7[b]	Eg
14	GGAG	rpl23[b]	L			rps2[b]	L
	GGAGG	ndh1	L			rps14[b]	L
						rpl21[b]	L
15	GGAG	atpE	T,B,P,R			frxC	L
	GGA	psbA	Ah			psaC	T
	GGA	petA	P			rbcL	Eg
	GGAG	infA	T				
16	GGAGG	psaB	T,L				
	GGAGG	psaA$_2$	P,M				
	GGAG	ndhA	T				
	GAGG	atpF	T				
	GGA	petB	L				
	GGA	rpl14[b]	L				
17	GGA	rps7[b]	SB				
18	AGG	rps18[b]	L				

Note: A — alfalfa (*Medicago entiva*);[178,179] Ah — *Amaranthus hybridus*;[180] B — barley (*Hordeum vulgare*);[181,182] BB — broad bean (*Vicia faba*);[183,184] C — *Chlamydomonas reinhardii*;[68,85,185,186] Cm — *Chlamydomonas moewusii*;[187] Eg — *Euglena gracilis*;[84,188-191] L — Liverwort (*Marchantia polymorpha*);[148] M — Maize (*Zea mays*);[82,176,192-200] O = *Oenothera hookeri*;[201] P = pea (*Pisum sativum*);[177,202-209] PT — Petunia (*Petunia hybrida*);[179,210] R — rice (*Oryza sativa*);[211] S — spinach (*Spinacia oleracea*);[132,212-217] SB — soybean (*Glycine max*);[83] S6803 — *Synechocystis 6803*;[218,219] Sn — *Solanum nigrum*;[220] So — *Spirodela oligorhiza*;[221] T — Tobacco (*Nicotiana tabacum*);[39,222] W — wheat (*Triticum aestivum*).[223-227]

[a]   N = the number of nucleotides between the A of the initiation codon AUG and the central A of the orthodox (GGAGG) Shine-Dalgarno sequence.
[b]   Indicates chloroplast ribosomal protein genes.
[c]   The *secX* gene is thought to encode protein L36 of the chloroplast ribosome large subunit.[176,177]

Data are from Bonham-Smith, P. C. and Bourque, D. P., *Nucleic Acid Res.*, 17, 2057, 1989.

variant of at least three consecutive nucleotides of a consensus SD sequence, capable of pairing with the chloroplast 16S rRNA anti-SD sequence, occurs within the first 25 nucleotides 5' to the initiation codon. Of the remaining sequences, 20% contain a SD sequence (or variant trinucleotide) in the sequence -26 to -100 from the initiation codon. It is possible, as postulated for the bacteriophage T4 gene 38,[165] that secondary structure, involving bases in the -100 to -26 sequence, could bring a distant SD sequence into closer proximity with the initiation codon. No consensus or variant SD sequence was located in 10% of the plant chloroplast sequences. A previous survey showed that 13 of 14 chloroplast-encoded thylakoid membrane protein genes possess one identifiable SD sequence.[50] By our criteria, four of these genes lack a SD sequence in the 20 bases examined upstream of the AUG initiation codon. The *dna*G gene of *E. coli* also appears to lack a SD sequence.[228] The absence of a SD sequence might correlate with a coding sequence which is translated without the formation of a new initiation complex, as might occur internally on a polycistronic mRNA. In prokaryotes, although each cistron of a polycistronic mRNA is preceded by its own SD sequence,[229,230] whether new initiation complexes are formed at these sites is unresolved.

Analysis of prokaryotic ribosome binding to synthetic AUG-containing oligonucleotides[231] shows that a U at position -1 (the AUG codon is $+1$ to $+3$) and a purine at position $+4$[232] result in a much higher efficiency of ribosome binding than do other nucleotides in these positions. Furthermore, mutational analysis of ribosome-binding sites suggests an A at position -3 enhances translation.[162] Nucleotide frequency analysis of the 25 bases 5' and 3' to the initiation codon in 175 plant chloroplast genes shows a preference for T at positions -1 (54%) and -2 (43%) and A at -3 (45%). Purines at position $+4$ are found in 74% of the sequences (Table 2). An A or T is the preferred nucleotide at every position except $+5$ (45% C) in chloroplasts. In prokaryotes, A or T is most frequent except for positions -12 to -8 (the location of many SD sequences), and positions $+5$ (45% C), $+21$ (30% C), and $+22$ (30% G).[161] Among the chloroplast genes, A is the preferred nucleotide at positions -12 to -9, (Table 1) however if the AT bias of the chloroplast genome (62% in tobacco) is corrected for, then G becomes the most common nucleotide at these positions (Table 4). Although the position of SD sequences in the 5' untranslated region of chloroplast genes is not as conserved as that found in prokaryotes, approximately 27% of the chloroplast SD sequences are located within the -12 to -9 region. The high conservation of C at nucleotide $+5$ in both prokaryotic and chloroplast genomes suggests that this nucleotide may be important in regulation of translation initiation.

The conserved nucleotide sequences (corrected for AT bias) immediately flanking the initiation codon of chloroplast-encoded mRNAs appear to be prokaryotic in nature, when compared to the consensus sequence proposed to flank eukaryotic (vertebrate) initiation codons (Table 4).[174] The consensus sequence of nuclear-encoded genes as well as the family of maize zein genes shows very little sequence identity to chloroplast sequences in this 5'-flanking region.[234]

Many as yet unidentified factors may act together with the proposed "required" mRNA sequences to influence the binding and proper alignment of ribosomes in the formation of translation initiation complexes with chloroplast mRNA. Recent comparisons of prokaryotic gene sequences[161] suggest that base-paired secondary structures surrounding internal methionine AUG codons help to distinguish them from initiation AUG sites. Diminished ribosome binding at these internal AUG sites would favor ribosome binding at the correct initiation sites. Other secondary structural features of mRNA, which are functionally equivalent to orthodox SD sequences, may also be required for initiation complex formation. Botterman and co-workers[235] have shown that, in the case of the EcoRV endonuclease mRNA, efficient initiation of translation requires a long distance, intramolecular, base-pairing interaction between sequences in the 3' untranslated region of the mRNA and the ribosomal-binding

TABLE 4
Comparison of the Consensus Nucleotide Sequences from −12 to +4 Relative to the Initiation Codon in Prokaryote, Chloroplast, and Eukaryote Genes

Genome	5′-Flanking sequence[a]	Coding sequence
	−12	+1
C[b] Chloroplast (56 genes, 196 sequences)	G G G G A G A T C A T T	A T G G
C:P Homology score[c]	* * * * o o o * *	* * * o
P Prokaryotic (123 sequences)	G G G G G A T A A A T A A A A A A T A T	A T G A G
C:PN Homology score	* *	* * * *
PN Plant nuclear (79 sequences)	T A A T A A T A A A C A A T T T	A T G G
C:Z Homology score	*	* * * o
Z Zein gene family (23 sequences)	A N C N T A N C A A C A T	A T G G
C:E Homology score	* * *	* * * *
E Eukaryotic (vertebrate-699 sequences)	C C C G C C G C C A C C G	A T G G
C:D Homology score	* * o * *	* * * o

D Drosophila (100 sequences)	? ? A A A A A C A A A
	      C T A C
		A T G A
		      G

[a] The chloroplast consensus sequence was determined by normalizing the sequence of each nucleotide in order to compensate for the high A/T content (62%) of tobacco chloroplast DNA. The top line of each sequence is the predominant nucleotide at this position. Bases indicated on the second line of a sequence are included when the difference in frequency between the two indicated nucleotides is small. The homology score for each pairwise comparison is given above the sequence to which the chloroplast consensus sequence is being compared.

[b] Consensus sequences compared are C (chloroplast), P (procaryotic, Reference 161), PN (plant nuclear, Reference 245), Z (zein, Reference 234), E (eucaryotic, Reference 174, and D (*Drosophila*, Reference 233).

[c] Identity between the predominant nucleotide of the comparison sequence with that of the chloroplast sequence is indicated by and asterisk (*). Identity between the second most common nucleotide of either sequence being compared is indicated by a circle (o). Undetermined nucleotides are indicated by a question mark.

Data are from Bonham-Smith, P. C. and Bourque, D. P., *Nucleic Acids Res.*, 17, 2057, 1989.

site. These features would allow discrimination of *bona fide* initiation sites from many possible SD sequences which are located throughout chloroplast gene coding sequences (see below). The above possibilities have yet to be evaluated experimentally for chloroplast mRNA.

Calame and Ihler[236,237] found that *E. coli* ribosomes form stable "pseudo-initiation"

TABLE 5
Requirements for *E. Coli* Ribosome Binding to Single-Stranded Chloroplast DNA

Reaction[a] mixture	cpm bound ($\times$ 10³)	% cpm bound
Complete mixture	25.0	8.30
Minus ssDNA	3.7	1.20
Minus ribosomes	4.7	1.50
Minus GTP	2.0	0.50
Minus ssDNA plus dsDNA	1.4	0.40
Minus initiation factors	1.6	0.15
Minus fMet-tRNA^fMet plus Met-tRNA^fMet	1.7	0.15

[a] Ribosomes were prepared according to Calame and Ihler.[236] Acylation and formylation of *E. coli* tRNA was performed as described by Stanley.[246] Quantitation of DNA-ribosome complex formation was by a modification of the filter-binding assay of Nirenberg as utilized by Calame and Ihler.[236] Data are from Jurgenson.[238]

complexes with single-stranded DNA (ϕX174 and bacteriophage lambda) in the presence of GTP, initiation factors, and fMet-tRNA^fMet. These complexes appeared to occur on noncoding strand sequences corresponding to predicted ribosome binding sites of mRNA transcripts of these DNA sequences. The homologies between chloroplast and *E. coli* genes, mRNA sequences, and ribosomes permit speculation that *E. coli* ribosomes might bind to single-stranded chloroplast DNA, producing "pseudo-initiation" complexes at sequences which could represent ribosome binding sites on mRNA transcripts. With this method, we carried out an indirect titration of potential initiation sites for protein synthesis from tobacco chloroplast DNA-encoded mRNAs. One hundred and eighty possible ribosome binding sites were detected.[238,239] The requirements for ssDNA, GTP, *E. coli* ribosomes, and fMet-tRNA^fMet for "pseudo-initiation" complex formation are indicated in Table 5. Distinct optima for ribosome concentration, ionic strength, and initiation factors were observed (Figure 3) at which complex formation was linearly dependent upon single-stranded DNA concentration (Figure 3d). At constant single-stranded chloroplast DNA concentration, increasing amounts of fMet-tRNA^fMet titrated the ribosome binding sites on chloroplast DNA (Figure 4). Complex formation at saturation represented about 180 ribosome binding sites per chloroplast genome.[238,239] These pseudo-initiation complexes were stable, sedimenting with ribosomes on sucrose gradients. Electron microscopy[237] of fractions containing DNA-ribosome complexes revealed structures in which the density of ribosomes (per unit length of DNA) was consistent with the binding site titration data in Figure 4.[238]

The ribosome binding assay uses all the components which are required for true translation initiation complex formation *in vitro*, with the exception being replacement of mRNA with single-stranded DNA. This assay is a reconstruction using heterologous components. However, ribosomes conceivably could bind to chloroplast non-coding strand DNA at sequences proximal to the 5'-end of polypeptide-encoding open reading frames. These binding sites may correspond to authentic ribosome binding sites on chloroplast mRNA. Synthesis of a functional chloroplast polypeptide is likely to require the presence of a ribosome binding sequence on the mRNA. Although numerous chloroplast transcripts contain multiple coding sequences, it is not known whether each coding sequence has its own ribosome binding site. Ribosome protection and footprinting studies using SP6 or T7 vectors to synthesize chloroplast protein-encoding RNA sequences would help to resolve this question. Experiments to measure homologous complex formation, with chloroplast ribosomes, chloroplast DNA, and mRNA from a single plant species, have yet to be carried out.

Of the 180 ribosome binding sites on tobacco chloroplast DNA, several may correspond

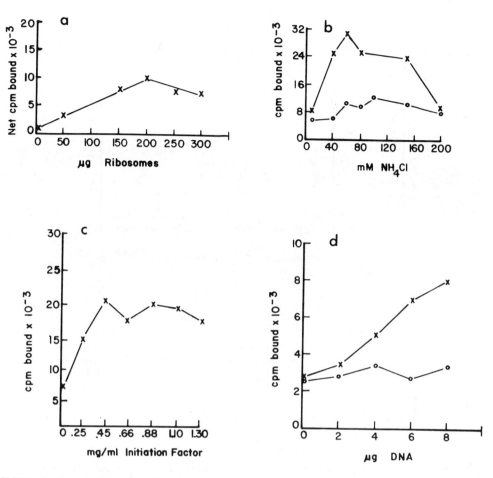

FIGURE 3.   Optimization of *E. coli* ribosome binding to denatured tobacco chloroplast DNA. Each assay contained the complete mixture as described in Table 5 except the ribosome concentration varied in (a), the ionic strength varied in (b), crude initiation factors prepared from salt washed ribosomes varied in (c), and the DNA concentration varied in (d). [x—x, ssDNA; ○—○, dsDNA]. Data are from Jurgenson.[238]

to genes (five protein-coding genes and ten URFs) present in both copies of the inverted repeats. Four ribosomal protein genes are located completely (excluding exon 1 of the split *rps*12 gene) within these repeats in tobacco chloroplast DNA.[222] In addition, some false binding signals may result from (1) AUG codons located within structural genes, (2) randomly located Shine-Dalgarno type sequences,[240,241] or (3) initiation AUG triplets of short open reading frames which are not structural genes. A computer-assisted[175] search of the tobacco chloroplast genome for the possible translation initiation sequence, GGANATG (where N is any sequence of 2 to 10 nucleotides), located 338 possible initiation sites. Due to their location within coding sequences of various protein-coding genes, 304 of these sites were determined to be "false-sites". Thirty-four sites were situated 5' to protein-coding sequences and therefore could be legitimate translation initiation sites. As mentioned above, applying other criteria to the identity of SD sequences, 149 chloroplast protein-coding genes possess plausible ribosome binding sites. DNA-ribosome complexes at false sites might be unstable due to an absence of other required components for prokaryotic ribosome binding at true translation initiation sites. Such requirements, e.g., the combination of a Shine-Dalgarno sequence in close proximity to an AUG initiation codon, have already been mentioned earlier in this section.

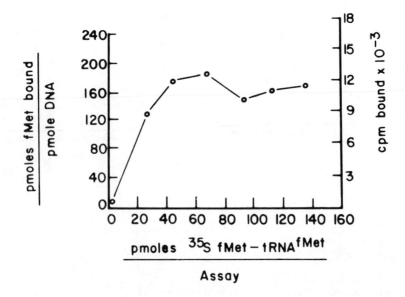

FIGURE 4.   Titration of ribosome binding sites on tobacco chloroplast DNA. Standard assay mixtures containing increasing amounts of ^{35}S-fMet-tRNA[fMet] were incubated with 4.6 μg denatured DNA. The counts retained by filters are displayed on the right ordinate. The pmoles of tRNA bound/pmole DNA molecule are shown on the left ordinate. Data are from Jurgenson.[238]

2. Codon Usage

In accord with the "wobble hypothesis" of Crick,[242] 32 tRNA genes occur in *Marchantia*[148] chloroplast DNA. However, in the chloroplast genome of tobacco only 30 tRNA species are encoded.[222] If no cytoplasmic tRNAs are transported into the tobacco chloroplast, a degenerate code, such as that observed in mitochondria where only 24 tRNAs are required,[243] may be utilized. As would be expected from the high AT content of chloroplast DNA,[148,222] there is a bias towards the use of these two bases in codons, especially in the third (wobble) position.[11] Within the chloroplast there appears to be a correlation between the relative abundance of the codons for each amino acid and the abundance of the corresponding tRNAs.[244] Furthermore, in bean (*Phaseolus vulgaris*) the cognate tRNA pool for amino acids coded by multiple codons corresponds to the frequency of codon usage within that pool.[244]

3. Termination

All three translational termination codons occur in chloroplast genes (Table 6), with UAA most frequent (70% of 169 plant sequences), followed by UAG (21%), and UGA (9%). Analysis of tobacco (*Nicotiana tabacum*) and liverwort (*Marchantia polymorpha*) protein-coding sequences (not including URFs) shows that UAA is the preferred stop codon (Table 6). However, while the other two stop codons (UAG>UGA) occur in 40% of the 49 tobacco genes examined, the *Marchantia* chloroplast genes only use the UAA (93%) and UAG (7%) stop codons. No evidence exists, in chloroplasts, for the use of UGA to code for tryptophan as has been found in mitochondria.[243]

IV. CHLOROPLAST RIBOSOMAL PROTEIN GENE STRUCTURE AND ORGANIZATION

A. CHLOROPLAST-ENCODED RIBOSOMAL PROTEINS
1. Identified Genes

Two-dimensional PAGE studies show that chloroplast ribosomes of tobacco[247-250] and other species[251-257] contain similar numbers of proteins as the *E. coli* ribosome. This typical

TABLE 6
Translation Termination Codon Usage of Various Plant Protein Coding Genes

	Total number of sequences	Stop codon	Number of occurrences
Various plant species	169	UAA	118
		UAG	36
		UGA	15
Tobacco[222] (*Nicotiana tabacum*)	48	UAA	29
		UAG	13
		UGA	6
Liverwort[148] (*Marchantia polymorpha*)	55	UAA	51
		UAG	4
		UGA	0

prokaryotic feature, along with others mentioned in Section II, suggest that each chloroplast ribosomal protein may share homology with an *E. coli* ribosomal protein. Nineteen chloroplast ribosomal protein genes are found in the tobacco chloroplast genome.[1] Genes for chloroplast ribosomal proteins have also been identified in spinach, maize, *Spirodela*, *Euglena*, soybean, wheat, pea, *Chlamydomonas*, and *Marchantia* (Table 3). All of these genes have significant translated amino acid sequence homology with corresponding *E. coli* ribosomal proteins. Chloroplast ribosomal protein gene nomenclature is that of the *E. coli* ribosomal protein to which the translated amino acid sequence of the chloroplast gene corresponds.[41]

The identified sequences which code for polypeptides homologous to *E. coli* ribosomal proteins are the same in tobacco and *Marchantia* with the exception of *rps*16 (present only in tobacco) and *rpl*21 (present only in *Marchantia*). A sequence homologous to the *sec*X gene of *E. coli* has been identified in tobacco,[1] *Marchantia*,[2] maize,[176] and pea.[177] In maize and pea, the predicted primary sequence of the protein product of this gene shows 62% identity[176,177] to its *E. coli* counterpart. Consequently, this gene has been identified as the ribosomal large subunit protein L36.[258] The tobacco and *Marchantia* *sec*X genes share 89% and 76% homology, respectively, with both the maize and pea genes. The remainder of this discussion will focus on tobacco chloroplast ribosomal protein genes.

2. Gene Organization
There are several distinguishing characteristics (relative to *E. coli* ribosomal protein genes) of tobacco chloroplast ribosomal protein genes described below:

1. Many are arranged in clusters which are similar in gene content and polarity with the *E. coli* str (*rps*12*-*rps*7), S10 (*rpl*23-*rpl*2*-*rps*19-*rpl*22-*rps*3-*rpl*16*), and *spc* (*rpl*14-*rps*8) operons[1-3] (Figure 5). Genes named are those in tobacco chloroplast DNA and those marked with an asterisk contain introns.
2. Other genes, *rps*4 and *rps*11 (from the α operon in *E. coli*), *rps*14 (from the *spc* operon in *E. coli*), *rps*2, *rps*15, *rps*16, *rps*18, *rpl*20, and *rpl*33 are found throughout the chloroplast genome.[1]
3. The operon-like gene clusters have examples of protein genes which are missing, relative to the *E. coli* operons. These genes (except for *rps*14) are not found elsewhere on the chloroplast genome and are probably nuclear genes. Furthermore, genes (S4, S11, *rpo*A) of an incomplete α-type operon are dispersed throughout the chloroplast genome.[1]
4. Some of the protein products of the homologous *E. coli* genes which are rRNA binding proteins (S4, S8, and S7) are important in *E. coli* 30S ribosome subunit assembly[259]

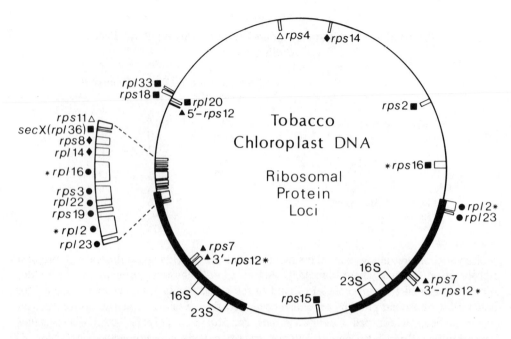

FIGURE 5. Circular map of the tobacco chloroplast genome showing the position of ribosomal protein genes and their relationship to *E. coli* ribosomal protein operons. More than one of the genes in each of four *E. coli* ribosomal protein gene operons (*str, spc, α*, S10) occur in the tobacco chloroplast genome. Genes shown on the outside of the map are coded on the A strand and those on the inside of the map on the B strand. An asterisk indicates an intron-containing gene. Relationship to *E. coli* operons is indicated by the following symbols: ▲, *str*; ♦, *spc*; △, α; ●, S10; ■, single genes from other operons.

and are also involved in feedback inhibition of their own translation.[260-261]

5.   The introns in chloroplast ribosomal protein genes and many other chloroplast genes can be classified as group II introns.[2,262]

6.   As described in section I, portions of the coding sequences of the *rps*12 gene are separated so that at least two different transcripts must be joined, by an unknown mechanism, during the maturation of the *rps*12 mRNA.[79-81]

B. NUCLEAR-ENCODED CHLOROPLAST RIBOSOMAL PROTEINS

1. Evidence for Nuclear Genes

The majority of chloroplast proteins, including those of the chloroplast ribosome, must be coded by nuclear genes. Evidence from genetic analysis in tobacco,[1,247,249] *Marchantia*[2] and *Chlamydomonas*,[263-269] as well as from experiments to determine the intracellular sites of synthesis,[253,257,270-273] shows that both chloroplast and nuclear genomes code for chloroplast ribosomal proteins. The elegant results from Gillham's laboratory[253] indicate that 43 or 44 *Chlamydomonas* chloroplast ribosomal proteins must be coded by nuclear genes, while only 20 are synthesized in the chloroplast.

Translation products of some nuclear genes are precursors[274] containing a transit peptide at the N-terminus. The transit peptide is recognized by an integral membrane protein of the chloroplast envelope[275,276] and cleaved[277] by a chloroplast stromal enzyme[276,278] during or after transport into the chloroplast. Transport and proteolytic cleavage are energy-requiring, and the transport process can be driven by exogenous ATP or photophosphorylation.[277,279] Chloroplast ribosomal proteins are synthesized as precursors in a manner consistent with these observations.

2. cDNA Identification

The first direct evidence of a nuclear origin for some chloroplast ribosomal proteins was published by Gantt and Key,[280] who found the DNA sequences of six cDNA clones for pea chloroplast ribosomal proteins. These cloned sequences were expressed in bacteriophage lambda gt11 recombinants as fusion polypeptides which reacted with antibodies to pea chloroplast ribosomal proteins. *In vitro* translation of hybrid selected mRNA from six full-length cDNA clones produced precursor polypeptides which, after *in vitro* transport into and processing by isolated chloroplasts, co-migrated on 2D-PAGE gels with pea chloroplast ribosomal proteins S16, L6, L12, L13, L18, and L25. Four of these sequences (S16, L6, L12, and L13) have translated homology with *E. coli* ribosomal proteins S17, L15, L24, and L9, respectively. The other two sequences are not homologous to known bacterial ribosomal protein sequences.[281]

3. Protein L24

A cDNA expression library to tobacco poly(A)-mRNA, cloned into lambda gt11, has been screened in our laboratory, using polyclonal antibodies to tobacco 50S or 30S chloroplast ribosomal protein subunits.[282] Plaques producing immunoreactive fusion polypeptides were bound to nitrocellulose filters and used to purify monospecific antibodies.[283] The sequence of one positive clone translates into a 69 amino-acid-long polypeptide having homology to the C-terminus of ribosomal protein L24 of *E. coli* (42%), *B. stearothermophilus* (49%), and pea chloroplast ribosomal protein L24 (78%).[279] The pea and tobacco L24 sequences are almost identical at the 3'-end of the homology with *E. coli* L24, but each has a C-terminal polypeptide extension of over 20 amino acids. A C-terminal extension occurs in the gene for *Neurospora* mitochondrial ribosomal protein S24 (which is homologous to *E. coli* ribosome protein S16), but the polypeptide lacks an N-terminal transit sequence.[284] No information on homologous L24 polypeptides from eukaryotes is published.

4. Protein L12

Other clones isolated from the tobacco cDNA library[282] encode a fusion polypeptide with a predicted amino acid sequence characteristic of *E. coli* ribosomal protein L12.[285] In *E. coli,* L12 is a 50S ribosomal subunit protein and it is the only protein present in the ribosome in multiple copies. A tetramer of two L12 and two L7 proteins[286] (L7 results from an acetylation of L12 at its N-terminal serine) constitutes the morphologically distinct L7/L12 stalk on the 50S subunit.[287] The tobacco L12 amino acid sequence has 46% identity with *E. coli* L12. The amino acid sequence of spinach L12 protein,[288] determined from manual micro-Edman degradation of the purified protein, is 79% identical to the predicted tobacco L12 sequence. While the N-terminus of the spinach L12 sequence is four amino acids longer than that of the *E. coli* L12 protein, the tobacco L12 sequence is at least twenty two amino acids longer than the *E. coli* protein. The differences between the spinach and tobacco L12 sequences may be resolved when the DNA sequence of the spinach *rpl*12 gene is determined. Since *E. coli* L7/L12 is important for GTPase activity,[289] it will be of interest to experimentally assess whether the chloroplast L12 is functionally homologous to the *E. coli* L12 and whether it is localized in a morphologically distinct L7/L12 stalk on chloroplast 50S ribosome subunits.

V. REGULATION OF CHLOROPLAST RIBOSOMAL PROTEIN GENE EXPRESSION

A. REGULATION BY CHLOROPLAST-ENCODED RIBOSOMAL PROTEINS

As a result of knowing the complete nucleotide sequence of the tobacco[1] and *Marchantia*[2] chloroplast DNA and the sophisticated arsenal of modern techniques for molecular biological research, experimental strategies to explore mechanisms of regulation of chloroplast ribosomal protein gene expression are now possible. Regulation of expression could occur at several levels — transcription, mRNA processing, and translation. Since chloroplast DNA genes have many prokaryotic properties, the logical paradigms for regulatory mechanisms are those which are well established for *E. coli*. However, several chloroplast ribosomal protein genes possess introns, indicating that they have eukaryotic as well as prokaryotic characteristics. Furthermore, the expression of one chloroplast ribosomal protein gene (*rps*12) is unique in requiring a *trans*-splicing event to join two separate transcripts during the maturation of its mRNA (see Section I). The regulatory mechanisms governing chloroplast ribosomal protein gene expression are unknown.

The simplest models to propose for regulation of expression of chloroplast genes are (1) that transcriptional regulation might utilize regulatory protein systems such as those which operate in the lactose operon of *E. coli*[290] and/or (2) that translational regulation might be a result of translational feedback mechanisms similar to those which occur in *E. coli* ribosomal protein operons.[291] Existing data suggest that differential promoter strengths determine the relative levels of chloroplast transcripts (see Section II). The expectation is that transcriptional regulation will play a minor role, relative to translational feedback mechanisms or control of transcript processing. However, in the S10 ribosomal protein operon of *E. coli,* the regulatory protein L4 modulates both the levels of transcription[292] and translation.[293] Site-directed mutagenesis of the 5'-leader sequence of the S10 polycistronic mRNA has shown that transcriptional control results from L4-mediated, premature transcription termination (attenuation) within this leader sequence.[292] Furthermore, a stem-loop structure which can be drawn immediately upstream of the premature termination site was found to be essential for L4 feedback control of both transcription and translation. Since transcriptional and translational control by L4 can be separated genetically, it is likely that different sequence features within the same region of the S10 leader are required for each type of control.[292]

The potential for feedback control of chloroplast ribosomal protein translation is interesting in that products of chloroplast as well as nuclear genes are required, if the mechanisms are related to those of the *E. coli* feedback mechanism. Several *E. coli* operons (*spc, str,* α, S2, and S15) code for autogenous regulatory proteins (S8, S4, S7, S2, and S15, respectively)[294] for which a homologous polypeptide is coded by chloroplast DNA in tobacco[1] and *Marchantia*.[2] Other chloroplast ribosomal proteins, homologous to *E. coli* proteins L1, L10, L4, S1 and S20, which regulate translation of mRNA transcribed from the L1, L10, L4, S1, and S20 operons, respectively, are likely to be products of plant nuclear genes. Possibilities exist that chloroplast ribosomal proteins which regulate translation may be the same ones found in the homologous *E. coli* operons. However, different ribosomal or nonribosomal proteins (chloroplast or nuclear-coded) may also act as translational regulators.

One common feature of *E. coli* ribosomal repressor proteins is their involvement in binding to rRNA during ribosome assembly. It has been postulated that there is a competition between rRNA and mRNA for the regulatory proteins and that the proteins have a higher affinity for rRNA than for mRNA. Therefore, if rRNA is in excess, these proteins would enter the ribosomal assembly pathway without inhibiting translation (reviewed in Reference 291). In support of this hypothesis are reports that the presumed binding sites for ribosomal proteins S4, S7,[261] and S8[295] (among others[294]) on mRNA transcripts have secondary structural homologies to their respective binding sites on 16S rRNA. Since the tobacco and *E. coli* 16S rRNA genes share high sequence identity, we were able to use the *E. coli* binding sequences as a guide to conduct a search (with the DM sequence analysis programs[175]) of the tobacco 16S rRNA.[296] The search revealed potential stem-loop structures, similar to those required for S4, S7, and S8 binding in *E. coli* (Figures 6 and 7). Furthermore, a possible binding site for the tobacco S7 protein occurs in exon 1 of the *rps*12 mRNA (Figure 7). In *E. coli* this binding site is located in the intercistronic region between *rps*12 and *rps*7.[261] As yet we have been unable to identify any possible mRNA binding sites for S4 or S8. Experimental evaluation of the role of tobacco S7 protein in the regulation of S12 expression is in progress.

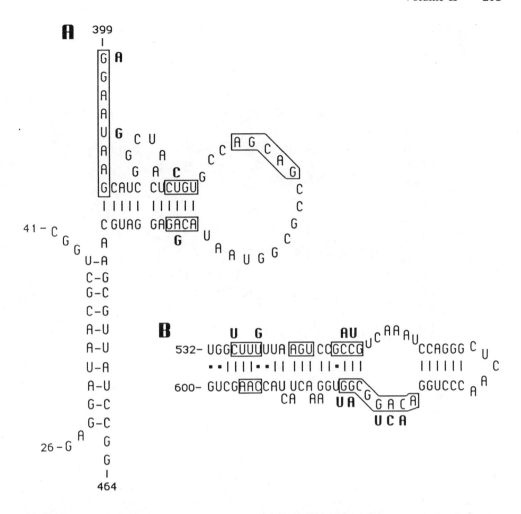

FIGURE 6.   Predicted binding sites, on tobacco 16S rRNA, for the ribosomal proteins S4 (A) and S8 (B). The boxed sequences are conserved between *E. coli* and tobacco chloroplast DNA and are thought to be important in recognition and binding of proteins S4 and S8, respectively, in *E. coli*.[261,295] Nucleotides in bold face are those found within the conserved boxes in *E. coli*. (From Bonham-Smith, P.C. and Bourque, D. P., unpublished results, 1988.)

B. REGULATION BY NUCLEAR-ENCODED CHLOROPLAST RIBOSOMAL PROTEINS

Evidence from our laboratory[282] shows that a protein homologous to *E. coli* ribosomal protein L24 is nuclear-encoded in tobacco (Figure 8). In *E. coli*, L24 is a primary RNA binding protein important early in the assembly of the 50S ribosome subunit.[297,398] The *E. coli* gene (*rplX*) for L24 is a component of the *spc* operon which is adjacent to and downstream from the gene for protein L14. Synthesis of other proteins in this operon is regulated by protein S8, but translation of L14 and L24 is not affected by S8.[295,299] Results of gene dosage experiments suggest that *E. coli* L24 might regulate the translation of its own mRNA[260] as well as that of several other ribosomal proteins.[300] Two mutants, including one lacking L24, show overexpression of different ribosomal protein gene transcription units.[300] An intriguing possibility exists that the nuclear-encoded L24 polypeptide of tobacco could be involved in feedback control of translation of the chloroplast-encoded ribosomal protein L14 or other chloroplast ribosomal proteins. Demonstration of this type of translational regulation would represent a clear case of regulation of chloroplast gene expression by a functional chloroplast protein which is the product of a nuclear gene.

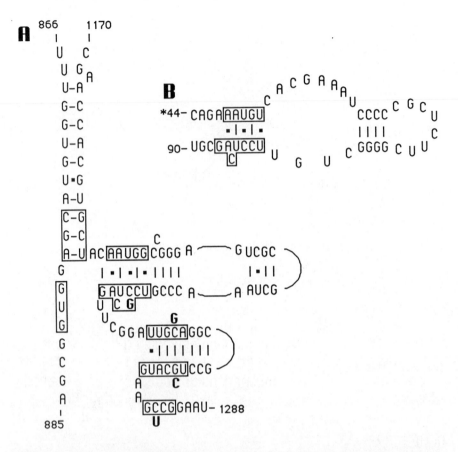

FIGURE 7.   Predicted binding sites on tobacco 16S rRNA (A) and S12-S7 mRNA (B) for ribosomal protein S7. The boxed sequences in (A) are conserved between *E. coli*[261] and tobacco chloroplast DNA, and the boxed sequences in (B) are conserved between the tobacco 16S rRNA and the S12-S7 mRNA. Nucleotides in bold face are those found within the conserved boxes in *E. coli*. (Bonham-Smith, P. C. and Bourque, D. P., unpublished results, 1988.)

These prokaryotic models for potential regulation of chloroplast gene expression do not acknowledge the eukaryotic nature of some chloroplast DNA-encoded genes and that most chloroplast ribosomal protein genes must be nuclear-encoded. Thus, it is reasonable to consider eukaryotic systems in which gene expression is regulated by ribosomal proteins. Two models are suggested by results concerning yeast cytoplasmic ribosomal proteins. In one instance, there is evidence that ribosomal protein YL3 can regulate expression of yeast 5S rRNA by a mechanism which couples 5S rRNA gene transcription to a step in ribosome assembly (binding of YL3 to 5S rRNA). YL3 competes with an analog to the presumptive transcription factor IIA which binds to and activates transcription of the yeast 5S rRNA gene.[301] A second example of modulation of ribosomal protein gene expression in yeast is the regulation by YL32 of splicing of its own transcript.[302] Whether similar mechanisms, mediated by nuclear-encoded chloroplast ribosomal proteins, operate to regulate the expression of chloroplast and/or nuclear-encoded genes remains to be tested.

VI. WHAT NEXT?

Although the complete nucleotide sequences of *Marchantia* and tobacco chloroplast DNA are known, the details of chloroplast DNA replication are unknown. Further characterization of replication origins[303-307] and of chloroplast DNA polymerase are necessary.[307,308]

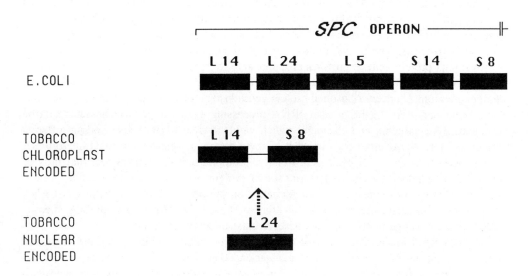

FIGURE 8.   Ribosomal protein gene arrangements of the *E. coli* S10 and *spc* operons which correspond to tobacco chloroplast ribosomal protein genes. Note the location of L24 in *E. coli* and its absence in tobacco chloroplast DNA. L24 is a nuclear-encoded tobacco chloroplast ribosomal protein whose sequence was deduced from clones isolated from a cDNA library derived from tobacco poly (A+)-mRNA.[282]

The molecular mechanisms which produce documented examples of chloroplast DNA rearrangements and the existence of potential recombination systems to achieve these DNA rearrangements deserve experimental attention. From these avenues of inquiry, strategies may develop for rational approaches to constructing vectors and systems for use in heritable genetic transformation of higher plant chloroplasts.

With regard to the genetic function of chloroplast DNA, several unidentified polypeptide coding sequences (URFs) exist. Determination of the identity and function of the products of these URFs are an immediate challenge to chloroplast molecular biologists. An urgent need is the purification and measurement of the functional properties of the chloroplast RNA polymerase which may contain the chloroplast DNA-encoded *rpo*A, *rpo*B, and *rpo*C gene products. This will help to resolve the question of uniqueness or multiplicity of RNA polymerases required to transcribe all classes of chloroplast genes.

Experimental definition of the mechanisms of chloroplast intron splicing and *rps*12 *trans*-splicing are necessary to evaluate the role of these processes in mRNA maturation. Further studies on transcription of chloroplast genes will yield information regarding the function of promoter and termination sequence elements. Attention should also be given to the potential occurrence and function of enhancer and/or other regulatory sequence elements of chloroplast genes. Functional analysis of translation initiation sequences on chloroplast mRNA is required. Evaluation of whether chloroplast polycistronic messages are translated without processing to monocistronic messages is necessary. This point is relevant for designing experimental approaches to examine the roles of RNA processing and translational control mechanisms in regulation of chloroplast gene expression.

Although chloroplast genes possess many prokaryotic properties, it is still an open question whether regulation of chloroplast gene expression follows expectations based on the mechanisms which operate in bacteria. Certainly the possibility of modulation of transcription by specific macromolecular or metabolite factors deserves further attention. Perhaps some of the chloroplast-encoded URFs may code for regulatory polypeptides. Expression of some chloroplast genes is stimulated by light. Whether specific macromolecular or other factors are involved in this type of regulation is unresolved.

Definitive experiments to elucidate potential mechanisms of chloroplast gene regulation require easily testable models to evaluate. At least in the case of the chloroplast ribosomal protein genes, several possibilities may be examined with a specific class of genes whose functions are related, i.e., some involvement in the structure of ribosomes and their function in protein synthesis. The *E. coli* models for both transcriptional and translational inhibition of ribosomal protein gene expression are amenable to experimental testing. The ribosomal protein genes also provide an opportunity to evaluate the role of less specific modulation of their expression. There are ribosomal protein gene clusters for which polycistronic transcripts have been identified. In these cases, RNA processing events occur which include intron excision, *trans*-splicing, and cleavages which yield separate RNA species coding different proteins. Thus, the relative rates of each of these events in determining functional mRNA levels for these genes can be evaluated. Relative promoter strength comparisons may provide evidence of whether all chloroplast ribosomal protein gene transcriptional units possess similar potential for expression based primarily on promoter affinity for chloroplast RNA polymerase. Similarly, measurements of relative efficiencies of ribosome binding sites on chloroplast ribosomal mRNA species deserve experimental attention.

As described in the discussion of chloroplast ribosomal protein L24, there exists the intriguing and testable possibility that nuclear-encoded ribosomal proteins, or other chloroplast proteins, may be regulators of chloroplast gene expression. In this area of investigation, the results will have broad implications for the understanding of regulatory interactions between nuclear and organelle genomes. It remains to be shown whether, as in the case of mammalian mitochondrial RNase MRP,[62] RNA transcripts of nuclear genes are transported into chloroplasts, where they may function to regulate chloroplast gene expression in a general or specific manner.

Clearly, continued research on chloroplast gene expression is likely to yield exciting results, particularly when specific genes and testable models are examined. Although complete chloroplast DNA nucleotide sequences are known, in many ways these landmark results only provide the keys to doors behind which are found the secrets of regulation of chloroplast gene expression. Ultimately, this exploration will uncover the details of the mechanisms of interactions between nuclear and chloroplast gene expression which engender fully functional, photosynthetically competent chloroplasts.

ACKNOWLEDGMENTS

The authors wish to express their thanks to Mark Dubnick for his help in searching the GenBank52 library. Also we thank C. W. Birky, B. Tharp, and F. Thomas for their suggestions and critical reviews of this manuscript. This work was supported by NIH grant GM26937 to D.P.B.

REFERENCES

1. **Shinozaki, D., Ohme, M., Tanaka, M., Wakasugi, T., Hahashida, N., Matsubayashi, T., Zaita, N., Chunwongse, J., Obokata, J., Yamaguchi-Shinozaki, K., Ohto, C., Torozawa, K., Meng, B. Y., Sugita, M., Deno, H., Kamogashira, T., Yamada, K., Kususa, J., Takaiwa, F., Kato, A., Tohdoh, N., Shimada, H., and Sugiura, M.,** The complete nucleotide sequence of the tobacco chloroplast genome: its organization and expression, *EMBO J.,* 5, 2043, 1986.

2. **Ohyama, K., Fukuzawa, H., Kohchi, T., Shirai, H., Sano, S., Umesono, K., Shiki, Y., Takeuchi, M., Chang, Z., Aota, S., Inokuchi, H., and Ozeki, H.,** Chloroplast gene organization deduced from complete sequence of liverwort *Marchantia polymorpha* chloroplast DNA, *Nature (London),* 322, 572, 1986.

3. **Whitfeld, P. R. and Bottomley, W.**, Organization and structure of chloroplast genes, *Annu. Rev. Plant Physiol.*, 34, 279, 1983.
4. **Weil, J. H.**, Organization and expression of the chloroplast genome, *Plant Sci.*, 49, 149, 1987.
5. **Umesono, K. and Ozeki, H.**, Chloroplast gene organization in plants, *Trends Genet.*, 3, 281, 1987.
6. **Palmer, J. D.**, Comparative organization of chloroplast genomes, *Annu. Rev. Genet.*, 19, 325, 1985.
7. **Thompson, J. A.**, Isolation and characterization of DNA from different plastid types of *Tropaesolum majus, Eur. J. Biol.*, 21, 37, 1980.
8. **Macherel, D., Kobayashi, H., Akazawa, T., Kawano, S., and Kuroiwa, T.**, Amyloplast nucleoids in sycamore cells and presence in amyloplast DNA of homologous sequences to chloroplast genes, *Biochem. Biophys. Res. Commun.*, 133, 140, 1985.
9. **Iwatsuki, N., Hirai, A., and Asahi, T.**, A comparison of tomato fruit chloroplast and chromoplast DNAs as analyzed with restriction endonucleases, *Plant Cell Physiol.*, 26, 599, 1985.
10. **Hansmann, P.**, Daffodil chromoplast DNA: comparison with chloroplast DNA, physical map and gene localization, *Z. Naturforsch.*, 42c, 118, 1987.
11. **Bohnert, H. J., Crouse, E. J., and Schmitt, J. M.**, Organization and expression of plastid genomes, in *Encyclopedia of Plant Physiology: Nucleic Acids and Proteins in Plants,* Boulter, D. and Parthier, B., Eds., Berlin, Springer Verlag, 1982, 475.
12. **Koller, B. and Delius, H.**, *Vicia faba* chloroplast DNA has only one set of ribosomal RNA as shown by partial denaturation mapping and R-loop analyses, *Mol. Gen. Genet.*, 178, 261, 1980.
13. **Kolodner, R. and Tewari, K. K.**, Inverted repeats in chloroplast DNA from higher plants, *Proc. Natl. Acad. Sci. U.S.A.*, 76, 41, 1979.
14. **Stummann, B. M., Lehmbeck, J., Bookjans, G., and Henningsen, K. W.**, Nucleotide sequence of the single ribosomal RNA operon of pea chloroplast DNA, *Physiol. Plant.*, 72, 139, 1988.
15. **Li, N. and Cattolico, R. A.**, Chloroplast genome characterization in the red alga *Griffithsia pacifica, Mol. Gen. Genet.*, 209, 343, 1987.
16. **Palmer, J. D. and Thompson, W. F.**, Rearrangements in the chloroplast genomes of mung bean and pea, *Proc. Natl. Acad. Sci. U.S.A.*, 78, 5533, 1981.
17. **Palmer, J. D. and Thompson, W. F.**, Chloroplast DNA rearrangements are more frequent when a large inverted repeat sequence is lost, *Cell,* 29, 537, 1982.
18. **Gray, P. W. and Hallick, R. B.**, Physical mapping of the *Euglena gracilis* chloroplast DNA and ribosomal RNA gene region, *Biochemistry,* 17, 284, 1978.
19. **Tomioka, N., Shinozaki, K., and Sugiura, M.**, Molecular cloning and characterization of ribosomal rRNA genes from a blue-green alga, *Anacystis nidulans, Mol. Gen. Genet.*, 184, 359, 1981.
20. **Machatt, M. A., Ebel, J. P., and Branlant, C.**, The 3′ terminal region of bacterial 23S ribosomal RNA: structure and homology with the 3′ terminal region of eukaryotic 28S rRNA and with chloroplast 4.5S rRNA, *Nucleic Acids Res.*, 9, 1533, 1981.
21. **Edwards, K., Bedbrook, J., Dyer, T. A., and Kossel, H.**, 4.5S rRNA from *Zea mays* chloroplasts shows structural homology with the 3′ end of prokaryotic 23S rRNA, *Biochem. Int.*, 2, 533, 1980.
22. **Kumagi, I., Pieler, T., Subramanian, A. R., and Erdmann, V. A.**, Nucleotide sequence and secondary structure analysis of spinach chloroplast 4.5S RNA, *J. Biol. Chem.*, 257, 12924, 1982.
23. **Strittmatter, G. and Kossel, H.**, Cotranscription and processing of 23S, 4.5S and 5S rRNA in chloroplasts from *Zea mays, Nucleic Acids Res.*, 12, 7633, 1984.
24. **McKay, R. M.**, The origin of plant chloroplast 4.5S ribosomal RNA, *FEBS Lett.*, 123, 17, 1981.
25. **Rochaix, J.-D. and Darlin, J.-D.**, Composite structure of the chloroplast 23S ribosomal RNA genes of *Chlamydomonas reinhardii.* Evolutionary and functional implications, *J. Mol. Biol.*, 159, 383, 1982.
26. **Schneider, M. and Rochaix, J.-D.**, Sequence organization of the chloroplast ribosomal spacer of *Chlamydomonas reinhardii:* uninterrupted tRNA[ile] and tRNA[ala] genes and extensive secondary structure, *Plant Mol. Biol.*, 6, 265, 1986.
27. **Allet, B. and Rochaix, J.-D.**, Structure analysis at the ends of the intervening DNA sequences in the chloroplast 23S ribosomal genes of *C. reinhardii, Cell,* 18, 55, 1979.
28. **Yamada, T. and Shimaji, M.**, Splitting of the ribosomal RNA operon on chloroplast DNA from *Chlorella ellipsoidea, Mol. Gen. Genet.*, 208, 377, 1987.
29. **Palmer, J. D.**, Chloroplast DNA exists in two orientations, *Nature (London),* 301, 92, 1983.
30. **Wolfe, K. H., Li, W.-H., and Sharp, P. A.**, Rates of nucleotide substitution vary greatly among plant mitochondrial, chloroplast and nuclear DNAs, *Proc. Natl. Acad. Sci. U.S.A.*, 84, 9054, 1987.
31. **Palmer, J. D.**, Evolution of chloroplast and mitochondrial DNA in plants and algae, in *Monographs in Evolutionary Biology: Molecular Evolutionary Genetics,* MacIntyre, R. J., Ed., New York, Plenum, 1985, 131.
32. **Vacek, A. T. and Bourque, D. P.**, Mode of inheritance and evidence for cistron heterogeneity of chloroplast 16S ribosomal RNA genes in *Nicotiana, Plasmid,* 4, 205, 1980.
33. **Moon, E., Kao, T.-H., and Wu, R.**, Rice chloroplast DNA molecules are heterogeneous as revealed by DNA sequences of a cluster of genes, *Nucleic Acids Res.*, 15, 611, 1987.

34. **Spreitzer, R. J., Chastain, C. J., and Ogren, W. L.,** Chloroplast gene suppression and defective ribulosebisphosphate carboxylase/oxygenase in *Chlamydomonas reinhardii:* evidence for stable heteroplasmic genes, *Curr. Genet.,* 9, 83, 1984.

35. **Spreitzer, R. J. and Chastain, C. J.,** Heteroplasmic suppression of an amber mutation in the *Chlamydomonas* chloroplast gene that encodes the large subunit of ribulosebisphosphate carboxylase/oxygenase, *Curr. Genet.,* 11, 611, 1987.

36. **Lonsdale, D. M., Hodge, T. P., and Fauron, C. M. R.,** The physical map and organization of the mitochondrial genome from the fertile cytoplasm of maize, *Nucleic Acids Res.,* 12, 9249, 1984.

37. **Leaver, C. J. and Gray, M. W.,** Mitochondrial genome organization and expression in higher plants, *Annu. Rev. Plant Physiol.,* 33, 373, 1984.

38. **Lonsdale, D. M.,** A review of the structure and function of the mitochondrial genome of higher plants, *Plant Mol. Biol.,* 3, 201, 1984.

39. **Hayashida, N., Matsubayashi, T., Shinozaki, K., Sugiura, M., Inou, K., and Hiyama, T.,** The gene for the 9 kd polypeptide, a possible apoprotein for the iron-sulfur centers A and B of the photosystem I complex in tobacco chloroplast DNA, *Curr. Genet,* 12, 247, 1987.

40. **Matsubayashi, T., Wakasugi, T., Shinozaki, K., Yamaguch-Shinozaki, K., Zaita, N., Hidaka, T., Meng, B. Y., Ohto, C., Tanaka, M., Kato, A., Maruyama, T., and Sugiura, M.,** Six chloroplast genes (ndhA-F) homologous to human mitochondrial genes encoding components of the respiratory chain NADH dehydrogenase are actively expressed: determination of the splice sites in *ndh*A and *ndh*B premRNAs, *Mol. Gen. Genet.,* 210, 385, 1987.

41. **Hallick, R. B. and Bottomley, W.,** Proposals for the naming of chloroplast genes, *Plant Mol. Biol. Rep.,* 1, 38, 1983.

42. **Poulsen, C.,** The barley chloroplast genome: physical structure and transcription activity *in vivo, Carlsberg Res. Commun.,* 48, 57, 1983.

43. **Oishi, K., Sumnicht, T., and Tewari, K. K.,** Messenger ribonucleic acid transcripts of pea chloroplast deoxyribonucleic acid, *Biochemistry,* 20, 5110, 1981.

44. **Koller, B. and Delius, H.,** Intervening sequences in chloroplast genomes, *Cell,* 36, 613, 1984.

45. **Schmidt, G. and Mishkind, M. L.,** The transport of proteins into chloroplasts, *Annu. Rev. Biochem.,* 55, 879, 1986.

46. **Capel, M. and Bourque, D. P.,** Characterization of *Nicotiana tabacum* chloroplast and cytoplasmic ribosomal proteins, *J. Biol. Chem.,* 257, 7746, 1982.

47. **Wasmann, C. C., Loeffelhardt, W., and Bohnert, H. J.,** Cyanelles: organization and molecular biology, in *Cyanobacteria,* Fay, P. and Ballen, C., Eds., Elsevier Science Publ., Amsterdam, 1986, 303.

48. **Steinmuller, D., Kaling, M., and Zetsche, K.,** *In vitro* synthesis of phycobiliproteins and ribulose 1,5-bisphosphate carboxylase by non-polyadenylated RNA of *Cyanidium caldarium* and *Prophyridium aerogineum, Planta,* 159, 308, 1983.

49. **Schmidt, G. M. and Mishkind, M. L.,** Rapid degradation of unassembled ribulose 1,5-bisphosphate carboxylase small subunits in chloroplasts, *Proc. Natl. Acad. Sci. U.S.A.,* 80, 2632, 1983.

50. **Herrmann, R. G., Westhoff, P., Alt, J., Tittgen, J., and Nelson, N.,** Thylakoid membrane proteins and their genes, in *Molecular Form and Function of the Plant Genome,* Vloten-Doting, L., Groot, G. S. P., and Hall, T. C., Eds., Plenum, New York, 1985, 233.

51. **Ohyama, K., Kohchi, T., Sano, T., and Yamada, Y.,** Newly identified groups of genes in chloroplasts, *Trends Biochem. Sci.,* 13, 19, 1988.

52. **Meng, B. Y., Matsubayashi, T., Wakasugi, T., Shinozaki, K., Hirai, A., Mikami, T., Kishima, Y., and Kinoshita, T.,** Ubiquity of the genes for components of a NADH dehydrogenase in higher plant chloroplast genomes, *Plant Sci.,* 47, 181, 1986.

53. **Bennoun, P.,** Evidence for a respiratory chain in the chloroplast, *Proc. Natl. Acad. Sci. U.S.A.,* 79, 4352, 1982.

54. **Godde, D.,** Evidence for membrane bound NADH-plastoquinone-oxidoreductase in *Chlamydomonas reinhardtii* CW-15, *Arch. Microbiol.,* 131, 197, 1982.

55. **Lemaire, C., Wollman, F. A., and Bennoun, P.,** Restoration of phototrophic growth in a mutant of *Chlamydomonas reinhardtii* in which the chloroplast *atp*B gene of the ATP synthase has a deletion: an example of mitochondria-dependent photosynthesis, *Proc. Natl. Acad. Sci. U.S.A.,* 85, 1344, 1988.

56. **Stern, D. B. and Lonsdale, D. A.,** Mitochondria and chloroplast genomes of maize have a 12 kilobase DNA sequence in common, *Nature (London),* 299, 698, 1982.

57. **Stern, D. B. and Palmer, J. D.,** Tripartite mitochondrial genome of spinach: physical structure, mitochondrial gene mapping and locations of transposed chloroplast DNA sequences, *Nucleic Acids Res.,* 14, 5651, 1986.

58. **Baldauf, S. L., Manhart, J. R., and Palmer, J. D.,** personal communication, 1986.

59. **Montandon, P. E. and Stutz, E.,** Nucleotide sequence of a *Euglena gracilis* chloroplast genome region for the elongation factor Tu; Evidence for a spliced mRNA, *Nucleic Acids Res.,* 11, 5877, 1983.

60. **Timmis, J. N. and Scott, N. S.,** Sequence homology between spinach nuclear and chloroplast genomes, *Nature (London),* 305, 65, 1983.
61. **Scott, N. S. and Timmis, J. N.,** Homologies between nuclear and plastid DNA in spinach, *Theor. Appl. Genet.,* 67, 279, 1984.
62. **Chang, D. D. and Clayton, D. A.,** A mammalian mitochondrial RNA processing activity contains nucleus-encoded RNA, *Science,* 235, 1178, 1987.
63. **Cozens, A. L. and Walker, J. E.,** The organization and sequence of the genes for ATP synthase subunits in the cyanobacterium *Synechococcus* 6301. Support for an endosymbiotic origin of chloroplasts, *J. Mol. Biol.,* 194, 359, 1987.
64. **Tanaka, M., Wakasugi, T., Sugita, M., and Sugiura, M.,** Genes for eight ribosomal proteins are clustered on the chloroplast genome of tobacco (*Nicotiana tabacum*): similarity to the S10 and *spc* operons of *Escherichia coli, Proc. Natl. Acad. Sci. U.S.A.,* 83, 6030, 1986.
65. **Bonnard, G., Michel, F., Weil, J. H., and Steinmetz, A.,** Nucleotide sequence of the split tRNA^{Leu} (UAA) gene from *Vicia faba* chloroplasts: evidence for structural homologies of the chloroplast tRNA^{Leu} intron with the intron of the autospliceable *Tetrahymena* ribosomal RNA precursor, *Mol. Gen. Genet.,* 194, 330, 1984.
66. **Michel, F. and Dujon, B.,** Conservation of RNA secondary structures in two intron families including mitochondrial-, chloroplast- and nuclear-encoded members, *EMBO J.,* 2, 33, 1983.
67. **Kruger, K., Grabowski, P. J., Zaug, A. J., Sands, J., Gottschling, D. E., and Cech, T. R.,** Self-splicing RNA: autoexcision and autocyclization of the ribosomal RNA intervening sequence of *Tetrahymena, Cell,* 31, 147, 1982.
68. **Erickson, J. M., Rahire, M., and Rochaix, J.-D.,** *Chlamydomonas reinhardii* gene for 32000 mol. wt. protein of photosystem II contains four large introns and is located entirely within the chloroplast inverted repeat, *EMBO J.,* 3, 2753, 1984.
69. **Rochaix, J.-D., Rahire, M., and Michel, M.,** The chloroplast ribosomal intron of *Chlamydomonas reinhardii* for a polypeptide related to mitochondrial maturases, *Nucleic Acids Res.,* 13, 975, 1985.
70. **Carignani, G., Groudinsky, O., Frezza, D., Schiavon, E., Bergantino, E., and Slonimski, P. P.,** An mRNA maturase is encoded by the first intron of the mitochondrial gene for the subunit 1 of cytochrome oxidase in *S. cerevisiae, Cell,* 35, 733, 1983.
71. **Koch, W., Edwards, K., and Kossel, H.,** Sequence of the 16S-23S spacer in a ribosomal RNA operon of *Zea mays* chloroplast DNA reveals two split tRNA genes, *Cell,* 25, 203, 1981.
72. **Takaiwa, F. and Sugiura, M.,** Nucleotide sequence of the 16S-23S spacer region in an rRNA gene cluster from tobacco chloroplast DNA, *Nucleic Acids Res.,* 10, 2665, 1982.
73. **Massenet, O., Martinez, P., Seyer, P., and Briat, J.-F.,** Sequence organization of the chloroplast ribosomal spacer of *Spinacia oleracea* including the 3' end of the 16S rRNA and the 5' end of the 23S rRNA, *Plant Mol. Biol.,* 10, 53, 1987.
74. **Michel, F., Jacquier, A., and Dujon, B.,** Comparison of fungal mitochondrial introns reveals intensive homologies in RNA secondary structure, *Biochimie,* 64, 867, 1982.
75. **Sugita, M., Shinozaki, K., and Sugiura, M.,** Tobacco chloroplast tRNA^{Lys} (UUU) gene contains a 2.5-kilobase-pair intron: an open reading frame and a conserved boundary sequence in the intron, *Proc. Natl. Acad. Sci. U.S.A.,* 82, 3557, 1985.
76. **Shinozaki, K., Deno, H., Sugita, M., Kuramitsu, S., and Sugiura, M.,** Intron in the gene for the ribosomal protein S16 of tobacco chloroplast and its conserved boundary sequences, *Mol. Gen. Genet.,* 202, 1, 1986.
77. **Hallick, R. B., Gingrich, J. C., Johanningmeier, U., and Passavant, C.,** Introns in *Euglena* and *Nicotiana* chloroplast protein genes, in *Molecular Form and Function of the Plant Genome,* Vloten-Doting, L., Groot, G. S. P., and Hall, T. C., Eds., Plenum Press, New York, 1985, 211.
78. **Cech, T.,** RNA splicing: three themes with variations, *Cell,* 34, 713, 1983.
79. **Hildebrand, M. M., Hallick, R. B., Passavant, C., and Bourque, D. P.,** *Trans*-splicing in chloroplasts: the RPS 12 loci of *Nicotiana tabacum, Proc. Natl. Acad. Sci. U.S.A.,* 84, 372, 1988.
80. **Koller, B., Fromm, H., Galun, E., and Edelman, M.,** Evidence for *in vivo trans*-splicing of pre-mRNAs in tobacco chloroplasts, *Cell,* 48, 111, 1987.
81. **Zaita, N., Torazawa, K., Shinozaki, K., and Sugiura, M.,** *Trans* splicing *in vivo*: joining of transcripts from the 'divided' gene for ribosomal protein S12 in the chloroplasts of tobacco, *FEBS Lett.,* 210, 153, 1987.
82. **Griese, K., Subramanian, A. P., Larrinua, I. M., and Bogorad, L.,** Nucleotide sequence, promoter analysis and linkage mapping of the unusually organized operon encoding ribosomal proteins S7 and S12 in maize chloroplasts, *J. Biol. Chem.,* 262, 15251, 1987.
83. **von Allmen, J. M. and Stutz, E.,** Complete sequence of 'divided' *rps*12 (r-protein S12) and *rps*7 (r-protein S7) gene in soybean chloroplast DNA, *Nucleic Acids Res.,* 15, 2387, 1987.

84. **Montandon, P. E. and Stutz, E.,** The genes for the ribosomal proteins S12 and S7 are clustered with the gene for the EF -Tu protein on the chloroplast genome of *Euglena gracilis, Nucleic Acids Res.,* 12, 2851, 1984.

85. **Kuck, U., Choquet, Y., Schneider, M., Dron, M., and Bennoun, P.,** Structural and transcription analysis of two homologous genes for the P700 chlorophyll *a*-apoproteins in *Chlamydomonas reinhardii*: evidence of *in vivo trans*-splicing, *EMBO J.,* 6, 2185, 1987.

86. **Crouse, E. J., Schmitt, J. M., and Bohnert, H. J.,** Chloroplast and cyanobacterial genomes, gene and RNAs: a compilation, *Plant Mol. Biol. Rep.,* 3, 43, 1985.

87. **Bedbrook, J. R., Link, G., Coen, D. M., Bogorad, L., and Rich, A.,** Maize plastid gene expressed during photoregulated development, *Proc. Natl. Acad. Sci. U.S.A.,* 75, 3060, 1978.

88. **Apel, K.,** The protochlorophyllide holochrome of barley (*Hordeum vulgare* L.). Phytochrome-induced decrease of translatable mRNA coding for the NADPH:protochlorophyllide oxidoreductase, *Eur. J. Biochem.,* 120, 89, 1981.

89. **Cuming, A. C. and Bennett, J.,** Biosynthesis of the light-harvesting chlorophyll a/b protein. Control of messenger RNA activity by light, *Eur. J. Biochem.,* 118, 71, 1981.

90. **Santel, H. and Apel, K.,** The protochlorophyllide holochrome of barley (*Hordeum vulgare* L.). The effect of light on the NADPH:protochlorophyllide oxireductase, *Eur. J. Biochem.,* 120, 95, 1981.

91. **Gallagher, T. F. and Ellis, R. J.,** Light-stimulated transcription of genes for two chloroplast polypeptides in isolated pea leaf nuclei, *EMBO J.,* 1, 1493, 1982.

92. **Crossland, L. D., Rodermel, S. R., and Bogorad, L.,** The single gene for the large subunit of ribulosebisphosphate carboxylase in maize yields two differently regulated mRNAs, *Proc. Natl. Acad. Sci. U.S.A.,* 81, 4060, 1984.

93. **Rodermel, S. R. and Bogorad, L.,** Maize plastid photogenes: mapping and photoregulation of transcript levels during light-induced development, *J. Cell Biol.,* 100, 463, 1985.

94. **Glick, R. E., McCauley, S. W., Gruissem, W., and Melis, A.,** Light quality regulates expression of chloroplast genes and assembly of photosynthetic membrane complexes, *Proc. Natl. Acad. Sci. USA,* 83, 4287, 1986.

95. **Link, G.,** Phytochrome control of plastid mRNA in mustard (*Sinapis alba* L.), *Planta,* 154, 81, 1982.

96. **Thompson, W. F., Everett, M., Polans, N. O., Jorgensen, R. A., and Palmer, J. D.,** Phytochrome control of RNA levels of developing pea and mung-bean leaves, *Planta,* 158, 487, 1983.

97. **Apel, K.,** Phytochrome-induced appearance of mRNA activity for the apoprotein of the light-harvesting chlorophyll a/b protein of barley (*Hordeum vulgare*), *Eur. J. Biochem.,* 97, 183, 1979.

98. **Tobin, E. M.,** Phytochrome-mediated regulation of messenger RNAs for the small subunit of ribulose 1,5-bisphosphate carboxylase and the light-harvesting chlorophyll a/b-protein in *Lemna gibba, Plant Mol. Biol.,* 1, 35, 1981.

99. **Steikma, W. J., Wimpee, C. F., Silverthorne, J., and Tobin, E. M.,** Phytochrome control of the expression of two nuclear genes encoding chloroplast proteins in *Lemna gibba* L. G-3, *Plant Physiol.,* 72, 717, 1983.

100. **Gollmer, I. and Apel, K.,** The phytochrome-controlled accumulation of mRNA sequences encoding the light-harvesting chlorophyll a/b protein of barley (*Hordeum vulgare* L.), *Eur. J. Biochem.,* 133, 309, 1983.

101. **Silverthorne, J. and Tobin, E. M.,** Demonstration of transcriptional regulation of specific genes by phytochrome action, *Proc. Natl. Acad. Sci. U.S.A.,* 81, 1112, 1984.

102. **Fromm, H., Devie, M., Fluhr, R., and Edelman, M.,** Control of *psb*A gene expression: in mature *Spirodela* chloroplasts light regulation of 32 kd protein synthesis is independent of transcript level, *EMBO J.,* 4, 291, 1985.

103. **Klein, R. R. and Mullet, J. E.,** Regulation of chloroplast-encoded biogenesis, *J. Biol. Chem.,* 261, 11138, 1986.

104. **Hawley, D. K. and McClure, W. R.,** Compilation and analysis of *Escherichia coli* promoter DNA sequences, *Nucleic Acids Res.,* 11, 2237, 1983.

105. **Kung, S. D. and Lin, C. M.,** Chloroplast promoters from higher plants, *Nucleic Acids Res.,* 13, 7543, 1985.

106. **Link, G.,** DNA sequence requirements for the accurate transcription of a protein-coding plastid gene in a plastid *in vitro* system from mustard (*Sinapis alba*), *EMBO J.,* 3, 1697, 1984.

107. **Gruissem, W. and Zurawski, G.,** Identification and mutational analysis of the promoter for a spinach chloroplast transfer RNA gene, *EMBO J.,* 4, 1637, 1985.

108. **Gruissem, W. and Zurawski, G.,** Analysis of promoter regions for the spinach chloroplast *rbc*L, *atp*B and *psb*A genes, *EMBO J.,* 4, 3375, 1985.

109. **Delp, G., Igloi, G. L., Beck, C. F., and Kossel, H.,** Functional *in vivo* verification in *E. coli* of promoter activities from the rDNA/tDNA^{Val} (GAC) leader region of *Zea mays* chloroplasts, *Curr. Genet.,* 12, 241, 1987.

110. **Simons, R. W., Hoopes, B. C., McClure, W. R., and Kleckner, N.,** Three promoters near the termini of IS 10: pIN, pOUT and pIII, *Cell,* 34, 673, 1983.

111. **Russell, D. and Bogorad, L.**, Transcription analysis of the maize chloroplast gene for the ribosomal protein S4, *Nucleic Acids Res.*, 15, 1853, 1987.

112. **Ryals, J., Little, R., and Bremmer, H.**, Control of rRNA and tRNA synthesis in *Escherichia coli* by guanosine tetraphosphate, *J. Bacteriol.*, 151, 1261, 1982.

113. **Little, R., Ryals, R., and Bremer, H.**, *rpo*B mutation in *Escherichia coli* alters control of ribosome synthesis by guanosine tetraphosphate, *J. Bacteriol.*, 154, 787, 1983.

114. **Jolly, S. O. and Bogorad, L.**, Preferential transcription of cloned maize chloroplast DNA sequences by maize chloroplast RNA polymerase, *Proc. Natl. Acad. Sci. U.S.A.*, 77, 822, 1980.

115. **Jolly, S. O., McIntosh, L., Link, G., and Bogorad, L.**, Differential transcription *in vivo* and *in vitro* of two adjacent maize chloroplast genes: the large subunit of ribulose bisphosphate carboxylase and the 2.2-kilobase gene, *Proc. Natl. Acad. Sci. U.S.A.*, 78, 6821, 1981.

116. **Hallick, R. B., Lipper, C., Richards, O. D., and Rutter, W. J.**, Isolation of a transcriptionally active chromosome from chloroplasts of *Euglena gracilis, Biochemistry*, 15, 3039, 1976.

117. **Rushlow, K. E., Orozco, E. M., Jr., Lipper, C., and Hallick, R. B.**, Selective *in vitro* transcription of *Euglena* chloroplast ribosomal RNA genes by a transcriptionally active chromosome, *J. Biol. Chem.*, 255, 3786, 1980.

118. **Greenberg, B. M., Narita, J. O., DeLuca-Flaherty, C., Gruissem, W., Rushlow, K. A., and Hallick, R. B.**, Evidence for two RNA polymerase activities in *Euglena gracilis* chloroplasts, *J. Biol. Chem.*, 259, 14880, 1984.

119. **De Luca-Flaherty, C.**, Messenger RNA transcription and co-transcript processing in *Euglena gracilis* chloroplast soluble extract, *Ph.D. dissertation*, University of Colorado, Boulder, 1985.

120. **Gauss, D. H. and Sprinzl, M.**, Compilation of tRNA sequences, *Nucleic Acids Res.*, 9, r1, 1981.

121. **Briat, J. F. and Mache, R.**, Properties and characterization of a spinach chloroplast RNA polymerase isolated from a transcriptionally active DNA-protein complex, *Eur. J. Biochem.*, 111, 503, 1980.

122. **Tewari, K. K. and Goel, A.**, Solubilization and partial purification of RNA polymerase from pea chloroplasts, *Biochemistry*, 22, 2142, 1983.

123. **Polya, G. M. and Jagendorf, A. T.**, Kinetic characterization and template specificities of nuclear, chloroplast and soluble enzymes, *Arch. Biochem. Biophys.*, 146, 649, 1971.

124. **Link, G.**, DNA sequence requirements for the accurate transcription of a protein-coding plastid gene in a plastid *in vitro* system from mustard (*Sinapis alba* L.), *EMBO J.*, 3, 1697, 1984.

125. **Tewari, K. K. and Wildman, S. G.**, Function of chloroplast DNA. II. Studies on DNA-dependent RNA polymerase activity of tobacco chloroplasts, *Biochem. Biophys. Acta*, 186, 358, 1969.

126. **Little, M. and Hallick, R. B.**, personal communication, 1988.

127. **Von Hippel, P. H., Bear, D. G., Morgan, W. D., and McSwiggen, J. A.**, Protein-nucleic acid interactions in transcription: a molecular analysis, *Annu. Rev. Biochem.*, 53, 389, 1984.

128. **Zacharias, M. and Wagner, R.**, Deletions in the t_L structure upstream of the rRNA genes in the *E. coli* rrnB operon cause transcription polarity, *Nucleic Acids Res.*, 15, 8235, 1987.

129. **Lindahl, L., Archer, R., and Zengel, J. M.**, Transcription of the S10 ribosomal protein operon is regulated by an attenuator in the leader, *Cell*, 33, 241, 1983.

130. **Briat, J. F., Dron, M., and Mache, R.**, Is transcription of higher plant chloroplast ribosomal operons regulated by an attenuator in the leader?, *Cell*, 33, 241, 1983.

131. **Rosenberg, M. and Court, D.**, Regulatory sequences involved in the promotion and termination of RNA transcription, *Annu. Rev. Genet.*, 13, 319, 1979.

132. **Zurawski, G., Perrot, B., Bottomley, W., and Whitfield, P. R.**, The structure of the gene for the large subunit of ribulose 1,5-bisphosphate carboxylase from spinach chloroplast DNA, *Nucleic Acids Res.*, 9, 3251, 1981.

133. **Akada, S., Xu, Y. Q., Machii, H., and Kung, S. D.**, personal communication, 1987.

134. **Wong, H. C. and Chang, S.**, Identification of a positive retroregulator that stabilizes mRNAs in bacteria, *Proc. Natl. Acad. Sci. U.S.A.*, 83, 3223, 1987.

135. **Birchmeier, C., Schumperli, D., Sconzo, G., and Birnstiel, M. L.**, 3' editing of mRNAs: sequence requirements and involvement of a 60-nucleotide RNA in maturation of histone mRNA precursors, *Proc. Natl. Acad. Sci. U.S.A.*, 81, 1057, 1984.

136. **Stern, D. B., and Gruissem, W.**, Control of plastid gene expression: 3' inverted repeats act as mRNA processing and stabilizing elements, but do not terminate transcription, *Cell*, 51, 1145, 1987.

137. **Gruissem, W., Prescott, D. M., Greenberg, B. M., and Hallick, R. B.**, Transcription of *E. coli* and *Euglena* chloroplast tRNA gene clusters and processing of polycistronic transcripts in a HeLa cell-free system, *Cell*, 30, 81, 1982.

138. **Gruissem, W., Prescott, D. M., Greenberg, B. M., and Hallick, R. B.**, Accurate processing and pseudo-uridylation of chloroplast transfer RNA in a chloroplast transcription system, *Plant Mol. Biol.*, 3, 97, 1984.

139. **Tanaka, M., Obokata, J., Chunwongse, J., Shinozaki, K., and Sugiura, M.**, Rapid splicing and stepwise processing of a transcript from the *psb*B operon in tobacco chloroplasts: determination of the intron sites in *pet*B and *pet*D, *Mol. Gen. Genet.*, 209, 427, 1987.

140. **Yamaguchi-Shinozaki, K., Shinozaki, K., and Sugiura, M.,** Processing of precursor tRNAs in a chloroplast lysate, *FEBS Lett.*, 215, 132, 1987.
141. **Berends, T., Gamble, P. E., and Mullet, J. E.,** Characterization of the barley chloroplast transcription units containing *psa*A-*psa*B and *psb*D-*psb*C, *Nucleic Acids Res.*, 15, 5217, 1987.
142. **Ohme, M., Kamogashira, T., Shinozaki, K., and Sugiura, M.,** Structure and cotranscription of tobacco chloroplast genes for tRNA^{Glu} (GUA) and tRNA^{ASP} (GUC), *Nucleic Acids Res.*, 13, 1045, 1985.
143. **Fukuzawa, H., Yoshida, T., Kohchi, T., Okumura, T., Sawano, Y., and Ohyama, K.,** Splicing of group II introns in mRNAs coding for cytochrome b$_6$ and subunit IV in the liverwort *Marchantia polymorpha* chloroplast genome, *FEBS Lett.*, 220, 61, 1987.
144. **Barkan, A., Miles, D., and Taylor, W. C.,** Chloroplast gene expression in nuclear, photosynthetic mutants of maize, *EMBO J.*, 5, 1421, 1986.
145. **Brosius, J., Palmer, M. L., Kennedy, P. J., and Noller, H. F.,** Complete nucleotide sequence of a 16S ribosomal RNA gene from *Escherichia coli*, *Proc. Natl. Acad. Sci. U.S.A.*, 75, 4801, 1978.
146. **Tohdoh, N. and Sugiura, M.,** The complete nucleotide sequence of a 16S ribosomal RNA gene from tobacco chloroplasts, *Gene*, 17, 213, 1982.
147. **Schwarz, Z. and Kossel, H.,** The primary structure of 16S rDNA from *Zea mays* chloroplast is homologous to *E. coli* 16S rRNA, *Nature (London)*, 283, 739, 1980.
148. **Ohyama, K., Fukuzawa, H., Kohchi, T., Shirai, H., Sano, T., Sano, S., Umesono, K., Shiki, Y., Takeuchi, M., Chang, Z., Aota, S., Inokuchi, H., and Ozeki, H.,** Complete nucleotide sequence of liverwort *Marchantia polymorpha* chloroplast DNA, *Plant Mol. Biol. Rep.*, 4, 148, 1986.
149. **Looman, A. C., Badlaender, J., de Gruyter, M., Vagelaar, A., and van Knippenberg, P. H.,** Secondary structure as primary determinant of the efficiency of ribosomal binding sites in *Escherichia coli*, *Nucleic Acids Res.*, 14, 5481, 1986.
150. **Chua, N.-H. and Luck, D.,** Biosynthesis of organelle ribosomes, in *Ribosomes*, Nomura, M., Tissieres, A., and Lengyel, P., Eds., Cold Spring Harbor Laboratory, Cold Spring Harbor, NY, 1974, 519.
151. **Ledoigt, G. and Freyssinet, G.,** Plastid ribosome, *Biol. Cell.*, 46, 215, 1982.
152. **Lake, J. A., Henderson, E., Clark, M. W., and Matheson, A. T.,** Mapping evolution with ribosome structure: intralineage constancy and interlineage variation, *Proc. Natl. Acad. Sci. U.S.A.*, 79, 5948, 1982.
153. **Schwarz, J. H., Meyer, R., Eisenstadt, J. M., and Brawermann, G.,** Involvement of N-formyl-methionine in initiation of protein synthesis in cell-free extracts of *Euglena gracilis*, *J.Mol. Biol.*, 25, 571, 1967.
154. **Lucas-Lenard, J. and Lipmann, F.,** Protein biosynthesis, *Annu. Rev. Biochem.*, 40, 409, 1971.
155. **Ozaki, M., Mizushima, S., and Nomura, M.,** Identification and functional characterization of the protein controlled by the streptomycin-resistant locus in *E. coli*, *Nature (London)*, 222, 333, 1969.
156. **Etzold, T., Fritz, C. C., Schell, J., and Schreier, P. H.,** A point mutation in the chloroplast 16S rRNA gene of a streptomycin resistant *Nicotiana tabacum*, *FEBS Lett.*, 219, 343, 1987.
157. **Hildebrand, M. M.,** Molecular characterization of streptomycin resistance and the trans-splicing *rps*12 gene in *Nicotiana tabacum* chloroplasts, *Ph.D. dissertation*, University of Arizona, Tucson, 1987.
158. **Anderson, P.,** Sensitivity and resistance to spectinomycin in *Escherichia coli*, *J. Bacteriol.*, 100, 939, 1969.
159. **Fromm, H., Edelman, M., Aviv, D., and Galun, E.,** The molecular basis for rRNA-dependent spectinomycin resistance in *Nicotiana* chloroplasts, *EMBO J.*, 6, 3233, 1987.
160. **Harris, E. H., Boynton, J. E., Gillham, N. W., Tingle, C. L., and Fox, S. B.,** Mapping of chloroplast genes involved in chloroplast ribosome biogenesis in *Chlamydomonas reinhardii*, *Mol. Gen. Genet.*, 155, 249, 1977.
161. **Ganoza, M. C., Kofoid, E. C., Marlier, P., and Louis, G. G.,** Potential secondary structure at translation-initiation sites, *Nucleic Acids Res.*, 15, 345, 1987.
162. **Childs, J., Villanueba, K., Barrick, D., Schneider, T. D., Stormo, G. D., Gold, L., Leitner, M., and Caruthers, M.,** Ribosome binding site sequences and function, in *Sequence Specificity in Transcription and Translation*, Calandar, R. and Gold, L., Eds., Alan R. Liss, New York, 1985, 341.
163. **Volckaert, G. and Fiers, W.,** Studies on the bacteriophage MS2. G-U-G as the initiation codon of the A-protein cistron, *FEBS Lett.*, 35, 91, 1973.
164. **Shine, J. and Dalgarno, L.,** The 3′ terminal sequence of *E. coli* 16S ribosomal RNA: complementarity to nonsense triplets and ribosome binding sites, *Proc. Natl. Acad. Sci. U.S.A.*, 71, 1342, 1974.
165. **Gold, L., Pribnow, D., Schneider, T., Shinedling, S., Singer, B. S., and Stormo, G.,** Translational initiation in prokaryotes, *Annu. Rev. Microbiol.*, 35, 365, 1981.
166. **Steitz, J. A.,** RNA:RNA interactions during polypeptide chain initiation, in *Ribosomes: Structure, Function and Genetics*, Chambliss, G., Craven, G. R., Davies, J., Davis, K., Kahan, L., and Nomura, M., Eds., University Park Press, Baltimore, 1980, 479.
167. **Steitz, J. A. and Jakes, K.,** How ribosomes select initiator regions in mRNA: base pair formation between the 3′ terminus of 16S rRNA and the mRNA during initiation of protein synthesis in *Escherichia coli*, *Proc. Natl. Acad. Sci. U.S.A.*, 72, 4734, 1975.

168. **de Lanversin, G., Pilay, D. T. N., and Jacq, B.**, Sequence studies on the soybean chloroplast 16S-23S rDNA spacer region, *Plant Mol. Biol.*, 10, 65, 1987.

169. **Janssen, I., Mucke, H., Loffelhardt, W., and Bohnert, H.-J.**, The central part of the cyanelle rDNA unit of *Cyanophora paradoxa*: sequence comparison with chloroplasts and cyanobacteria, *Plant Mol. Biol.*, 9, 478, 1987.

170. **Keus, R. J. A., Dekker, A. F., van Roon, M. A., and Groot, G. S. P.**, The nucleotide sequences of the regions flanking the genes coding for 23S, 16S and 4.5S ribosomal RNA on chloroplast DNA from *Spirodel oligorhiza*, *Nucleic Acids Res.*, 11, 6465, 1983.

171. **Steege, D. A., Graves, M. C., and Spremulli, L. L.**, *Euglena gracilis* chloroplast small subunit rRNA, *J. Biol. Chem.*, 257, 10430, 1982.

172. **Stormo, G., Schneider, T., and Gold, L.**, Characterization of translation initiation sites in *E. coli*, *Nucleic Acids Res.*, 10, 2971, 1982.

173. **Scherer, G. M., Walkinshaw, M., Arnott, S., and Moore, D. J.**, The ribosome binding sites recognized by *E. coli* ribosomes have regions with signal character in both the leader and protein coding segments, *Nucleic Acids Res.*, 8, 3895, 1980.

174. **Kozak, M.**, Comparison of initiation of protein synthesis in prokaryotes, eukaryotes, and organelles, *Micro. Rev.*, 47, 1, 1983.

175. **Mount, D. W. and Conrad, B.**, Improved programs for DNA and protein sequence analysis on the IBM personal computer and other standard computer systems, *Nucleic Acids Res.*, 14, 443, 1986.

176. **Markmann-Mulisch, U., vonKnoblauch, K., Lehmann, A., and Subramanian, A. R.**, Nucleotide sequence and linkage map position of the *secX* gene in maize chloroplast and evidence that it encodes a protein belonging to the 50S ribosomal subunit, *Biochem. Int.*, 15, 1057, 1987.

177. **Purton, S. and Gray, J. C.**, Nucleotide sequence of the gene for ribosomal protein L36 in pea chloroplast DNA, *Nucleic Acids Res.*, 15, 9080, 1987.

178. **Aldrich, J., Cherney, B. and Merlin, E.**, Sequence of the chloroplast-encoded *psb*A gene for the Qb polypeptide of alfalfa, *Nucleic Acids Res.*, 14, 9537, 1986.

179. **Aldrich, J., Cherney, B., Merlin, E. and Palmer, J.**, Sequence of the *rbc*L gene for the large subunit of ribulose bisphosphate carboxylase-oxygenase from alfalfa, *Nucleic Acids Res.*, 14, 9535, 1986.

180. **Herschberg, J. and McIntosh, L.**, Molecular basis of herbicide resistance in *Amaranthus hybridus*, *Science*, 222, 1346, 1983.

181. **Zurawski, G. and Clegg, M. T.**, The barley chloroplast DNA *atpBE*, *trnM*2 and *trnV*1 loci, *Nucleic Acids Res.*, 12, 2549, 1984.

182. **Zurawski, G., Clegg, M. and Brown, A. H. D.**, The nature of nucleotide sequence divergence between barley and maize chloroplast DNA, *Genetics*, 106, 735, 1984.

183. **Ko, K. and Straus, N. A.**, Sequence of the apocytochrome *f* gene encoded by the *Vicia faba* chloroplast genome, *Nucleic Acids Res.*, 15, 2391, 1987.

184. **Shinozaki, K., Sun, C. R., and Sugiura, M.**, Gene organization of chloroplast DNA from the broadbean *Vicia faba*, *Mol. Gen. Genet.*, 197, 363, 1984.

185. **Erickson, J. M., Rahire, M., Malnoe, P., Girard-Bascou, J., Pierre, Y., Bennoun, P., and Rochaix, J.-D.**, Lack of the D2 protein in a *Chlamydomonas reinhardtii psb*D mutant affects photosystem II stability and D1 expression, *EMBO J.*, 5, 1745, 1986.

186. **Woessner, J. P., Gillham, N. W., and Boyton, J. E.**, The sequence of the chloroplast *atp*B gene and its flanking regions in *Chlamydomonas reinhardtii*, *Gene*, 44, 17, 1986.

187. **Yong, R. C. A., Dove, M., Selgy, V. L., Lemieux, C., Turmel, M., and Narang, S. A.**, Complete nucleotide sequence and mRNA-mapping of the large subunit gene of ribulose-1,5-bisphosphate carboxylase/oxygenase (RuBisCo) from *Chlamydomonas moewusii*, *Gene*, 50, 259, 1986.

188. **Manzara, T., Hu, J.-X., Price, C. A., and Hallick, R. B.**, Characterization of the *Trn*D, *Trn*K, *Psa*A locus of *Euglena gracilis* chloroplast DNA, *Plant Mol. Biol.*, 8, 327, 1987.

189. **Manzara, T. and Hallick, R. B.**, Nucleotide sequence of the *Euglena gracilis* chloroplast gene for ribosomal protein L20, *Nucleic Acids Res.*, 15, 3927, 1987.

190. **Karabin, G. D., Farley, M. and Hallick, R. B.**, Chloroplast gene for Mr 32000 polypeptide of photosystem II in *Euglena gracilis* is interrupted by four introns with conserved boundary sequences, *Nucleic Acids Res.*, 12, 5801, 1984.

191. **Gingrich, J. C. and Hallick, R. B.**, The *Euglena gracilis* chloroplast ribulose 1,5-bisphosphate carboxylase gene, *J. Biol. Chem.*, 260, 16156, 1985.

192. **McLaughlin, W. E. and Larrinua, I. G.**, The sequence of the maize *rps*19 locus and of the inverted repeat/unique region junctions, *Nucleic Acids Res.*, 15, 3932, 1987.

193. **McLaughlin, W. E. and Larrinua, I. M.**, The sequence of the maize plastid encoded *rpl* 22 locus, *Nucleic Acids Res.*, 15, 4356, 1987.

194. **Srinivasa, B. R. and Subramanian, A. R.**, Nucleotide sequence and linkage map position of the gene for maize chloroplast ribosomal protein S14, *Biochemistry*, 20, 3188, 1987.

195. **McLaughlin, W. E. and Larrinua, I. M.,** The sequence of the first exon and part of the intron of the maize plastid encoded *rpl*16 locus, *Nucleic Acids Res.*, 15, 5896, 1987.

196. **McLaughlin, W. E. and Larrinua, I. M.,** The sequence of the maize plastid encoded *rps*3 locus, *Nucleic Acids Res.*, 15, 4689, 1987.

197. **McIntosh, L., Poulsen, C., and Bogorad, L.,** Chloroplast gene sequence for the large subunit of ribulose bisphosphatecarboxylase of maize, *Nature (London)*, 288, 556, 1980.

198. **Bradley, D. and Gatenby, A. A.,** Mutational analysis of the maize ATPase-*b* subunit gene promoter: the isolation of promoter mutants in *E. coli* and their characterization in a chloroplast *in vitro* transcription system, *EMBO J.*, 4, 3641, 1985.

199. **Steinmetz, A. A., Castroviejo, M., Sayre, R. T., and Bogorad, L.,** Protein PSII-G, *J. Biol. Chem.*, 261, 2485, 1986.

200. **Fish, L. E., Kueck, U., and Bogorad, L.,** Two partially homologous adjacent light-inducible maize chloroplast genes encoding polypeptides of the P700 chlorophyll *a*-apoprotein complex of photosystem I, *J. Biol. Chem.*, 260, 1413, 1985.

201. **Carrillo, N., Seyer, P., Tyagi, A., and Herrmann, R. G.,** Cytochrome *b*-559 genes from *Oenothera hookeri* and *Nicotiana tabacum* show a remarkably high degree of conservation as compared to spinach, *Curr. Genet.*, 10, 619, 1986.

202. **Lehmbeck, J., Rasmussen, O. F., Bookjans, G. B., Jepsen, B. R., Stummann, B. M., Henningsen, K. W.,** Sequence of two genes in pea chloroplast DNA coding for 84 and 82 kD polypeptides of the photosystem I complex, *Plant Mol. Biol.*, 7, 3, 1986.

203. **Zurawski, G., Bottomley, W., and Whitfeld, P. R.,** Sequence of the genes for the *b* and *e* subunits of ATP synthase from pea chloroplasts, *Nucleic Acids Res.*, 14, 3974, 1986.

204. **Zurawski, G., Whitfeld, P. R., and Bottomley, W.,** Sequence of the gene for the large subunit of ribulose 1,5-biphosphate carboxylase from pea chloroplasts, *Nucleic Acids Res.*, 14, 3975, 1986.

205. **Purton, S. and Gray, J. C.,** Nucleotide sequence of the gene for ribosomal protein S11 in pea chloroplast DNA, *Nucleic Acids Res.*, 15, 1873, 1987.

206. **Willey, D. L., Auffret, A. D., and Gray, J. C.,** Pea (*P. sativum*) chloroplast cytochrome *f* gene and flanks, *Cell*, 36, 555, 1984.

207. **Rasmussen, O. F., Bookjans, G., Stutmann, B. M., and Henningsen, K. W.,** Localization and nucleotide sequence of the gene for the membrane polypeptide D₂ from pea chloroplast DNA, *Plant Mol. Biol.*, 3, 191, 1984.

208. **Phillips, A. L. and Gray, J. C.,** Location and nucleotide sequence of the gene for the 15.2 kDa polypeptide of the cytochrome b-f complex from pea chloroplasts, *Mol. Gen. Genet.*, 194, 477, 1984.

209. **Cozens, A. L., Walker, J. E., Phillips, A. L., Huttly, A. K., and Gray, J. C.,** A sixth subunit of ATP synthase, an Fo component is encoded in the pea chloroplast genome, *EMBO J.*, 5, 217, 1986.

210. **Aldrich, J., Cherney, B., Merlin, E., Christopherson, L. A., and Williams, C.,** Sequence of the chloroplast-encoded *psb*A gene for the QB polypeptide of petunia, *Nucleic Acids Res.*, 14, 9536, 1986.

211. **Moon, E., Kao, T., and Wu, R.,** Sequence of the chloroplast-encoded *atp*B-*atp*E-*trn*M gene clusters from rice, *Nucleic Acids Res.*, 15, 4358, 1987.

212. **Tahar, S. B., Bottomley, W., and Whitfeld, P. R.,** Characterization of the spinach chloroplast genes for the S4 ribosomal protein, tRNA[Thr] (UGU) and tRNA[Ser] (GGA), *Plant Mol. Biol.*, 7, 63, 1986.

213. **Sijben-Muller, G., Hallick, R. B., Alt, J., Westhoff, P., and Herrmann, R. G.,** Spinach plastid genes coding for initiation factor IF-1, ribosomal protein S11 and RNA polymerase *a*-subunit, *Nucleic Acids Res.*, 14, 1029, 1986.

214. **Morris, J. and Herrmann, R. G.,** Nucleotide sequence of the gene for the P-680 chlorophyll *a* apoprotein of the photosystem II reaction centre from spinach, *Nucleic Acids Res.*, 12, 2837, 1984.

215. **Zurawski, G., Bohnert, H. J., Whitfeld, P. R., and Bottomley, W.,** Nucleotide sequence of the gene for the Mr 32000 thylakoid membrane protein from *Spinacia oleracea* and *Nicotiana debneyi* predicts a totally conserved primary translation product of Mr 38950, *Proc. Natl. Acad. Sci. U.S.A.*, 79, 7699, 1982.

216. **Zurawski, G., Bottomley, W., and Whitfeld, P. R.,** Junctions of the large single copy region and the inverted repeats in *Spinacia oleracea* and *Nicotiana debneyi* chloroplast DNA sequence of the genes for tRNA[His] and the ribosomal proteins S19 and L2, *Nucleic Acids Res.*, 12, 6547, 1984.

217. **Holschuh, K., Bottomley, W., and Whitfeld, P. R.,** Structure of the spinach chloroplast genes for the D2 and 44 kd reaction-centre proteins of photosystem II and for tRNA[Ser] (UGA), *Nucleic Acids Res.*, 12, 8819, 1984.

218. **Vermaas, W. F. J., Williams, J. G. K., and Arntzen, C. J.,** Sequencing and modification of *psb*B, the gene encoding the CP-47 protein of photosystem II, in the cyanobacterium *Synechocystis* 6803, *Plant Mol. Biol.*, 8, 317, 1987.

219. **Osiewacz, H. D. and McIntosh, L.,** Nucleotide sequence of a member of the psbA multigene family from the unicellular cyanobacterium *Synechocystis* 6803, *Nucleic Acids Res.*, 15, 10585, 1987.

220. **Goloubinoff, P., Edelman, M., and Haliick, R. B.,** Chloroplast-coded atrazine resistance in *Solanum nigrum psa*A loci from susceptible and resistant biotypes are isogenic except for a single codon change, *Nucleic Acids Res.,* 12, 9489, 1985.

221. **Posno, M., van Vliet, A., and Groot, G. S. P.,** The gene for *Spirodela oligorhiza* chloroplast ribosomal protein homologous to *E. coli* ribosomal protein L16 is split by a large intron near its 5' end: structure and expression, *Nucleic Acids Res.,* 14, 3181, 1986.

222. **Shinozaki, K., Ohme, M., Tanaka, M., Wakasugi, T., Hayshida, N., Matsubayasha, T., Zaita, N., Chunwongse, J., Obokata, J., Yamaguchi-Shinozaki, K., Ohto, C., Torazawa, K., Meng, B. Y., Sugita, M., Deno, H., Kamagashire, T., Yamada, K., Kususa, J., Takaiwa, F., Kata, A., Tohdoh, N., Shimada, H., and Sugiura, M.,** The complete sequence of the tobacco chloroplast genome, *Plant Mol. Biol. Rep.,* 4, 111, 1986.

223. **Dunn, P. P. J. and Gray, J. C.,** Nucleotide sequence of the frxB gene in wheat chloroplast DNA, *Nucleic Acids Res.,* 16, 348, 1988.

224. **Hoglund, A. S. and Gray, J. C.,** Nucleotide sequence of the gene for ribosomal protein S2 in wheat chloroplast DNA, *Nucleic Acids Res.,* 15, 1987.

225. **Willey, D. L., Howe, C. J., Auffret, A. D., Bowman, C. M., Dyer, T. A., and Gray, J. C.,** Location and nucleotide sequence of the gene for cytochrome *f* in wheat chloroplast DNA, *Mol. Gen. Genet.,* 194, 416, 1984.

226. **Howe, C. J., Auffret, A. D., Doherty, A., Bowman, C. M., Dyer, T. A., and Gray, J. C.,** Location and nucleotide sequence of the gene for the proton-translocating subunit of wheat chloroplast ATP synthase, *Proc. Natl. Acad. Sci. U.S.A.,* 79, 6903, 1982.

227. **Hird, S. M., Willey, D. L., Dyer, T. A., and Gray, J. C.,** Location and nucleotide sequence of the gene for cytochrome *b*-559 in wheat chloroplast DNA, *Mol. Gen. Genet.,* 203, 95, 1986.

228. **Smiley, B., Lupaki, J., Svec, P., McMacken, R., and Godson, G. N.,** Sequences of the *Escherichia coli dna* G primase gene and regulation of its expression, *Proc. Natl. Acad. Sci. U.S.A.,* 79, 4550, 1982.

229. **Post, L. E. and Nomura, M.,** DNA sequences from the *str* operon of *Escherichia coli, J. Biol. Chem.,* 255, 4660, 1980.

230. **Yanofsky, C., Platt, T., Crawford, I., Nichols, B., Christie, G., Horowitz, H., van Cleemput, M., and Wu, A. M.,** The complete nucleotide sequence of the tryptophan operon of *Escherichia coli, Nucleic Acids Res.,* 9, 6647, 1981.

231. **Eckhardt, H. and Luhrman, R.,** Recognition by initiator tRNA of a uridine 5' adjacent to the AUG codon, *Biochemistry,* 20, 2075, 1981.

232. **Schmitt, M., Mandershied, U., Kryiatsaulis, A., Brinckman, U., and Gassen, H. G.,** Tetranucleotides as effectors for the binding of initiator tRNA to *Escherichia coli* ribosomes, *Eur. J. Biochem.,* 109, 291, 1980.

233. **Cavener, D. R.,** Comparison of the consensus sequence flanking translational start sites in *Drosophila* and vertebrates, *Nucleic Acids Res.,* 15, 1353, 1987.

234. **Heidecker, G. and Messing, J.,** Structural analysis of plant genes, *Annu. Rev. Plant Physiol.,* 37, 439, 1986.

235. **Botterman, J., de Almeida, L., Bouguelert, L., and Zabeau, M.,** Regulation of translation initiation by an intermolecular long distance interaction within the mRNA, in *Sequence Specificity in Transcription and Translation,* Calandar, R. and Gold, L., Eds., Alan R. Liss, New York, 1985, 397.

236. **Calame, K. and Ihler, G.,** Visualization of ribosome-single stranded DNA complexes in the electron microscope, *Biochemistry,* 16, 964, 1977.

237. **Calame, K. and Ihler, G.,** Physical location of ribosome binding sites on lambda DNA, *J. Mol. Biol.,* 116, 841, 1978.

238. **Jurgenson, J. E.,** *Nicotiana tabacum* chloroplast DNA, structure and gene content, *Ph.D. dissertation,* University of Arizona, Tucson, 1980.

239. **Jurgenson, J. E. and Bourque, D. P.,** unpublished data, 1980.

240. **Bahramian, M. B.,** How bacterial ribosomes select translation initiation sites, *J. Theor. Biol.,* 84, 103, 1980.

241. **Scherer, G. F. E., Walkinshaw, M. D., Arnott, S., and Morre, D. J.,** The ribosome binding sites recognized by *E. coli* ribosomes have regions with signal character in both the leader and protein coding segments, *Nucleic Acids Res.,* 8, 3895, 1980.

242. **Crick, F. H. C.,** Codon-anticodon pairing: the wobble hypothesis, *J. Mol. Biol.,* 19, 548, 1966.

243. **Bonitz, S. G., Berlani, R., Coruzzi, G., Li, M., Macino, G., Nobrega, F. G., Nobrega, M. P., Thalenfeld, B. E., and Tzagaloff, A.,** Codon recognition rules in yeast mitochondria, *Proc. Natl. Acad. Sci. U.S.A.,* 77, 3167, 1980.

244. **Pfitzinger, H., Guillemaut, P., Weil, J. H., and Pillay, D. T. N.,** Adjustment of the tRNA population to the codon usage in chloroplasts, *Nucleic Acids Res.,* 15, 1377, 1987.

245. **Joshi, C. P.,** An inspection of the domain between putative TATA box and translation start site in 79 plant genes, *Nucleic Acids Res.,* 15, 6643, 1987.

246. **Stanley, W. M.,** Specific aminoacylation of the methionine-specific tRNAs of eukaryotes, in *Methods in Enzymology,* Vol. 29, part E, Moldave, K. and Grossman, L., Eds., Academic Press, New York, 1974, 530.

247. **Bourque, D. P. and Wildman, S. G.,** Evidence that nuclear genes code for several chloroplast ribosomal proteins, *Biochem. Biophys. Res. Commun.,* 50, 532, 1973.

248. **Capel, M. S., Redman, B., and Bourque, D. P.,** Quantitative comparative analysis of complex two dimensional electropherograms, *Anal. Biochem.,* 97, 210, 1979.

249. **Capel, M. S.,** Characterization and genomic partitioning of chloroplast ribosomal proteins from higher plants, *Ph.D. dissertation,* University of Arizona, 1982.

250. **Capel, M. S. and Bourque, D. P.,** Characterization of *Nicotiana tabacum* chloroplast and cytoplasmic ribosomal proteins, *J. Biol. Chem.,* 257, 7746, 1982.

251. **Eneas-Filho, J., Hartlet, M. R., and Mache, R.,** Pea chloroplast ribosomal protein: characterization and site of synthesis, *Mol. Gen. Genet.,* 184, 484, 1982.

252. **Spiess, H.,** Analysis of the chloroplast ribosomal proteins from *Chlamydomonas reinhardii,* streptomycin-resistant and dependent mutants by two-dimensional gel electrophoresis, *Plant Sci. Lett.,* 10, 103, 1977.

253. **Schmidt, R. F., Richardson, C. B., Gillham, N. W., and Boynton, J. E.,** Sites of synthesis of chloroplast ribosomal proteins in *Chlamydomonas reinhardii, J. Cell Biol.,* 96, 1451, 1983.

254. **Freyssinet, G.,** The protein synthesizing system of *Euglena:* synthesis of ribosomal proteins *in vivo* and their characterization, *Physiol. Veg.,* 15, 519, 1977.

255. **Gualerzi, C., Janda, H. G., Passow, H., and Stoffler, F.,** Studies on the protein moiety of plant ribosomes, *J. Biol. Chem.,* 249, 347, 1974.

256. **Mache, R., Dorne, A. M., and Batlle, R. M.,** Characterization of spinach plastid ribosomal proteins by two dimensional gel electrophoresis, *Mol. Gen. Genet.,* 177, 333, 1980.

257. **Posno, M., vanNoort, M., Debise, R., and Groot, G. S. P.,** Isolation, characterization, phosphorylation and site and synthesis of *Spinacia* chloroplast ribosomal proteins, *Curr. Genet.,* 8, 147, 1984.

258. **Wada, A. and Sako, T.,** Primary structures of and genes for new ribosomal proteins A and B in *Escherichia coli, J. Biochem.,* 101, 817, 1987.

259. **Nomura, M. and Held, W. A.,** Reconstitution of ribosomes: studies of ribosome structure, function and assembly, in *Ribosomes,* Nomura, M., Tissieres, A., and Lengyel, P., Eds., Cold Spring Harbor Laboratory, Cold Spring Harbor, NY, 1974, 193.

260. **Nomura, M., Gourse, R., and Baughman, G.,** Regulation of the synthesis of ribosomes and ribosomal components, *Annu. Rev. Biochem.,* 53, 75, 1984.

261. **Nomura, M., Yates, J. L., Dean, D., and Post, L. E.,** Feedback regulation of ribosomal protein gene expression in *Escherichia coli:* structural homology of ribosomal RNA and ribosomal protein mRNA, *Proc. Natl. Acad. Sci. U.S.A.,* 77, 7084, 1980.

262. **Hildebrand, M. M. and Bourque, D. P.,** unpublished results, 1987.

263. **Mets, L. J. and Bogorad, L.,** Mendelian and uniparental alterations in erythromycin binding by plastid ribosomes, *Science,* 174, 707, 1971.

264. **Davidson, J. N., Hanson, M. R., and Bogorad, L.,** An altered chloroplast ribosomal protein in *ery*-M1 mutants of *Chlamydomonas reinhardii, Mol. Gen. Genet.,* 132, 119, 1974.

265. **Davidson, J. N., Hanson, M. R., and Bogorad, L.,** Erythromycin resistance and the chloroplast ribosome in *Chlamydomonas reinhardii, Genetics,* 8, 281, 1978.

266. **Schlanger, G. and Sager, R.,** Mutation of a cytoplasmic gene in *Chlamydomonas reinhardii* alters chloroplast ribosome function, *Proc. Natl. Acad. Sci. U.S.A.,* 69, 3551, 1972.

267. **Schlanger, G. and Sager, R.,** Localization of five antibiotic resistances at the subunit level in chloroplast ribosomes of *Chlamydomonas reinhardii, Proc. Natl. Acad. Sci. U.S.A.,* 71, 1715, 1974.

268. **Ohta, N., Sager, R., and Inouye, M.,** Identification of a chloroplast ribosomal protein altered by chloroplast mutation in *Chlamydomonas reinhardii, J. Biol. Chem.,* 250, 3655, 1975.

269. **Conde, M. E., Boynton, J. E., Gillham, N. W., Haris, E. H., Tingle, C. L., and Wang, W. L.,** Chloroplast genes in *Chlamydomonas* affecting organelle ribosomes, *Mol. Gen. Genet.,* 140, 183, 1975.

270. **Dorne, A. M., Lescure, A. M., and Mache, R.,** Site of synthesis of spinach chloroplast ribosomal proteins and formation of incomplete ribosomal particles in isolated chloroplasts, *Plant Mol. Biol.,* 3, 83, 1984.

271. **Freyssinet, G.,** Determination of the site of synthesis of some *Euglena* cytoplasmic and chloroplast ribosomal proteins, *Exp. Cell Res.,* 115, 207, 1978.

272. **Kloppstech, K. and Schweiger, H. G.,** Nuclear genome codes for chloroplast ribosomal proteins in *Acetabularia, Exp. Cell Res.,* 80, 69, 1973.

273. **Kloppstech, K. and Schweiger, H. G.,** The site of synthesis of chloroplast ribosomal proteins, *Plant Sci. Lett.,* 2, 101, 1974.

274. **Chua, N.-H. and Schmidt, G. W.,** Post-translational transport into intact chloroplasts of a precursor to the small subunit of ribulose-1,5-bisphosphate carboxylase, *Proc. Natl. Acad. Sci. U.S.A.,* 75, 6110, 1978.

275. **Pfisterer, J. P., Lachmann, P., and Kloppstech, K.,** Transport of proteins into chloroplasts. Binding of nuclear-encoded chloroplast proteins to the chloroplast envelope, *Eur. J. Biochem.,* 126, 143, 1982.

276. **Pain, D., Kanwar, Y. S., and Blobel, G.,** Identification of a receptor for protein import into chloroplasts and its localization to envelope contact zones, *Nature (London),* 331, 232, 1988.
277. **Grossman, A., Bartlett, S., and Chua, N.-H.,** Energy-dependent uptake of cytoplasmically synthesized polypeptides by chloroplasts, *Nature (London),* 285, 625, 1980.
278. **Lamppa, G. K. and Abad, M. S.,** Processing of a wheat light-harvesting chlorophyll a/b protein precursor by a soluble enzyme from higher plant chloroplasts, *J. Cell Biol.,* 105, 2641, 1987.
279. **Grossman, A. R., Bartlett, S. G., Schmidt, G. W., Mullet, J. P., and Chua, N.-H.,** Optimal conditions for post-translational uptake of proteins by isolated chloroplasts, *J. Biol. Chem.,* 257, 1558, 1982.
280. **Gantt, J. S. and Key, J. L.,** Isolation of nuclear encoded plastid ribosomal protein cDNAs, *Mol. Gen. Genet.,* 202, 186, 1986.
281. **Gantt, S.,** personal communication, 1987.
282. **El Hag, G. and Bourque, D. P.,** unpublished data, 1988.
283. **Johnson, L. M., Snyder, M., Chang, L. M. S., Davis, R. W., and Campbell, J. L.,** Isolation of the gene encoding yeast DNA polymerase I, *Cell,* 43, 369, 1985.
284. **Kuiper, M. T. R., Akins, R. A., Holtrop, M., deVries, H., and Lambowitz, A. M.,** Isolation and analysis of the *Neurospora crassa* cyt-21 gene. A nuclear gene encoding a mitochondrial ribosomal protein, *J. Biol. Chem.,* 263, 2840, 1988.
285. **Terhorst, C. P., Moller, W., Laursen, R., and Wittman-Liebold, B.,** Amino acid sequence of 50S ribosomal proteins involved in both EFG and EFT dependent GTP-hydrolysis, *FEBS Lett.,* 28, 325, 1972.
286. **Subramanian, A. R.,** Copies of proteins L7 and L12 and heterogeneity of the large subunit of *Escherichia coli* ribosomes, *J. Mol. Biol.,* 95, 1, 1975.
287. **Strycharz, W. A., Nomura, M., and Lake, J. A.,** Ribosomal proteins L7/L12 localized at a single region of the large subunit by immune electron microscopy, *J. Mol. Biol.,* 126, 123, 1978.
288. **Bartsch, M., Kimura, M., and Subramanian, A. R.,** Purification, primary structure and homology relationships of a chloroplast ribosomal protein, *Proc. Natl. Acad. Sci. U.S.A.,* 79, 6871, 1982.
289. **Lake, J. A.,** Evolving ribosome structure: domains in archaebacteria, eubacteria, eocytes and eukaryotes, *Annu. Rev. Biochem.,* 54, 507, 1985.
290. **Miller, J. H. and Reznikoff, W. S.,** *The Operon,* Cold Spring Harbor Laboratory, Cold Spring Harbor, NY, 1978.
291. **Nomura, M.,** Regulation of the synthesis of ribosomes and ribosomal components in *Escherichia coli;* translational regulation and feedback loops, in *Regulation of Gene Expression,* Booth, I. and Higgins, C., Eds., Cambridge University Press, Cambridge, 1986, 199.
292. **Freedman, L. P., Zengel, J. M., Archer, R. H., and Lindahl, L.,** Autogenous control of the S10 ribosomal protein operon of *Escherichia coli:* genetic dissection of transcriptional and posttranscriptional regulation, *Proc. Natl. Acad. Sci. U.S.A.,* 84, 6516, 1987.
293. **Yates, J. and Nomura, M.,** *E. coli* ribosomal protein L4 is a feedback regulatory protein, *Cell,* 21, 517, 1980.
294. **Lindahl, L. and Zengel, J. M.,** Ribosomal genes in *Escherichia coli, Annu. Rev. Genet.,* 20, 297, 1986.
295. **Olins, P. O. and Nomura, M.,** Translational regulation by ribosomal protein S8 in *Escherichia coli:* structural homology between rRNA binding site and feedback target on mRNA, *Nucleic Acids Res.,* 9, 1757, 1981.
296. **Bonham-Smith, P. C. and Bourque, D. P.,** unpublished data, 1988.
297. **Leibold-Wittman, B.,** Ribosomal proteins: their structure and evolution, in *Structure, Function and Genetics of Ribosomes,* Hardesty, G. and Kramer, G., Eds., Springer-Verlag, New York, 1986, 326.
298. **Nowotny, V. and Nierhaus, K. H.,** Initiator proteins for the assembly of the 50S subunit from *Escherichia coli* ribosomes, *Proc. Natl. Acad. Sci. U.S.A.,* 79, 7238, 1982.
299. **Dean, D., Yates, J. L., and Nomura, M.,** *Escherichia coli* ribosomal protein S8 feedback regulates part of the *spc* operon, *Nature (London),* 289, 89, 1981.
300. **Dabbs, E. R.,** Organization and regulation of ribosomal protein genes in bacteria, in *Genetics: New Frontiers,* IBH Publishing, New Delhi, 1984, 35.
301. **Braw, D. A. and Geiduschek, E. P.,** Modulation of yeast 5S rRNA synthesis *in vitro* by ribosomal protein YL3, *J. Biochem.,* 262, 13953, 1987.
302. **Dohera, M. D. and Warner, J. R.,** The yeast ribosomal protein L32 and its gene, *J. Biol. Chem.,* 262, 16055, 1987.
303. **Waddell, J., Wang, X.-M., and Wu, M.,** Electron microscopic localization of the chloroplast DNA replication origins in *Chlamydomonas reinhardii, Nucleic Acids Res.,* 12, 3843, 1984.
304. **Wang, X.-M., Chang, C. H., Waddell, J., and Wu, M.,** Cloning and delimiting one chloroplast DNA replicative origin of *Chlamydomonas, Nucleic Acids Res.,* 12, 3857, 1984.
305. **Wu, M., Lou, J. K., Chang, D. Y., Chang, C. H., and Nie, Z. Q.,** Structure and function of a chloroplast DNA replication origin of *Chlamydomonas reinhardii, Proc. Natl. Acad. Sci. U.S.A.,* 83, 676, 1986.

306. **Gold, B., Carillo, N., Tewari, K. K., and Bogorad, L.,** Nucleotide sequence of a preferred maize chloroplast genome template for *in vitro* DNA synthesis, *Proc. Natl. Acad. Sci. U.S.A.,* 84, 194, 1987.
307. **Meeker, R., Nielsen, B., and Tewari, K. K.,** Localization of replication origins in pea chloroplast DNA, *Mol. Cell. Biol.,* 8, 1216, 1988.
308. **McKown, R. L. and Tewari, K. K.,** Purification and properties of a pea chloroplast DNA polymerase, *Proc. Natl. Acad. Sci. U.S.A.,* 81, 2354, 1984.

Index

INDEX